Android 应用开发教程（下册）

——基于Android Studio的案例开发全析

张冬玲　张光显　编著

清华大学出版社
北京

内容简介

本书以Android 11为系统平台，以Studio 4.0.1为开发环境，全面介绍Android应用开发的相关知识和技术。全书共15章，分上、下两册。上册主要涉及Android入门级基础内容：第1～3章，主要介绍Android平台概述及基本概念；第4～8章，主要介绍Android应用项目页面的常见布局管理器、控件的使用及事件处理等技术。上册内容覆盖了Android应用的用户界面编程全部内容。下册主要涉及Android进阶技术：第9～14章分别介绍Android的数据存储、后台处理、多媒体应用、手机基本功能、网络通信和第三方开发包应用开发，覆盖了Android应用开发中涉及的数据处理技术和逻辑控制技术；第15章介绍"我的音乐盒"实战项目的完整开发过程，对实际应用开发极具参考价值。本书精心设计出各章后面的练习题，汇合集成之后便是下册最后的实例项目的主要功能模块。本书内容全面，案例丰富，实践性强。各章节内容讲述透彻，注重知识的来龙去脉，案例解析清晰。章与章之间环环相扣，内容由浅入深，引导读者逐步步入Android应用开发的奇妙世界。

本书不仅可作为本科院校、大中专院校、IT技能开发培训机构的相关课程的教材，也可作为移动应用开发设计人员的参考用书。

图书在版编目(CIP)数据

Android应用开发教程：基于Android Studio的案例开发全析. 下册/张冬玲，张光显编著. —北京：清华大学出版社，2021.6
ISBN 978-7-302-57913-7

Ⅰ. ①A… Ⅱ. ①张… ②张… Ⅲ. ①移动终端—应用程序—程序设计—教材 Ⅳ. ①TP929.53

中国版本图书馆CIP数据核字(2021)第060932号

责任编辑：刘向威　常晓敏
封面设计：文　静
责任校对：李建庄
责任印制：刘海龙

出版发行：清华大学出版社
网　　址：http://www.tup.com.cn，http://www.wqbook.com
地　　址：北京清华大学学研大厦A座　　**邮　　编**：100084
社 总 机：010-62770175　　**邮　　购**：010-83470235
投稿与读者服务：010-62776969，c-service@tup.tsinghua.edu.cn
质量反馈：010-62772015，zhiliang@tup.tsinghua.edu.cn
课件下载：http://www.tup.com.cn，010-83470236
印 装 者：三河市铭诚印务有限公司
经　　销：全国新华书店
开　　本：185mm×260mm　　**印　　张**：26.75　　**字　　数**：654千字
版　　次：2021年7月第1版　　**印　　次**：2021年7月第1次印刷
印　　数：1～1500
定　　价：69.00元

产品编号：091884-01

前言

在移动应用开发中，Android仍是一个优秀的开源开发平台。本书以Android 11为系统平台，使用Android Studio 4.0.1为开发集成工具，介绍在Android平台上进行原生开发的知识和技术。Android由于其开源性特点，版本的升级十分频繁，每年皆有数次版本更新，Android的API支持库也更新到了AndroidX。本书以2013年出版的《Android应用开发教程》为基础，引入当前最新的Android版本和最新的Android Studio开发环境，全面介绍Android的原生开发知识与技术，包括当前流行的较新技术。本书依据实际开发中经常使用的应用技术，吸纳Android开发设计类书籍的优点，从教学的角度全面介绍Android应用程序的开发设计，深浅适宜，实例丰富，不仅可作为本科院校、大中专院校、IT培训机构相关课程的教材，而且也可作为Android系统开发人员的参考用书。

全书共15章，分上、下两册。上册包括第1～8章，下册包括第9～15章。

第1～3章介绍Android概述、Android项目的开发基础。主要内容涉及Android平台概要介绍，开发环境搭建，应用项目的目录结构，Android项目的生命周期，项目的控件机制及项目组件之间的联系。

第4～8章介绍项目用户界面的开发入门。主要内容涉及用户界面的布局管理器，以及布局在其上的各种控件的添加、设置属性、添加绑定数据、适配数据，对控件交互的监听及事件处理，包括在布局上设置标签栏、导航栏、菜单、对话框及绘制图形和动画技术。

第9～14章介绍Android项目的开发进阶。主要内容涉及数据存储、后台处理、多媒体应用、手机基本功能、网络通信及第三方SDK应用等内容。掌握这些技术就可以实现对应用项目中的页面内容进行数据处理和控制处理。

第15章讲述综合应用实例开发。该章以项目开发周期为主线，从需求分析开始，逐一对项目设计开发的步骤展开介绍。

Android课程内容十分丰富，实践性强，教学课时建议不低于100学时，并且需要保证充足的实践课时数，建议实践课时不低于50学时。

本书作者张冬玲从事计算机本科教学数十年，另一作者张光显从事Android项目开发数十年。教程内容凝聚了两位作者多年的教学与移动应用开发经验，讲解深入透彻，论述通俗易懂，注重知识的来龙去脉，案例解析清晰透彻。凡具备编程基础的人员，都可以通过本书的学习，掌握 Android 的应用编程。

本书的主要章节由张冬玲编写。张光显完成大部分案例的技术支持、第 15 章实例开发和主要内容的编写。全书由张光显统审，张冬玲统稿与定稿。在此还要感谢杨宁、张泽宾、刘涛涛等同事的支持和帮助，没有他们的鼎力相助，本书无法按期顺利完成。

由于作者水平有限，书中难免会有疏漏与错误，敬请各位读者与专家批评指正。

张冬玲

2020 年 12 月

目录

第 9 章 数据存储

程序本身就是数据的输入、处理和输出的过程，不管是操作系统还是应用项目都不可避免地要用到大量的数据。因此，数据存储是程序最基本的问题。在手机这种特殊设备里，也经常会进行存取数据处理，例如，存取通讯录、图片文件、音频文件和视频文件等数据的处理。

在 Android 系统中，所有应用项目的数据为该应用项目所私有，同时也提供了多个应用项目之间的数据通信标准方式。Android 作为一种手机操作系统，提供了以下几种数据存储和数据共享方式：SharedPreferences(配置)、File(文件)、数据库 SQLite、ContentProvider、SD Card 和网络。关于网络数据存储将在第 13 章中详细介绍，下面针对本地的几种存储方式进行介绍。

9.1 SharedPreferences 存储

SharedPreferences 提供了一种轻量级的数据存取方法，一般用于数据较少的配置信息的存储场合。例如，一些登录的用户名和密码，一些默认的应用项目问候词，程序关闭前界面的主要属性值等。

SharedPreferences 存储是以"键-值"对的方式将数据保存到一个内部的 XML 配置文件中。如果在 Android 的应用项目中使用了 SharedPreferences 保存数据，可以在 Android Studio 的 Device File Explorer 窗口区域内找到该文件。保存 SharedPreferences 键-值对信息的文件存放在/data/data/< package name >/shared_prefs 下。这里，< package name >为应用项目包名。注意，必须先启动 Android Studio 的模拟器，然后单击 Android Studio 编辑窗右下角的 Device File Explorer 标签，即可看到 Device File Explorer 区域中的内容，如图 9-1 所示。

9.1.1 SharedPreferences 接口

SharedPreferences 是一个接口，位于 android. content. SharedPreferences 包中。使用 SharedPreferences 存储时，首先需要获得 SharedPreferences 对象。

1. 获取 SharedPreferences 对象

在 Android 中，获取 SharedPreferences 对象有两种方式：一是调用 Context 对象的 getSharedPreferences()方法；二是调用 Activity 对象的 getPreferences()方法。两种方式

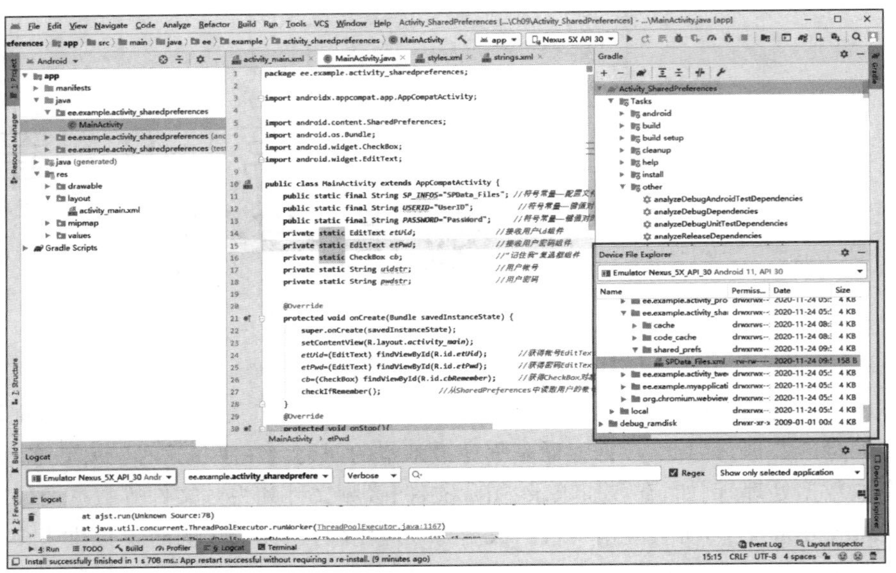

图 9-1　Device File Explorer 窗口区域

的区别是：调用 Context 对象的 getSharedPreferences()方法获得的 SharedPreferences 对象可以被同一应用项目下的其他组件共享；调用 Activity 对象的 getPreferences()方法获得的 SharedPreferences 对象只能在该 Activity 中使用。

下面着重对 getSharedPreferences()进行介绍。每个应用项目都有一个 SharedPreferences 对象。通过调用 Context 的 getSharedPreferences()方法可以获取该对象。调用 SharedPreferences 对象的语法如下。

```
Context.getSharedPreferences(String name,int mode);
```

参数说明如下。

Context：当前上下文，一般指当前 Activity。

name：本组件的配置文件名。

mode：操作模式。操作模式有三种：MODE_PRIVATE(值为 0,应用项目私有,常用)；MODE_WORLD_READABLE(值为 1,其他程序可读)；MODE_WORLD_WRITEABLE(值为 2,其他程序可写)。

2. SharedPreferences 对象获得数据的方法

在使用 SharedPreferences 对象存储数据时，还需要使用 SharedPreferences 的一个内部接口 SharedPreferences. Editor，以及相关的获得数据的方法。常用的方法见表 9-1。

表 9-1　SharedPreferences 常用的方法及说明

方　法	说　明
edit()	返回 SharedPreferences 的内部接口 SharedPreferences. Editor
contains (String key)	判断是否包含该键值
getAll()	从 SharedPreferences 中返回所有的信息

续表

方法	说明
getBoolean(String key,boolean defValue)	获得一个 boolean 值
getFloat(String key,float defValue)	获得一个 float 值
getInt(String key,int defValue)	获得一个 int 值
getLong(String key,long defValue)	获得一个 long 值
getString(String key,String defValue)	获得一个 String 值

调用 SharedPreferences 的 edit()方法返回 SharedPreferences. Editor 内部接口,该接口中提供了保存数据的方法,常用的方法见表 9-2。

表 9-2 SharedPreferences. Editor 常用方法及说明

方法	说明
clear()	清空 SharedPreferences 里的所有数据
commit()	当 Editor 编辑完成后,提交修改到 SharedPreferences 的 XML 配置文件中
putBoolean(String key,boolean value)	保存一个 boolean 值
putFloat(String key,float value)	保存一个 float 值
putInt(String key,int value)	保存一个 int 值
putLong(String key,long value)	保存一个 long 值
putString(String key,String value)	保存一个 String 值
remove(String key)	删除 SharedPreferences 里指定 key 对应的数据项

使用 SharedPreferences 对象保存信息,即向 SharedPreferences 对象写入数据,在编写代码时一般步骤如下。

(1) 获得 SharedPreferences 对象。

(2) 使 SharedPreferences 对象处于编辑状态。即调用 SharedPreferences 对象的 edit()方法得到内部接口 SharedPreferences. Editor。

(3) 保存键-值对。即使用相应的 put...()方法保存键-值对。例如保存字符串型使用 putString()方法。

(4) 调用 commit()方法提交数据,将键-值对数据写入 SharedPreferences 的 XML 配置文件中。

相对地,读取 SharedPreferences 对象数据的编程步骤比较简单:首先获取 SharedPreferences 对象的键 key,然后再由 key 获取相应的键值。在这里,对于不同类型的键值有不同的 get()方法。下面通过一个案例来说明 SharedPreferences 的具体用法。

9.1.2 SharedPreferences 应用案例

【案例 9.1】 在用户登录界面中,增加一个复选框“记住我”。当勾选了该复选框后,系统的用户登录 Activity 将保存最近的一次用户登录信息。

说明:使用 SharedPreferences 对账号和密码的输入信息进行存取。在 Activity 的构造方法中读取 XML 配置文件,如果有保存的登录信息则显示在相应位置;如果有勾选“记住我”，则在单击“登录”按钮或退出 Activity 时写入登录信息到 XML 配置文件中。

开发步骤及解析：过程如下。

1）创建项目

在 Android Studio 中创建一个名为 Activity_SharedPreferences 的项目。其包名为 ee.example.activity_sharedpreferences。

2）准备图片

将 back.jpg 作为背景图片资源复制到本项目的 res/drawable 目录中。

3）准备字符串资源

编写 res/values 目录下的 strings.xml 文件，代码大多数与 4.4 节中的案例 4.16 相似，在文件中多定义了一个“< string name="cbRemember">记住我</string >”，其余的同案例 4.16，在此不赘述。

4）准备样式资源

编写 res/values 目录下的样式描述文件 styles.xml，在< resources >元素中定义了< style name="button">、< style name="title">、< style name="text">和< style name="content">等样式，这几个样式都是定义文本的 textSize、textColor 和 textStyle 属性。

5）设计布局

编写 res/layout 目录下的布局文件 activity_main.xml。该布局文件与 4.4 节中的案例 4.16 布局文件相似，在按钮布局前增加了一个线性布局，在其内添加一个 CheckBox 控件声明，代码片段如下。

```
<!-- 新增加的 LinearLayout -->
<LinearLayout
    android:orientation = "horizontal"
    android:layout_gravity = "center_horizontal"
    android:layout_width = "wrap_content"
    android:layout_height = "wrap_content" >
    <CheckBox
        android:id = "@ + id/cbRemember"
        android:text = "@string/cbRemember"
        style = "@style/text"
        android:layout_width = "match_parent"
        android:layout_height = "wrap_content"
        android:layout_gravity = "center_vertical"
        android:checked = "false" />
 </LinearLayout>
```

6）开发逻辑代码

打开 java/ee.example.activity_sharedpreferences 包下的 MainActivity.java 文件，并进行编辑，代码如下。

```
package ee.example.activity_sharedpreferences;

import androidx.appcompat.app.AppCompatActivity;
import android.content.SharedPreferences;
import android.os.Bundle;
import android.view.View;
```

```
import android.widget.CheckBox;
import android.widget.EditText;

public class MainActivity extends AppCompatActivity{
    public static final String SP_INFOS = "SPData_Files";   //符号常量——配置文件名
    public static final String USERID = "UserID";           //符号常量——键-值对的账号键名
    public static final String PASSWORD = "PassWord";       //符号常量——键-值对的密码键名
    private static EditText etUid;                          //接收用户 id 组件
    private static EditText etPwd;                          //接收用户密码组件
    private static CheckBox cb;                             //"记住我"复选框组件
    private static String uidstr;                           //用户账号
    private static String pwdstr;                           //用户密码

    @Override
    protected void onCreate(Bundle savedInstanceState){
        super.onCreate(savedInstanceState);
        setContentView(R.layout.activity_main);
        etUid = (EditText) findViewById(R.id.etUid);              //获得账号 EditText
        etPwd = (EditText) findViewById(R.id.etPwd);              //获得密码 EditText
        cb = (CheckBox) findViewById(R.id.cbRemember);            //获得 CheckBox 对象
        checkIfRemember();                                        //从 SharedPreferences 中
                                                                  //读取用户的账号和密码

        findViewById(R.id.btnLogin).setOnClickListener(new View.OnClickListener() {
            @Override
            public void onClick(View v){
                uidstr = etUid.getText().toString().trim();       //获得输入的账号
                pwdstr = etPwd.getText().toString().trim();       //获得输入的密码
                if (!uidstr.isEmpty() && !pwdstr.isEmpty()     && cb.isChecked()){
                    rememberMe(uidstr,pwdstr);                    //将用户的账号与密码存
                                                                  //入 SharedPreferences

                }
                //TODO something for Login
            }
        })
    }
    @Override
    protected void onStop(){
        super.onStop();
        if(cb.isChecked()){
            uidstr = etUid.getText().toString().trim();       //获得输入的账号
            pwdstr = etPwd.getText().toString().trim();       //获得输入的密码
            rememberMe(uidstr,pwdstr);                        //将用户的账号与密码存
                                                              //入 SharedPreferences

        }
    }
    @Override
    protected void onDestroy(){
        super.onDestroy();
    }

    //方法：从 SharedPreferences 中读取用户的账号和密码
    public void checkIfRemember(){
```

```
57          SharedPreferences sp = getSharedPreferences(SP_INFOS,MODE_PRIVATE);
                                                            //获得 Preferences
58          uidstr = sp.getString(USERID, null);            //取键-值对中的账号值
59          pwdstr = sp.getString(PASSWORD, null);          //取键-值对中的密码值
60          if(uidstr != null && pwdstr!= null){
61              etUid.setText(uidstr);                      //给 EditText 控件赋账号
62              etPwd.setText(pwdstr);                      //给 EditText 控件赋密码
63              cb.setChecked(true);
64          }
65      }
66
67      //方法：将用户的 id 和密码存入 SharedPreferences
68      public void rememberMe(String uid,String pwd){
69          SharedPreferences sp = getSharedPreferences(SP_INFOS,MODE_PRIVATE);
70          SharedPreferences.Editor editor = sp.edit();    //获得 Editor
71          editor.putString(USERID, uid);     //将用户的账号存入 Preferences
72          editor.putString(PASSWORD, pwd);                //将密码存入 Preferences
73          editor.commit();
74      }
75
76  }
```

(1) 第 11～13 行定义了三个字符串常量，分别代表配置文件名、账号的键名、密码的键名。

(2) 第 27 行调用 checkIfRemember()自定义方法，它将完成从 SharedPreferences 中读取用户的账号和密码。

(3) 第 56～65 行，定义 checkIfRemember()方法。其中，第 57 行获取 SharedPreferences 对象 sp；第 58、59 行分别取得账号值和密码值；第 61、62 行分别向账号和密码的 EditText 控件中传入取得的值，第 63 行将“记住我”复选框设置为选中状态。

(4) 第 29～39 行实现登录按钮的单击监听。在重写 onClick()方法中，当用户账号、密码都已填写并且勾选了“记住我”复选框的情况下，调用 rememberMe(uidstr,pwdstr)自定义方法，它将完成把用户的账号与密码存入 SharedPreferences 的配置文件中。

(5) 第 42～49 行重写 onStop()方法。如果勾选了“记住我”复选框，调用 rememberMe(uidstr,pwdstr)自定义方法。

(6) 第 68～74 行，定义 rememberMe(String uid,String pwd)方法。其中，第 70 行使用 edit()方法获得一个 SharedPreferences 的 Editor 对象 editor；第 71、72 行将用户的账号值和密码值写到 editor 对象中；第 73 行将 editor 对象保存到 SharedPreferences 配置文件中。

运行结果：在 Android Studio 支持的模拟器上运行 Activity_SharedPreferences 项目，运行结果如图 9-2 所示。

在模拟器上运行并退出该案例程序后，本应用将回调 onStop()方法，此时会把输入的账号和密码写入 SharedPreferences 配置文件中，在重启模拟器后，可以在 Device File Explorer 窗格下查看路径/data/data/ee.example.activity_sharedpreferences/shared_prefs，其下有个名为 SPData_Files.xml 的文件，如图 9-3 所示。

SPData_Files.xml 文件内容如下。

```
<?xml version = '1.0' encoding = 'utf - 8' standalone = 'yes' ?>
< map >
```

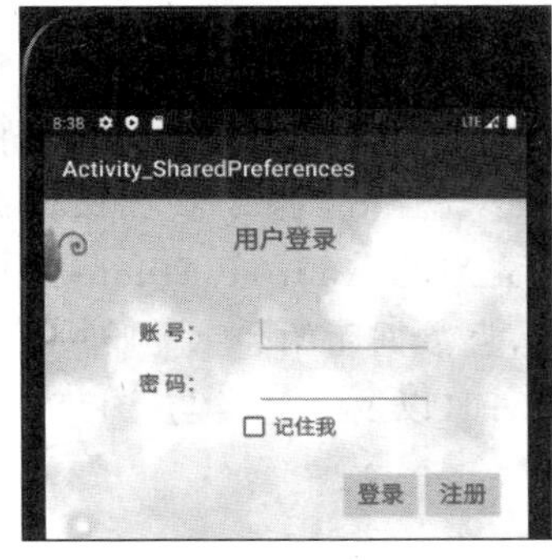

(a) 初始运行界面

(b) 选中“记住我”复选框，再次运行界面

图 9-2　用户登录案例运行效果

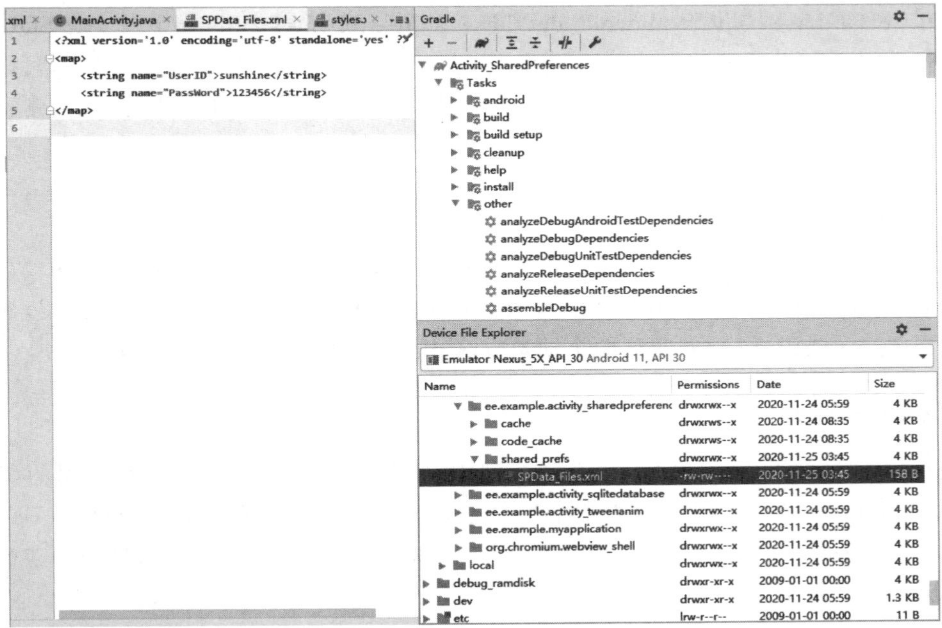

图 9-3　SPData_Files.xml 文件的存储位置

```
    <string name = "UserID"> sunshine </string>
    <string name = "PassWord"> 123456 </string>
</map>
```

使用 SharedPreferences 的存储方式非常方便，但是它只适合存储比较简单且内容少的数据，如果需要存储更多的数据，就必须使用其他的存储方式。

9.2　SQLite 数据库

Android 平台中内嵌一个轻量级的、功能强大的嵌入式关系数据库 SQLite，使用方便，可以用于完成各种复杂的数据处理，实现结构化的数据存储。

SQLite 数据库是一种关系数据库，支持多数的 SQL92 标准，最大支持数据库到 2TB。它没有服务进程，是一种嵌入应用项目内部的数据库，所包含的数据库、表等所有数据都存放在一个单一的文件中，文件名的扩展名为 db。可以在 Android Studio 的 Device File Explorer 窗口区域内的/data/data/< package name >/databases 下找到该数据库文件。注意，要查看 Device File Explorer 窗口内的内容必须先启动 Android Studio 的模拟器。在默认情况下，SQLite 数据库文件属于应用项目所私有，且数据库的名字是唯一的。

Android 提供了创建和使用 SQLite 数据库的 API，以及一些类与接口，下面分别介绍。

9.2.1 SQLite 数据库相关的类与接口

1. SQLiteDatabase 类

SQLiteDatabase 类位于 android. database. sqlite. SQLiteDatabase 包中，是 SQLite 的数据库管理类。一个 SQLiteDatabase 对象代表一个数据库。在 Android 平台下，可以通过 SQLiteDatabase 类的静态方法创建或打开数据库，以及对数据库的记录进行增、删、改、查等操作。

2. SQLiteOpenHelper 类

SQLiteOpenHelper 类位于 android. database. sqlite. SQLiteOpenHelper 包中，它是一个辅助类，主要用来管理数据库的创建和版本。SQLiteOpenHelper 是一个抽象类，使用时通常需要创建子类继承它，并实现两个抽象方法 onCreate()和 onUpgrade()。

3. Cursor 接口

Cursor 位于 android. database. Cursor 包中，它是 Android 的一个非常有用的游标接口，通过 Cursor 可以对数据库的查询结果集进行随机的读写访问。调用 SQLiteDatabase 的 query 和 rawQuery 方法时，返回的都是 Cursor 对象，可以通过一条一条地查询游标的内容来遍历整个结果集。

4. ContentValues 类

ContentValues 类位于 android. content. ContentValues 包中，它存储一些键-值对，提供数据库表的列名、数据映射信息。ContentValues 类似于映射 Map，不同的是，ContentValues 的键是一个 String 类型的数据，对应数据库表中的列名；而其相应的值是一个基本数据类型的数据，对应数据库表中的数据，所以一个 ContentValues 对象代表数据库表的一行数据。ContentValues 主要用于记录增加和更新操作。

9.2.2 管理 SQLite 数据库相关的方法及编程

在 Android 中，通过上述类的相关方法来对 SQLite 数据库进行创建和管理。下面分别介绍常用的方法及其用法。

1. 相关方法

SQLiteDatabase 中常用的方法见表 9-3。

表 9-3 SQLiteDatabase 常用的方法及说明

方 法	参 数	说 明
openDatabase(String path, SQLiteDatabase. CursorFactory factory, int flags)	path：指定数据库文件的路径 factory：构造查询时返回的 Cursor 对象 flags：打开模式，模式参数包括： OPEN_READONLY(只读方式)； OPEN_READWRITE(可读可写)； CREATE_IF_NECESSARY(当数据库文件不存在时，创建该数据库)	打开或创建数据库
openOrCreateDatabase(String path, SQLiteDatabase. CursorFactory factory)	path：指定数据库文件的路径 factory：构造查询时返回的 Cursor 对象	相当于上面的方法中第三个参数取 CREATE_IF_NECESSARY 的情形
create (SQLiteDatabase. CursorFactory factory)	factory：构造查询时返回的 Cursor 对象	创建一个内存数据库
insert(String table,String nullColumnHack, ContentValues values)	table：数据表名称 nullColumnHack：空列的默认值 values：封装了列名和列值的 Map	添加一条记录
delete(String table, String whereClause, String[] whereArgs)	table：数据表名称 whereClause：删除条件 whereArgs：删除条件值数组	删除一条记录
update(String table, ContentValues values, String whereClause, String[] whereArgs)	table：数据表名称 values：更新列 ContentValues 类型键-值对 whereClause：更新条件(即 where 子句) whereArgs：更新条件值数组	修改记录
query(String table, String[] columns, String selection, String[] selectionArgs, String groupBy, String having, String orderBy)	table：数据表名称 columns：列名数组 selection：条件子句，相当于 where selectionArgs：条件子句，where 的值数组 groupBy：分组的列名 having：分组条件 orderBy：排序的列名	查询记录，其返回值为一个 Cursor 实例
execSQL(String sql)	sql：SQL 语句	执行一条 SQL 语句
close()		关闭数据库

SQLiteOpenHelper 中常用的方法见表 9-4。

表 9-4 SQLiteOpenHelper 的常用方法及说明

方法	说明
onCreate (SQLiteDatabase db)	创建数据库时调用
onUpgrade(SQLiteDatabase db,int oldVersion, int newVersion)	版本更新时调用
getReadableDatabase()	创建或打开一个只读数据库
getWritableDatabase()	创建或打开一个读写数据库

在数据库第一次生成时会调用 onCreate()方法，一般在这个方法里写生成数据表的代码。当数据库需要升级时，Android 系统会自动调用 onUpgrade()方法，一般在这个方法里会删除数据表，并建立新的数据表，当然，还可以根据应用的需求，编写相关的代码完成其他的操作。

Cursor 中常用的方法见表 9-5。

表 9-5 Cursor 的常用方法及说明

方法	说明
getCount()	获取游标结果集的总记录条数
isFirst()	判断游标是否是第一条记录
isLast()	判断游标是否是最后一条记录
isClosed()	判断游标结果集是否关闭
move(int offset)	向后移动游标到指定记录
moveToFirst()	移动游标到第一条记录
moveToLast()	移动游标到最后一条记录
moveToNext()	移动游标到下一条记录
moveToPrevious()	移动游标到上一条记录
getColumnIndexOrThrow(String columnName)	根据列名获得列索引
getInt(int columnIndex)	根据列索引获得 int 类型值
getString(int columnIndex)	根据列索引获得 String 类型值

2. 相关操作

下面给出一些代码片段，说明如何使用这些方法对数据库、表进行操作。

1）创建数据库

如果要在/data/data/ee. example. activity_abc/databases 目录下创建一个名为 pocketblog. db 的数据库，可使用如下方法。

```
openOrCreateDatabase("/data/data/ee.example.activity_abc/databases/pocketblog.db",null);
```

2）创建数据表

在这里使用 execSQL()方法来创建一个数据表。首先，编写一条 SQL 语句存放到一个 String 变量中，然后再调用 execSQL()方法执行先前编写的 SQL 语句。例如，创建一个数据表 UserTb，其属性列分别为：uid(主键并且自动增加)、uname(用户名)、upsw(密码)，可使用代码如下。

```
String sql = "Create table UserTb(uid integer primary key autoincrement, uname text, upsw text)";
db.execSQL(sql);          //db是指定的数据库对象
```

3）插入数据

分别使用execSQL()方法和insert()方法来完成向数据表UserTb插入记录。下面给出代码片段，以便读者更好地理解insert()方法的使用，代码片段如下。

```
String sql = "insert into UserTb(uname,upsw) values('user01','123')";
db.execSQL(sql);          //执行SQL语句,向db的数据表UserTb中插入一条记录
```

或

```
ContentValues mycv = new ContentValues();
mycv.put("uname"," user01'");
mycv.put("upsw"," 123'");
db.insert("UserTb",null,mycv);   //向db的数据表UserTb中插入一条记录
```

4）删除数据

和插入数据类似，使用两种方法来实现删除数据表UserTb中的记录，代码片段如下。

```
String sql = "delete from UserTb where uid = 2";
db.execSQL(sql);          //执行SQL语句,在db的数据表UserTb中删除uid为2的记录
```

或

```
String whereClause = "uid = ?";
String[ ] whereArgs = {String.valueOf(2)};
db.delete("UserTb", whereClause, whereArgs);   //在db的数据表UserTb中删除uid为2的记录
```

这里，String. valueOf(2)是将数值2转换成字符串值“2”。String. valueOf()是将其他基本数据类型转换成String的静态方法，例如：String. valueOf(int i)将int变量i中的值转换成字符串，String. valueOf(boolean b)将boolean变量b中的值转换成字符串，String. valueOf(char c)将char变量c中的值转换成字符串，等等。

5）修改数据

和插入或删除数据类似，使用两种方法来实现修改数据表UserTb中的记录，代码片段如下。

```
String sql = "update UserTb set upsw = 666 where uid = 1";
db.execSQL(sql);          //执行SQL语句,在db的数据表UserTb中修改uid为1的记录
```

或

```
ContentValues newvalue = new ContentValues();
newvalue.put("upsw"," 666");
String whereClause = "uid = ?";
String[ ] whereArgs = {String.valueOf(1)};
db.update("UserTb",newvalue, whereClause, whereArgs);   //在db的数据表UserTb中修改
                                                        //uid为1的记录
```

6）查询数据

查询数据方法将返回一个Cursor对象，这里存放的是查询数据表的结果集。在应用编

程中,往往要使用 Cursor 的方法,才能在让用户看到查询结果集中的内容。例如要查询 UserTb 中 uid 大于 5 的用户信息,可使用代码段如下。

```
Cursor qc = db.query("UserTb",null,"uid=?",new String[] { String.valueOf(5)},null,null,null);
if(qc.moveToFirst()){                                    //判断游标是否为空
    for (int i = 0; i < qc.getCount(); i++){             //遍历游标
        qc.move(i);
        int userid = qc.getInt(0);                       //获得用户 ID
        String username = qc.getString(1);               //获得用户名
        String userpassword =  qc.getString(2);          //获得用户密码
        System.out.println(userid + ":" + username + "," + userpassword);
                                                         //在屏幕上显示用户 ID、用户名和密码
    }
}
```

以上主要是针对 SQLiteDatabase 类给出的对 SQLite 数据库进行操作的使用方法。在实际开发中,比较常见的是开发一个数据库辅助类来方便地对数据库进行操作。下来通过一个案例,来说明如何使用 SQLiteOpenHelper 辅助类来实现 SQLite 数据库的应用。

9.2.3 SQLite 应用案例

【案例 9.2】 对某商家顾客信息进行管理,要求把 VIP 用户信息存储到数据库中,并且实现对数据库中信息的浏览和删除。

说明: 本案例通过 SQLiteOpenHelper 来实现对数据库进行创建及增、删、改的操作,对数据结构及其初始化用一个独立的代码文件实现。使用 ListView 对数据库的信息进行展示,并在每一条信息后增加删除按钮。每个删除按钮对对应的一条数据记录做删除操作。

开发步骤及解析: 过程如下。

1) 数据库设计

在实际项目中需要保存到数据库中的信息不少,在此只设计与日志相关的数据表。在该项目中,数据库命名为 customerserve,数据表命名为 user_info。数据表的结构见表 9-6。

表 9-6 user_info 表结构及说明

列　名	类　型	默认值	是否可以为空	说　　明
u_id	integer	0	否	用户 ID,主键,自增
u_name	varchar	“”	否	用户姓名
u_age	integer	0	否	用户年龄
u_sex	integer	true	否	用户性别
buy_credits	long	0l	否	用户积分
buy_totals	float	0.0f	否	用户总消费金额
update_time	varchar	“”	否	更新时间
u_phone	varchar	“”	可为空,也可不为	用户电话号码
u_password	varcjar	“”	空	用户密码

2) 创建项目

在 Android Studio 中创建一个名为 Activity_SQLiteDatabase 的项目。其包名为

ee.example.activity_sqlitedatabases。

3）准备颜色资源

编写 res/values 目录下的 colors.xml 文件，分别声明 black、white、grey、darkgrey、red 和 blue 这些颜色。在前面案例中已经多次展示过 colors.xml 的代码，在此就不赘述了。

4）准备图片和背景资源

将 back.jpg 作为背景图片资源复制到本项目的 res/drawable 目录中。在该目录中，还使用 XML 文件定义了文本输入框的背景样式和 text_cursor.png。其中，edittext_selector.xml 声明文本框各种状态的样式，shape_edit_focus.xml 声明文本框获取焦点时的样式，shape_edit_normal.xml 声明文本框失去焦点后（即通常状态）的样式。

edittext_selector.xml 代码如下。

```
1    <?xml version = "1.0" encoding = "utf-8"?>
2    <selector xmlns:android = "http://schemas.android.com/apk/res/android">
3    <item
4        android:state_focused = "true"
5        android:drawable = "@drawable/shape_edit_focus"/>
6    <item
7        android:drawable = "@drawable/shape_edit_normal"/>
8    </selector>
```

第 4 行声明文本框获得焦点状态，第 5 行声明用 shape_edit_focus.xml 来定义样式。第 7 行声明用 shape_edit_normal.xml 来定义文本框其他状态时的样式。

shape_edit_focus.xml 代码如下。

```
1    <?xml version = "1.0" encoding = "utf-8"?>
2    <shape xmlns:android = "http://schemas.android.com/apk/res/android" >
3        <solid android:color = "@color/white" />
4        <stroke
5            android:width = "1dp"
6            android:color = "@color/blue" />
7        <corners
8            android:bottomLeftRadius = "5dp"
9            android:bottomRightRadius = "5dp"
10           android:topLeftRadius = "5dp"
11           android:topRightRadius = "5dp" />
12       <padding
13           android:bottom = "2dp"
14           android:left = "10dp"
15           android:right = "10dp"
16           android:top = "2dp" />
17   </shape>
```

第 2～17 行，声明一个圆角矩形，常用的属性有：solid 设置矩形框内的填充属性，stroke 设置边框属性，corners 用于设置圆角半径，padding 用于设置矩形框内的内容与四周边框的间距。

shape_edit_normal.xml 文件代码与 shape_edit_focus.xml 几乎一样，只是在第 6 行代码中设置的颜色是深灰色（android:color="@color/darkgrey" />），在此不赘述。

5）设计布局

本项目至少有三个 Activity。activity_main.xml 设计主 Activity，activity_sqlite_write.xml 设计写入数据库 Activity，activity_sqlite_read.xml 设计浏览和删除数据库信息的 Activity。

activity_main.xml，是应用被启动后的第一个可见 Activity（即主 Activity）的布局文件。在其中只声明了两个按钮，一个是写入数据库按钮，其 ID 为 btn_sqlite_write，另一个是读取数据库按钮，其 ID 为 btn_sqlite_read。

activity_sqlite_write.xml，是向数据库中写入信息的 Activity 布局文件。该布局文件的几点说明如下。

（1）在该布局中，其根布局要设置可获取焦点操作和通过触屏操作可获取焦点属性，设置这两个属性代码如下。

```
<LinearLayout xmlns:android="http://schemas.android.com/apk/res/android"
    ...
    android:focusable="true"
    android:focusableInTouchMode="true"
    ...>
```

（2）其中声明了 4 个 EditText 输入框：ID 为 et_name 的输入框用于输入姓名，ID 为 et_age 的输入框用于输入年龄，ID 为 et_credits 的输入框用于输入积分，ID 为 et_totals 的输入框用于输入购买总额数据。为了保证输入数据的有效性，这些输入框使用 android:inputType 属性来控制输入信息的类型，常用的有：text 表示只能输入文本字符，number 表示只能输入整数，numberDecimal 表示只能输入可以包含小数的数字，等等。

（3）输入性别部分使用了 Spinner 控件，其 ID 为 sp_bemale，提供男、女性别选项。设置 Spinner 的模式 spinnerMode 的值为 dialog。该模式有两个值：dialog（弹窗显示）和 dropdown（下拉显示），当 spinnerMode 的值是 dialog 时，弹出的对话框是下拉列表式的提示；如果 spinnerMode 的值是 dropdown 时不能立即有显示效果，必须设置下拉列表框的长度和宽度，使用引用数据源 ArrayAdapter，添加适配器监听才能正常显示，实现起来相对复杂，参见案例 6.2。如果只是 2～3 个选项，建议使用 dialog 模式。

该 Spinner 控件的选中状态和下拉状态样式由 item_select.xml 和 item_dropdown.xml 文件定义。

在 res/layout 目录下创建 item_select.xml 文件，编写代码如下。

```
<TextView xmlns:android="http://schemas.android.com/apk/res/android"
    android:id="@+id/tv_name"
    android:layout_width="match_parent"
    android:layout_height="wrap_content"
    android:singleLine="true"
    android:gravity="center"
    android:textSize="17sp"
    android:textColor="@color/blue" />
```

在 res/layout 目录下创建 item_dropdown.xml 文件，编写代码如下。

```
<TextView xmlns:android="http://schemas.android.com/apk/res/android"
    android:id="@+id/tv_name"
```

```
        android:layout_width = "match_parent"
        android:layout_height = "40dp"
        android:singleLine = "true"
        android:gravity = "center"
        android:textSize = "17sp"
        android:textColor = "@color/red" />
```

(4) 最后添加一个按钮，ID 为 btn_save，用于保存到数据库操作的触发控件。

activity_sqlite_read. xml，从数据库中读取信息并显示对于每一条信息可做删除操作的 Activity 布局文件。该布局文件的几点说明如下。

(1) 该布局的根布局也要设置可获取焦点操作和通过触屏操作可获取焦点属性。

(2) 该布局中只添加一个 ListView 列表控件，其 ID 为 ListView02。用于展示数据库中的所有记录信息。而 ListView 中的每一项由 sqlite_readlistitem. xml 文件给出布局。在 res/layout 目录下创建 sqlite_readlistitem. xml 文件，编写代码如下。

```
<?xml version = "1.0" encoding = "utf - 8"?>
< LinearLayout xmlns:android = "http://schemas.android.com/apk/res/android"
    android:orientation = "horizontal"
    android:padding = "10dp"
    android:layout_width = "match_parent"
    android:layout_height = "match_parent">

< TextView android:id = "@ + id/iteminfo"
    android:layout_width = "wrap_content"
    android:layout_height = "wrap_content"
    android:textColor = "@color/blue"
    android:layout_weight = "1"
    android:textSize = "13sp" />
< Button
    android:id = "@ + id/itembtn"
    android:layout_width = "52dp"
    android:layout_height = "wrap_content"
    android:text = "删除"
    android:layout_gravity = "center"
    android:gravity = "end"
    />
</LinearLayout >
```

6) 开发数据库管理操纵代码

本案例是数据库的应用，需要开发与数据库管理和操纵相关的类。这里需要编写一个类来描述数据结构，编写一个类来管理对数据库表的操纵方法，编写一个工具类来规范数据库中的各种数据格式的处理方法。

开发数据表结构的类代码。在项目包中创建一个 bean 目录，在其下创建一个名为 UserInfo. java 的代码文件。该代码文件主要是定义一个数据库表的数据结构类，并在构造方法中对数据进行初始化，代码如下。

```
package ee.example.activity_sqlitedatabase.bean;

public class UserInfo {
    public int userid;
```

```
5         public String name;
6         public int age;
7         public boolean sex;
8         public long credits;
9         public float totals;
10        public String update_time;
11        public String phone;
12        public String password;
13
14        public UserInfo() {
15            userid = 0;
16            name = "";
17            age = 0;
18            sex = true;
19            credits = 0l;
20            totals = 0.0f;
21            update_time = "";
22            phone = "";
23            password = "";
24        }
25    }
```

第 19 行为 credits 赋初值为 0，最后的“l”表示是长整型。第 20 行为 totals 赋初值为 0.0，最后的“f”表示是浮点型。

开发对数据库表进行管理的类代码。在项目包中创建一个 database 目录，创建并编写数据库辅助类子类代码文件 MyDBHelper.java，代码如下。

```
package ee.example.activity_sqlitedatabase.database;

import android.content.ContentValues;
import android.content.Context;
import android.database.Cursor;
import android.database.sqlite.SQLiteDatabase;
import android.database.sqlite.SQLiteOpenHelper;
import android.util.Log;

import java.util.ArrayList;

import ee.example.activity_sqlitedatabase.bean.UserInfo;

public class MyDBHelper extends SQLiteOpenHelper {
    private static final String TAG = "MyDBHelper";
    private static final String DB_NAME = "customerserve.db";
    private static final int DB_VERSION = 1;
    private static MyDBHelper mHelper = null;
    private SQLiteDatabase mDB = null;
    private static final String TABLE_NAME = "user_info";

    private MyDBHelper(Context context) {
        super(context, DB_NAME, null, DB_VERSION);
    }

    private MyDBHelper(Context context, int version) {
```

```
            super(context, DB_NAME, null, version);
        }

        public static MyDBHelper getInstance(Context context, int version) {
            if (version > 0 && mHelper == null) {
                mHelper = new MyDBHelper(context, version);
            } else if (mHelper == null) {
                mHelper = new MyDBHelper(context);
            }
            return mHelper;
        }

        public SQLiteDatabase openReadLink() {
            if (mDB == null || mDB.isOpen() != true) {
                mDB = mHelper.getReadableDatabase();
            }
            return mDB;
        }

        public SQLiteDatabase openWriteLink() {
            if (mDB == null || mDB.isOpen() != true) {
                mDB = mHelper.getWritableDatabase();
            }
            return mDB;
        }

        public void closeLink() {
            if (mDB != null && mDB.isOpen() == true) {
                mDB.close();
                mDB = null;
            }
        }

        public String getDBName() {
            if (mHelper != null) {
                return mHelper.getDatabaseName();
            } else {
                return DB_NAME;
            }
        }

        @Override
        public void onCreate(SQLiteDatabase db) {
            Log.d(TAG, "onCreate");
            String drop_sql = "DROP TABLE IF EXISTS " + TABLE_NAME + ";";
            Log.d(TAG, "drop_sql:" + drop_sql);
            db.execSQL(drop_sql);
            String create_sql = "CREATE TABLE IF NOT EXISTS " + TABLE_NAME + " ("
                + "u_id INTEGER PRIMARY KEY AUTOINCREMENT NOT NULL,"
                + "u_name VARCHAR NOT NULL,"
                + "u_age INTEGER NOT NULL,"
                + "u_sex INTEGER NOT NULL,"
                + "buy_credits LONG NOT NULL,"
                + "buy_totals FLOAT NOT NULL,"
```

```
                + "update_time VARCHAR NOT NULL,"
                + "u_phone VARCHAR,"
                + "u_password VARCHAR"
                + ");";
        Log.d(TAG, "create_sql:" + create_sql);
        db.execSQL(create_sql);
    }

    @Override
    public void onUpgrade(SQLiteDatabase db, int oldVersion, int newVersion) {
        …//写版本更新时数据库更新代码
    }

    public int delete(String whereClause,String[] whereArgs) {
        int count = mDB.delete(TABLE_NAME, whereClause, whereArgs);
        return count;
    }

    public int deleteAll() {
        int count = mDB.delete(TABLE_NAME, "1=1", null);
        return count;
    }

    public long insert(UserInfo info) {
        ArrayList<UserInfo> infoArray = new ArrayList<UserInfo>();
        infoArray.add(info);
        return insert(infoArray);
    }

    public long insert(ArrayList<UserInfo> infoArray) {
        long result = -1;
        for (int i = 0; i < infoArray.size(); i++) {
            UserInfo info = infoArray.get(i);
            ArrayList<UserInfo> tempArray = new ArrayList<UserInfo>();
            //如果存在同名记录,则更新记录
            if (info.name!= null && info.name.length()> 0) {
                String condition = String.format("u_name='%s'", info.name);
                tempArray = query(condition);
                if (tempArray.size() > 0) {
                    update(info, condition);
                    result = tempArray.get(0).userid;
                    continue;
                }
            }
            //不存在唯一性重复的记录,则插入新记录
            ContentValues cv = new ContentValues();
            cv.put("u_name", info.name);
            cv.put("u_age", info.age);
            cv.put("u_sex", info.sex);
            cv.put("buy_credits", info.credits);
            cv.put("buy_totals", info.totals);
            cv.put("update_time", info.update_time);
            cv.put("u_phone", info.phone);
            cv.put("u_password", info.password);
```

```
                result = mDB.insert(TABLE_NAME, "", cv);
                //添加成功后返回行号,失败后返回 - 1
                if (result == -1) {
                    return result;
                }
            }
            return result;
        }

        public int update(UserInfo info, String condition) {
            ContentValues cv = new ContentValues();
            cv.put("u_name", info.name);
            cv.put("u_age", info.age);
            cv.put("u_sex", info.sex);
            cv.put("buy_credits", info.credits);
            cv.put("buy_totals", info.totals);
            cv.put("update_time", info.update_time);
            cv.put("u_phone", info.phone);
            cv.put("u_password", info.password);
            int count = mDB.update(TABLE_NAME, cv, condition, null);
            return count;
        }

        public ArrayList<UserInfo> query(String condition) {
            String sql = String.format("select u_id,u_name,u_age,u_sex,buy_credits,buy_
                         totals,update_time," + "u_phone,u_password from %s where %s;",
                         TABLE_NAME, condition);
            Log.d(TAG, "query sql: " + sql);
            ArrayList<UserInfo> infoArray = new ArrayList<UserInfo>();
            Cursor cursor = mDB.rawQuery(sql, null);
            if (cursor.moveToFirst()) {
                for (;; cursor.moveToNext()) {
                    UserInfo info = new UserInfo();
                    info.userid = cursor.getInt(0);
                    info.name = cursor.getString(1);
                    info.age = cursor.getInt(2);
                    //SQLite没有布尔型,用0表示false,用1表示true
                    info.sex = (cursor.getInt(3) == 0)?false:true;
                    info.credits = cursor.getLong(4);
                    info.totals = cursor.getFloat(5);
                    info.update_time = cursor.getString(6);
                    info.phone = cursor.getString(7);
                    info.password = cursor.getString(8);
                    infoArray.add(info);
                    if (cursor.isLast() == true) {
                        break;
                    }
                }
            }
            cursor.close();
            return infoArray;
        }
    }
```

(1) 第 14～20 行以静态常量的方式定义数据库的 TAG、名称、表名、版本及数据库对象信息等。

(2) MyDBHelper 类继承自 SQLiteOpenHelper 类。第 22～24 行和第 26～28 行分别定义 MyDBHelper 类的两个构造方法。

(3) 第 30～37 行定义获取 MyDBHelper 类实例的方法。

(4) 第 39～44 行定义以只读权限打开数据库的方法。第 46～51 行定义以可写权限打开数据库的方法。第 53～58 行定义关闭数据库的方法。第 60～66 行定义获取数据库名的方法。

(5) 第 68～77 行定义 MyDBHelper 类的 onCreate()方法，在此方法中创建数据库表结构。其中，第 71 行执行在数据库中创建新表前先删除同名表，以确保所创建表的唯一性。

(6) 第 79～82 行定义 onUpgrade()写当数据库版本更新时需要执行的数据库更新代码。当数据库需要升级时，会调用这个方法。应该使用这个方法来实现删除表、添加表或者做一些需要升级新的策略版本等操作。如果没有任何操作，也需要重写该方法，否则会抛出异常。

(7) 第 84～87 行定义对数据表删除满足 where 子句的记录操作的方法，返回值为删除的记录条数。如果返回值为 0，则表示没有符合 where 子句条件的记录，删除记录数为 0。

(8) 第 89～92 行定义删除所有记录的方法。其中第 91 行中的“1＝1”表示 where 子句条件恒成立。

(9) 第 94～98 行定义插入数据库记录方法，其方法的参数是一个 UserInfo 类定义的对象，其对象的值由一组 ArrayList 值 infoArray 传入，而 insert(infoArray)方法由第 100～132 行定义。在这个方法中，如果检测到输入的记录的名字与数据库表的名字相同，则定位该记录做修改操作，第 110 行调用 update()方法对记录进行更新修改，更新操作的方法由第 134～146 行定义；否则插入新记录。第 116～124 行创建一个 ContentValues 的对象 cv，并逐一向 cv 中压入新记录的键-值对序列，第 125 行调用 SQLiteDatebase 的 insert()方法，把新记录插入数据库中。

(10) 第 148～174 行定义对数据库的查询方法。第 152 行创建一个游标对象 cursor，并将查询结果集保存到 cursor 中。第 154～170 行定义的循环，从游标 cursor 对象的第一行，一条一条地写入 infoArray 数组列表中。在 SQLite 中是没有布尔类型的，第 160 行定义用数字 0 表示 false，用 1 表示 true。

在这个 MyDBHelper 类中，包含了本项目对数据库的所有管理及操纵的方法。

开发格式化并获取当前日期时间数据的类代码。在项目包中创建一个 util 目录，创建并编写代码文件 DateUtil.java，代码如下。

```
package ee.example.activity_sqlitedatabase.util;

import java.text.SimpleDateFormat;
import java.util.Date;

public class DateUtil {
    public static String getNowDateTime(String formatStr) {
        String format = formatStr;
        if (format == null || format.length()<= 0) {
```

```
10                  format = "yyyy-MM-dd HH:mm:ss";
11              }
12              SimpleDateFormat s_format = new SimpleDateFormat(format);
13              return s_format.format(new Date());
14          }
15
16          public static String getNowTime() {
17              SimpleDateFormat s_format = new SimpleDateFormat("HH:mm:ss");
18              return s_format.format(new Date());
19          }
20      }
```

在 Java 中用 SimpleDateFormat 对当前日期进行格式化,第 12、17 行指定日期时间数据的格式。new Date()获取本机系统的日期时间。

7) 开发 Activity 类实现代码

本案例有三个 Activity,必须有三个 Java 实现代码。

MainActivity.java,实现主 Activity 类的代码。打开本案例包下的 MainActivity.java 文件,并进行编辑,代码如下。

```
package ee.example.activity_sqlitedatabase;

import androidx.appcompat.app.AppCompatActivity;
import android.content.Intent;
import android.os.Bundle;
import android.view.View;

public class MainActivity extends AppCompatActivity {

    @Override
    protected void onCreate(Bundle savedInstanceState) {
        super.onCreate(savedInstanceState);
        setContentView(R.layout.activity_main);
        findViewById(R.id.btn_sqlite_write).setOnClickListener(new BtnOnClick());
        findViewById(R.id.btn_sqlite_read).setOnClickListener(new BtnOnClick());
    }

    private class BtnOnClick implements View.OnClickListener{
        @Override
        public void onClick(View v) {
            if (v.getId() == R.id.btn_sqlite_write) {
                Intent intent = new Intent(MainActivity.this, SQLiteWriteActivity.class);
                startActivity(intent);
            } else if (v.getId() == R.id.btn_sqlite_read) {
                Intent intent = new Intent(MainActivity.this,SQLiteReadActivity.class);
                startActivity(intent);
            }
        }
    }
}
```

首页只有两个按钮:一个是跳转到向数据库写入信息的 SQLiteWriteActivity,另一个是跳转到从数据库中读取信息的 SQLiteReadActivity。

SQLiteWriteActivity.java，实现向数据库写入信息类的代码。在项目包下创建并编写SQLiteWriteActivity.java文件，代码如下。

```
package ee.example.activity_sqlitedatabase;

import androidx.appcompat.app.AppCompatActivity;
import android.os.Bundle;
import android.view.View;
import android.view.View.OnClickListener;
import android.widget.AdapterView;
import android.widget.AdapterView.OnItemSelectedListener;
import android.widget.ArrayAdapter;
import android.widget.EditText;
import android.widget.Spinner;
import android.widget.Toast;

import ee.example.activity_sqlitedatabase.bean.UserInfo;
import ee.example.activity_sqlitedatabase.database.MyDBHelper;
import ee.example.activity_sqlitedatabase.util.DateUtil;

public class SQLiteWriteActivity extends AppCompatActivity implements OnClickListener {

    private MyDBHelper mHelper;
    private EditText et_name;
    private EditText et_age;
    private EditText et_credits;
    private EditText et_totals;
    private boolean bemale = true;

    @Override
    protected void onCreate(Bundle savedInstanceState) {
        super.onCreate(savedInstanceState);
        setContentView(R.layout.activity_sqlite_write);
        et_name = (EditText) findViewById(R.id.et_name);
        et_age = (EditText) findViewById(R.id.et_age);
        et_credits = (EditText) findViewById(R.id.et_credits);
        et_totals = (EditText) findViewById(R.id.et_totals);
        findViewById(R.id.btn_save).setOnClickListener(this);

        ArrayAdapter<String> typeAdapter = new ArrayAdapter<String>(this,
                                                R.layout.item_select, typeArray);
        typeAdapter.setDropDownViewResource(R.layout.item_dropdown);
        Spinner sp_bemale = (Spinner) findViewById(R.id.sp_bemale);
        sp_bemale.setPrompt("请选择性别");
        sp_bemale.setAdapter(typeAdapter);
        sp_bemale.setSelection(0);
        sp_bemale.setOnItemSelectedListener(new TypeSelectedListener());
    }

    private String[] typeArray = {"男", "女"};
    class TypeSelectedListener implements OnItemSelectedListener {
        public void onItemSelected(AdapterView<?> arg0, View arg1, int arg2, long arg3) {
            bemale = (arg2 == 0)?true:false;
        }
```

```
            public void onNothingSelected(AdapterView<?> arg0) {
            }
        }

        @Override
        protected void onStart() {
            super.onStart();
            mHelper = MyDBHelper.getInstance(this, 2);
            mHelper.openWriteLink();
        }

        @Override
        protected void onStop() {
            super.onStop();
            mHelper.closeLink();
        }

        @Override
        public void onClick(View v) {
            if (v.getId() == R.id.btn_save) {
                String name = et_name.getText().toString();
                String age = et_age.getText().toString();
                String credits = et_credits.getText().toString();
                String totals = et_totals.getText().toString();
                if (name == null || name.length()<= 0) {
                    showToast("请先填写姓名");
                    return;
                }
                if (age == null || age.length()<= 0) {
                    showToast("请先填写年龄");
                    return;
                }
                if (credits == null || credits.length()<= 0) {
                    showToast("请先填写积分");
                    return;
                }
                if (totals == null || totals.length()<= 0) {
                    showToast("请先填写购买总额");
                    return;
                }

                UserInfo info = new UserInfo();
                info.name = name;
                info.age = Integer.parseInt(age);
                info.sex = bemale;
                info.credits = Long.parseLong(credits);
                info.totals = Float.parseFloat(totals);
                info.update_time = DateUtil.getNowDateTime("yyyy-MM-dd HH:mm:ss");
                mHelper.insert(info);
                showToast("数据已写入 SQLite 数据库");
            }
        }

    private void showToast(String desc) {
```

```
105            Toast.makeText(getApplicationContext(), desc, Toast.LENGTH_SHORT).show();
106        }
107    }
```

(1) 第 37～44 行创建一个 Spinner 对象 sp_bemale，并绑定适配器 typeAdapter。typeAdapter 的默认显示布局和下拉项的布局分别由 item_select.xml 和 item_dropdown.xml 定义。第 42 行定义 sp_bemale 的默认显示内容为下拉选项中的第 1 项。

(2) 第 46～53 行定义适配器的接口监听器。第 49 行定义当选择项是第 1 项(即 arg2==0)时，bemale 的值为 true，否则为 false。这就意味着选择第 1 项为男，选择第 2 项为女，这正是我们要的结果。

(3) 第 55～60 行重写类的 onStart()方法，第 62～66 行重写类的 onStop()方法，它们分别用于以可写方式打开数据库和关闭数据库。

(4) 第 69～102 行定义了按钮的 onClick()方法。在该方法中监听保存按钮是否被按下。如果被按下，则先检验姓名、年龄、积分和购买总额是否为空？若为空，则等待输入并给出提示；若不为空，则将每个输入框的输入值赋值给 UserInfo 类对象 info，第 99 行实现把 info 中的内容保存到数据库中。

SQLiteReadActivity.java，实现读取数据库中信息，并删除某些信息类的代码。在项目包下创建并编写 SQLiteReadActivity.java 文件，代码如下。

```
package ee.example.activity_sqlitedatabase;

import androidx.appcompat.app.AppCompatActivity;
import android.content.Context;
import android.os.Bundle;
import android.view.LayoutInflater;
import android.view.View;
import android.view.ViewGroup;
import android.widget.BaseAdapter;
import android.widget.Button;
import android.widget.ListView;
import android.widget.TextView;
import android.widget.Toast;

import ee.example.activity_sqlitedatabase.bean.UserInfo;
import ee.example.activity_sqlitedatabase.database.MyDBHelper;

import java.util.ArrayList;
import java.util.HashMap;

public class SQLiteReadActivity extends AppCompatActivity {

    private MyDBHelper myHelper;
    private ListView listv;

    @Override
    protected void onCreate(Bundle savedInstanceState) {
        super.onCreate(savedInstanceState);
        setContentView(R.layout.activity_sqlite_read);
        listv = (ListView) this.findViewById(R.id.ListView02); //初始化 ListView
```

```java
        }

        @Override
        protected void onStart() {
            super.onStart();

            myHelper = MyDBHelper.getInstance(this, 2);
            myHelper.openReadLink();
            if ((readSQLite() == null)|(readSQLite().size()<=0)) {
                showToast("数据库为空.");
                finish();
            } else {
                MyAdapter BAdapter = new MyAdapter(this);  //得到一个 MyAdapter 对象
                listv.setAdapter(BAdapter); //为 ListView 绑定 Adapter
            }
        }

        @Override
        protected void onStop() {
            super.onStop();
            myHelper.closeLink();
        }

        private void showToast(String desc) {
            Toast.makeText(this, desc, Toast.LENGTH_SHORT).show();
        }

        /** 读取数据库数据的方法 **/
        protected ArrayList<UserInfo> readSQLite() {
            if (myHelper == null) {
                showToast("数据库连接为空.");
                return null;
            }
            ArrayList<UserInfo> userArray = myHelper.query("1=1");
            return userArray;
        }

        /** 获得数据的方法 **/
        private ArrayList<HashMap<String, String>> getData() {
            ArrayList<HashMap<String, String>> listItem = new ArrayList<HashMap<String, String>>();
            String desc;
            ArrayList<UserInfo> uArray = readSQLite();
            setTitle(String.format("数据库查询到%d条记录,详情如下:", uArray.size()));
            /** 为动态数组添加数据 **/
            for (int i = 0; i < uArray.size(); i++) {
                HashMap<String, String> tvitem = new HashMap<String, String>();
                desc = "";
                UserInfo info = uArray.get(i);
                desc = String.format("第%d个用户:", i + 1);
                desc = String.format("%s\n  姓名:%s", desc, info.name);
                desc = String.format("%s\n  年龄:%d", desc, info.age);
                if (info.sex)
                    desc = String.format("%s\n  性别:男", desc);
```

```
                else
                    desc = String.format("%s\n 性别:女", desc);
                desc = String.format("%s\n 积分:%d", desc, info.credits);
                desc = String.format("%s\n 购买总额:%8.2f", desc, info.totals);
                desc = String.format("%s\n 更新时间:%s", desc, info.update_time);
                tvitem.put("ItemId", String.valueOf(info.userid));
                tvitem.put("ItemText", desc);
                listItem.add(tvitem);
            }
            return listItem;
        }

        /** 定义 ViewHolder 类,存放控件 **/
        private final class ViewHolder {
            private TextView text;
            private Button btn;
    }

    /** 定义适配器 **/
    private class MyAdapter extends BaseAdapter {
        private LayoutInflater mInflater;   //声明一个 LayoutInflater 对象用来导入布局
        private ArrayList<HashMap<String, String>> datas; //声明 HashMap 类型的数组

        /** 构造函数 **/
        public MyAdapter(Context context) {
            this.mInflater = LayoutInflater.from(context);
            this.datas = getData();          //为 HashMap 类型的数组赋初值
        }

        @Override
        public int getCount() {
            return datas.size();             //返回数组的长度
        }

        @Override
        public Object getItem(int position) {
            return null;
        }

        @Override
        public long getItemId(int position) {
            return 0;
        }

        /** 动态生成每个下拉项对应的 View,每个下拉项由一个 TextView 和一个按钮组成 **/
        @Override
        public View getView(int position, View convertView, ViewGroup parent) {
            ViewHolder holder;

            if (convertView == null) {
                convertView = mInflater.inflate(R.layout.sqlite_readlistitem, null);
                holder = new ViewHolder();
                /** 得到各个控件的对象 **/
                holder.text = (TextView) convertView.findViewById(R.id.iteminfo);
```

```
138                 holder.btn = (Button) convertView.findViewById(R.id.itembtn);
139                 convertView.setTag(holder);                        //绑定 ViewHolder 对象
140             } else {
141                 holder = (ViewHolder) convertView.getTag(); //取出 ViewHolder 对象
142             }
143 
144             HashMap< String, String > map = datas.get(position);
145 
146             /** 设置 TextView 显示的内容,即存放在动态数组中的数据 **/
147             holder.text.setText(map.get("ItemText"));
148             /** 为 Button 添加单击事件 **/
149             holder.btn.setFocusable(false);                        //释放焦点
150             holder.btn.setTag(String.valueOf(position));
151             holder.btn.setOnClickListener(
152                 new View.OnClickListener() {
153                     @Override
154                     public void onClick(View v) {
155                         int position = Integer.parseInt((String) v.getTag());
156                         String[] whereArgs = {datas.get(position).get("ItemId")};
157                         myHelper.closeLink();
158                         myHelper.openWriteLink();
159                         myHelper.delete("u_id = ?", whereArgs);
160                         /** 重新读取数据库 **/
161                         myHelper.closeLink();
162                         myHelper.openReadLink();
163                         datas = getData();                        //重新获取数据
164                         notifyDataSetChanged();                   //刷新 ListView
165                     }
166             });
167             return convertView;
168         }
169     }
170 }
```

(1) 第 34～46 行重写 onStart()方法,在该方法中以只读权限打开数据库,如果数据库为空给出提示,并返回到上级 Activity。当数据库不为空时创建一个自定义适配器对象 BAdapter,并将 ListView 绑定该对象。其中第 39 行中的 readSQLite()是一个自定义的读取数据库数据的方法,该方法定义在第 59～66 行,主要实现查询数据库中的全部记录,并返回到一个列表数组中,如果数据库为空则返回 null。

(2) 第 69～94 行定义 getData()方法,在该方法中首先回调 readSQLite()方法,获取一个列表数组(即数据库中的每条记录都保存在一个列表中)。第 79～88 行将记录的其他信息合并到 desc 字符串变量中,第 89 行将记录的用户 ID 压入键-值对 ItemId,第 90 行将 desc 的值压入键-值对 ItemText 中,第 91 行将保存两个键-值对的 tvitem 添加到列表项中。这样循环往复,直到数据库记录列表数组 uArray 取完为止。

(3) 第 97～100 行定义了 ViewHolder 类,用于存放 ListView 中每一项的控件构成,如此做可以缓存显示数据的视图(View),加快用户界面(UI)的响应速度。

(4) 第 103～169 行定义适配器类 MyAdapter 子类,MyAdapter 继承自 BaseAdapter 类,在其构造方法中为 HashMap 类型数组 datas 赋初值,此时调用了 getData()方法。作为一个适配器类,都会重写 getCount()、getItem()、getItemId()和 getView()四种方法。通常重点放在

getView()的编写中，当适配器的内容改变时需要强制调用 getView()来刷新每个 Item 的内容。

(5) 第 133～142 行从 sqlite_readlistitem. xml 布局文件中取得 listView 的每一项布局，然后将布局中每一控件绑定到 ViewHolder 对象上，从而实现对这些控件的监听。

(6) 第 149 行从按钮中释放焦点，这样，就可以把焦点先放在 ListView 上。否则用户将无法对整个 ListView 进行单击操作。

(7) 第 151～166 行实现对每一项中按钮的监听编程，当监听到按钮被按下后，第 156 行执行定位数据库的记录，第 157～158 行以可写权限重新打开数据库，第 159 行回调 myHelper 的 delete()方法删除指定记录，第 161～162 行以只读权限重新打开数据库，第 163 行重新获取 datas 数据，第 164 行刷新 ListView 的显示列表内容。

运行结果： 在 Android Studio 支持的模拟器上，运行 Activity_SQLiteDatabase 项目，运行结果如图 9-4 所示。

(a) 初始运行界面

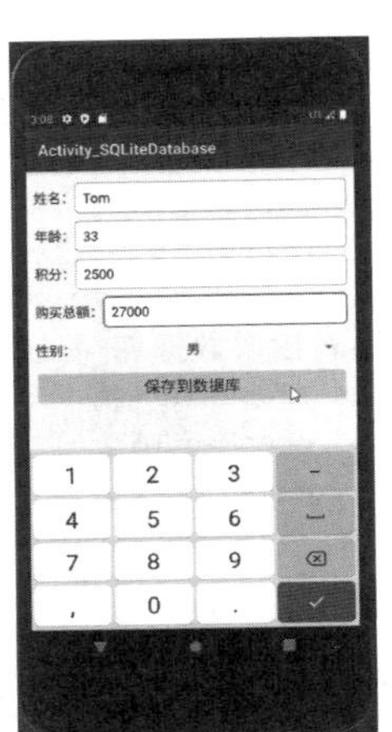

(b) “写入数据库”运行界面

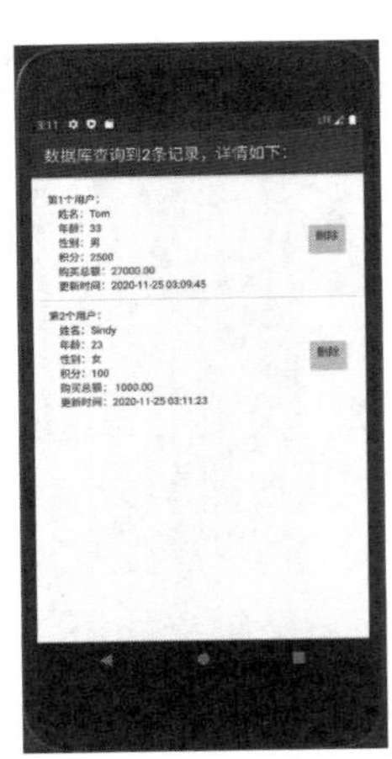

(c) “读取数据库”运行界面

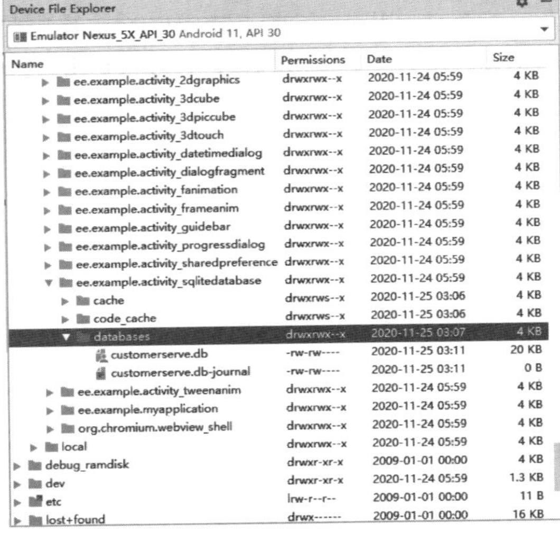

(d) 在Device File Explorer窗格内查看数据库文件

图 9-4 SQLite 应用案例运行效果

9.3　访问 SD 卡简介

手机的存储空间分为两部分，一部分称为本机内存，另一部分称为外部存储。外部存储空间是使用 SD 卡来扩展存储空间的，SD 卡（Secure Digital Memory Card，安全数码卡）是一种基于半导体快闪记忆器的新一代记忆便携式设备，拥有高记忆容量、快速数据传输率、极大的移动灵活性以及很好的安全性。早期的 SD 卡是可拔插的存储芯片，后来多数手机把 SD 卡固化在手机内部，但仍然称之为外部存储空间。

9.3.1　访问 SD 卡常用的方法及常量

在应用编程中，Android 通过 Environment 类来访问 SD 卡上的信息。Environment 类位于 android. os. Environment 包中，是专门提供访问环境变量的类。在应用编程中，Environment 类的常用方法见表 9-7。

表 9-7　Environment 类的常用方法及说明

方　　法	说　　明
getDataDirectory()	返回 File，获取 Android 数据目录
getDownloadCacheDirectory()	返回 File，获取 Android 下载/缓存内容目录
getExternalStorageDirectory()	返回 File，获取外部存储目录即 SD 卡目录，Android 10(API 29)废弃
getExternalStoragePublicDirectory(String type)	返回 File，取一个高端的公用的 SD 卡指定类型目录的路径，Android 10(API 29)废弃
getExternalStorageState()	返回 File，获取 SD 卡的当前状态，如是否存在、是否可读写
getRootDirectory()	返回 File，获取 Android 的根目录
getExternalStorageDirectory(). getFreeSpace()	判断 SD 卡剩余可用空间，Android 10(API 29)废弃

Environment 类提供访问环境变量，包含了大量的系统常量。getExternalStorageState()是获取 SD 卡存储状态的方法，由其返回的状态常量见表 9-8。

表 9-8　SD 卡存储状态常量、取值及说明

常　量　名	常　量　值	说　　明
MEDIA_BAD_REMOVAL	bad_removal	SD 卡被卸载前已被移除
MEDIA_CHECKING	checking	正在进行磁盘检查
MEDIA_MOUNTED	mounted	SD 卡已经挂载，并具有可读/写权限
MEDIA_MOUNTED_READ_ONLY	mounted_ro	SD 卡已经挂载，并具有只读权限
MEDIA_NOFS	nofs	SD 卡为空白或正在使用不支持的文件系统
MEDIA_REMOVED	removed	SD 卡已经移除
MEDIA_SHARED	shared	SD 卡未安装，但通过 USB 共享返回
MEDIA_UNKNOWN	unknown	未知
MEDIA_UNMOUNTABLE	unmountable	SD 卡无法挂载。即 SD 卡存在但不可以被安装
MEDIA_UNMOUNTED	unmounted	SD 卡已卸掉，即 SD 卡存在但是没有被安装

getExternalStoragePublicDirectory(String type)是获取 SD 卡指定类型目录路径的方法，相关的目录类型常量见表 9-9。

表 9-9　SD 卡目录类型常量、取值及说明

常　量　名	常　量　值	说　　明
MEDIA_BAD_REMOVAL	bad_removal	SD 卡被卸载前已被移除
DIRECTORY_DCIM	DCIM	相机拍摄照片存放的标准目录
DIRECTORY_DOCUMENTS	document	文档存放的标准目录
DIRECTORY_DOWNLOADS	download	下载文件存放的标准目录
DIRECTORY_MOVIES	movies	视频存放的标准目录
DIRECTORY_MUSIC	music	音乐存放的标准目录
DIRECTORY_PICTURES	picture	图片存放的标准目录

9.3.2　访问 SD 卡权限设置

1. 访问 SD 卡权限声明

为了正常访问 SD 卡，需要在 AndroidManifest. xml 文件中声明相关的访问权限。例如，要往 SD 卡中写入数据，需要添加 WRITE_EXTERNAL_STORAGE 权限。要从 SD 卡中读取数据，需要添加 READ_EXTERNAL_STORAGE 权限。一般在 AndroidManifest. xml 文件的< application >元素前添加权限，代码片段如下。

```
<?xml version = "1.0" encoding = "utf - 8"?>
< manifest xmlns:android = "http://schemas.android.com/apk/res/android"
    package = "com.example.ee.helloandroid">
    < uses - permission android:name = "android.permission.READ_EXTERNAL_STORAGE" />
    < uses - permission android:name = "android.permission.WRITE_EXTERNAL_STORAGE" />
    < application … >
    </application >
</manifest >
```

2. 为 App 运行时授权

从 Android 6.0(API 23)以后，系统在运行时权限上增加了对用户隐私权的保护功能。Google 将权限分为两类，一类是普通权限(Normal Permissions)，这类权限不涉及用户隐私，是不需要用户进行授权的，比如手机振动、获取 WiFi 连接状态、访问网络等；另一类是危险权限(Dangerous Permission)，这类权限会涉及用户隐私，需要用户进行授权，比如获取位置信息、读取 SD 卡、访问通讯录等。

对于 Android 6.0(API 23)之前的版本，只需要在 AndroidManifest. xml 中添加权限声明即可。但是在 Android 6.0(API 23)及更高版本上，除了需要在 AndroidManifest. xml 中添加相应的访问权限声明外，还需要在应用项目 App 运行时，为每个危险权限类向系统进行请求授权。从外部存储设备中读写文件属于危险权限，因此需要在 Java 代码中使用系统提供的 API 检查并请求权限动态授权。这部分相关的权限检查和请求授权方法定义在

ActivityCompat 类中，位于 androidx. core. app. ActivityCompat 包中。常用的方法如下。

1）检查权限方法

使用方法 ActivityCompat. checkSelfPermission(android. content. Context，java. lang. String)检查是否拥有指定的权限，返回类型为 int 类型，一般用符号常量表示。如果 App 已经拥有了该权限，则返回值为 PackageManager. PERMISSION_GRANTED。如果 App 没有该权限，返回值为 PERMISSION_DENIED，此时需要向用户请求授权。

2）展示需要授权理由的方法

```
shouldShowRequestPermissionRationale(Activity activity, String permission)
```

这个方法在用户拒绝权限后返回 true。也就是说在用户第一次单击一个需要权限的地方，该方法返回 false，如果当用户选择了拒绝该权限，下次单击此权限处，该方法会返回 true。可在里面进行对该权限需要的说明，然后弹出权限让用户选择，并且对话框增加了 don't ask again 选项框。当用户选择 don't ask again 后，该方法总是返回 false，并且不再弹出 ActivityCompat. requestPermissions 对话框，系统直接拒绝权限。

3）请求权限方法

使用方法 requestPermissions(Activity activity，String[] permissions，int requestCode)为该 App 请求指定的权限。其中参数说明如下。

第二个参数 permissions 是一个 String 数组，数组元素是 App 所需要申请的所有权限。这些权限常量可以在 Manifest 类中找到。

第三个参数 requestCode 标志该次授权的唯一请求码，当用户进行授权操作后在回调方法中可以根据这个标识符区分不同的授权操作。

4）请求权限后的回调方法

```
onRequestPermissionsResult(int requestCode, String[] permissions, int[] grantResults)
```

该方法是 OnRequestPermissionsResultCallback 类中的方法，当执行了 requestPermissions() 方法后，系统会自动回调 onRequestPermissionsResult()方法。

其中，第三个参数 grantResults 是 int 类型的数组，每个元素的值分别对应 permissions 的每个请求，取值由 PackageManager 的常量表示，为 PackageManager. PERMISSION_GRANTED 或 PackageManager. PERMISSION_DENIED。

例如，我们需要对 SD 卡上的文件进行读写操作，因此需要为应用项目 App 请求外存储器读写授权，如果用户开始时拒绝了授权，则给出提示授权信息，编写代码如下。

```
private static final int REQUEST_EXTSTO = 1;
private static String[] PERMISSIONS_STORAGE = { Manifest.permission.READ_EXTERNAL_STORAGE,
                                    Manifest.permission.WRITE_EXTERNAL_STORAGE};
int permission_read = ActivityCompat.checkSelfPermission(activity,
                                    Manifest.permission.READ_EXTERNAL_STORAGE);
int permission_write = ActivityCompat.checkSelfPermission(activity,
                                    Manifest.permission.WRITE_EXTERNAL_STORAGE);
//检查是否有读外存权限
if (permission_write != PackageManager.PERMISSION_GRANTED
                        || permission_read != PackageManager.PERMISSION_GRANTED) {
    if (ActivityCompat.shouldShowRequestPermissionRationale(activity,
                                    Manifest.permission.READ_EXTERNAL_STORAGE) ||
```

```
        ActivityCompat.shouldShowRequestPermissionRationale(activity,
                                Manifest.permission.WRITE_EXTERNAL_STORAGE) ) {
         AlertDialog dialog = new AlertDialog.Builder(this)
             .setTitle("该权限保证对 SD 卡上文件进行读写操作")
             .setPositiveButton("需要此权限!", new DialogInterface.OnClickListener() {
                 @Override
                 public void onClick(DialogInterface dialog, int which) {
                     ActivityCompat.requestPermissions(MainActivity.this,
                            PERMISSIONS_STORAGE, REQUEST_EXTSTO);
                 }
             }).setNegativeButton("不授权", new DialogInterface.OnClickListener() {
                 @Override
                 public void onClick(DialogInterface dialog, int which) {
                     Toast.makeText(MainActivity.this, "重新授权", Toast.LENGTH_SHORT).show();
                 }
             }).show();
} else {
    //没有权限,需要动态申请
    ActivityCompat.requestPermissions(activity, PERMISSIONS_STORAGE, REQUEST_EXTSTO);
}
```

注意：在完成权限请求方法之后，即执行了 ActivityCompat.requestPermissions()方法后，如果有一些需要设置的操作可以在 onRequestPermissionsResult()方法内编码实现，系统会自动回调 onRequestPermissionsResult()方法。

在此需要说明的是，请求授权时会弹出 Android 系统内置的对话框，例如应用项目 Activity_SurfaceView 需要读取 SD 卡中的文件，在请求授权后会弹出访问授权对话框，如图 9-5 所示。用户不能自定义该对话框。该授权对话框一经用户同意授权，以后运行该 App 都不会再出现。建议用户选择 Allow，同意该授权。

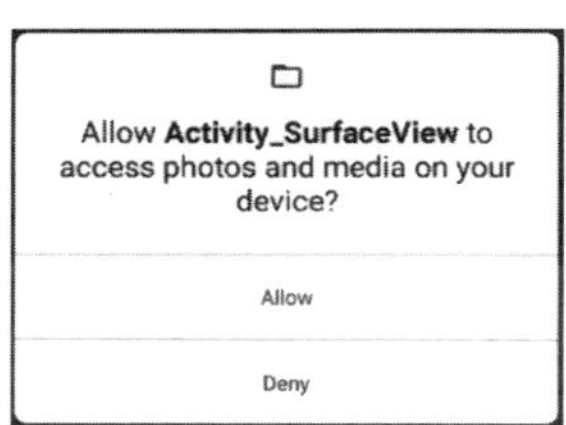

图 9-5　Activity_SurfaceView 项目的访问授权对话框

此时在 Device File Explorer 的 sdcard 目录下可以看到要访问的资源，如图 9-6 所示。当然，这个文件资源管理器必须在 Android 模拟器打开状态才可见。

9.3.3　关于 SD 卡的相关编程

在应用编程中，如果涉及 SD 卡的应用，可以使用 Environment 类中的上述方法、常量来实现。下面给出一些有关应用的代码片段。

1. 检测 SD 卡是否可用

编写一个 sdcardisavailable()方法来检测 SD 卡是否可用，该方法的代码如下。

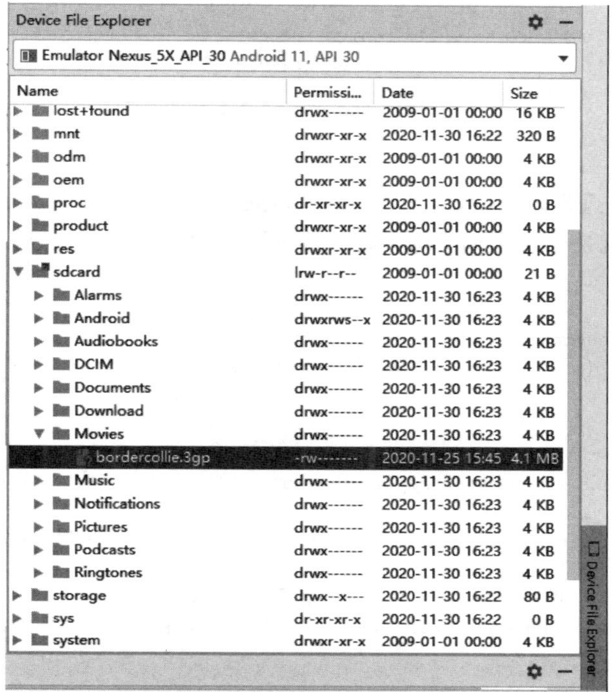

图 9-6　模拟器的设备文件管理器

```
public static boolean sdcardisavailable() {
    string status = environment.getexternalstoragestate();
    if (!status.equals(environment.media_mounted)) {
        return false;
    }
    return true;
}
```

2. 获取 SD 卡的实际空间大小

编写一个 getrealsizeonsdcard()方法来获取 SD 卡的实际空间大小，该方法的代码如下。

```
public static long getrealsizeonsdcard() {
    file path = new file(environment.getexternalstoragedirectory().getabsolutepath());
    statfs stat = new statfs(path.getpath());
    long blocksize = stat.getblocksize();
    long availableblocks = stat.getavailableblocks();
    return availableblocks * blocksize;
}
```

3. 向 SD 卡写入数据

编写一个 savetosdcard()方法来实现将数据写入 SD 卡的文件中，其中参数 filename 是文件名，content 是文件内容。

```
public void savetosdcard(String filename,String content) throws exception{
    file sdfile = new file(environment.getexternalstoragedirectory(), filename);
    outputstream out = new fileoutputstream(sdfile);
    out.write(content.getbytes());
    out.close();
}
```

对 SD 卡进行写操作时，如果不成功，有可能是当前 SD 卡没插入或 SD 卡为写保护状态。下列代码片段就是实现在向 SD 卡进行信息写入前，判断 SD 卡是否为已插入并且为可写状态。

```
string en = environment.getexternalstoragestate();
if(en.equals(environment.media_mounted)){   //获取 SD 卡状态，如果 SD 卡插入了手机且为
                                            //可写状态
    try {
        service.savetosdcard(filename, content);        //向 SD 卡中保存
        toast.maketext(getapplicationcontext(), "保存成功", 1).show();
    } catch (exception e) {
        toast.maketext(getapplicationcontext(), "保存失败", 1).show();
    }
}else{        //提示用户 SD 卡不存在或者为写保护状态
  toast.maketext(getapplicationcontext(), "SD 卡不存在或者为写保护状态", 1).show();
}
```

9.4 文件存储

可以将一些数据以文件的形式保存在设备中，例如，一些文本文件、PDF 文件、图片文件、音频文件和视频文件等。

使用文件存储，Java 的 I/O 包中的常用类也可以正常工作在 Android 平台上，例如：Java. io. BufferedReader、Java. io. BufferedWriter、Java. io. FileReader、Java. io. FileWriter、Java. io. FileOutputStream、Java. io. FileInputStream、Java. io. OutputStream、Java. io. InputStream、Java. io. OutputStreamWriter、Java. io. InputStreamReader、Java. io. PrintStream、Java. io. PrintWriter、Java. io. FileNotFoundException 和 Java. io. IOException 等。

在 Android 中，可以通过 openFileOutput()方法打开一个指定的文件，通过 load()方法来获取文件中的数据，通过 deleteFile()方法来删除一个指定的文件，通过 openFileInput()方法写入文件，等等。例如，使用下述文件代码段。

```
String FILE_NAME = "infofile.txt";                          //获取要操作文件的文件名
FileOutputStream fos = openFileOutput(FILE_NAME,Context.NODE_PRIVATE);
                                                            //初始化一个文件流
FileInputStream fis = openFileInput(FILE_NAME);             //创建写入文件流
```

注意：在使用上述文件流读写方法时，只能对当前应用项目所在目录下的文件进行操作，即指定的文件名中不能包含路径。如果调用 FileOutputStream 时指定的文件不存在，Android 系统会自动创建它。在默认情况下，写入操作是覆盖式的写入，如果需要把新的数据写入文件中，而不覆盖原文件内容，则要将 openFileOutput()方法的第二个参数设置为 Context. MODE_APPEND。

在 Android 中对文件的存储操作与传统的 Java 对文件实现 I/O 的程序类似，在此不做赘述。

到此为止，本章所介绍的应用项目数据存储方式，无论是使用配置文件存储还是使用数据库存储，数据都是应用项目私有的。在 Android 系统中，没有一个公共的内存区域，供多个应用共享存储数据，所以应用项目之间是相互独立的，它们分别运行在自己的进程中。而 ContentProvider 机制可以支持在多个应用项目中存储和读取数据，这个组件是解决跨应用共享数据的一种方式。

9.5 ContentProvider

ContentProvider 是所有应用项目之间数据存储和检索的桥梁，它的作用就是使得各个应用项目之间实现数据共享。在 Android 中，ContentProvider 是一种特殊的存储数据的类型，它提供了一套标准的接口用来获取、操作数据。例如音频、视频、图片和个人联系信息等几种常用的 ContentProvider。它们被定义在 android. provider 包下。通过这些定义好的 ContentProvider 可以方便地进行数据读取，也可以进行数据删除等操作。当然，执行这些操作必须拥有适当的权限，记住，要在应用项目的 AndroidManifest. xml 中把相关权限添加进去。

9.5.1 实现数据共享的相关类、接口与权限

1. ContentProvider

ContentProvider 类位于 android. content. ContentProvider 包中。一个程序可以通过实现一个 ContentProvider 的抽象方法接口将自己的数据暴露出去，外部程序可以通过一组标准的接口来和本程序内的数据打交道。无论数据的来源是什么，ContentProvider 都会认为是一种表，然后把数据组织成表格。就像前面介绍的 SQLite 数据库操作一样，所不同的是操作对象是本设备其他应用的内部数据。所以定义一个 ContentProvider 必须要实现几个抽象方法接口，这些常用抽象方法见表 9-10。

表 9-10 ContentProvider 的常用抽象方法及说明

方法名	说明
query(Uri uri, String[] projection, String selection, String[] selectionArgs, String sortOrder)	通过 URI 进行查询，返回一个 Cursor
insert(Uri uri,ContentValues values)	将一组数据插入 URI 指定的地方
delete(Uri uri,String where, String[] selectionArgs)	删除指定 URI 并且符合一定条件的数据
update(Uri uri, ContentValues values, String where, String[] selectionArgs)	更新 URI 指定位置并且符合一定条件的数据
getType(Uri uri)	获得 MIME 数据类型
onCreate()	当 ContentProvider 创建时调用

在上述方法中使用最多的是 query()方法，下面对其参数进行说明。

（1）参数 uri 为指定的 URI 地址。

（2）参数 projection 为指定的返回列名的列表。

（3）参数 selection 用于指定查询条件，确定返回的行内容。相当于 SQL 语句中的 WHERE 子句。

（4）参数 selectionArgs 指定查询条件中的参数取值的列表，对参数 selection 中出现的"?"进行替换。

（5）参数 sortOrder 指定返回结果的排序方式。

2. ContentResolver

ContentResolver 类位于 android. content. ContentResolver 包中，是内容解析器类。当外部应用项目需要对 ContentProvider 中的数据进行添加、删除、修改和查询操作时，就要通过 ContentResolver 来访问。ContentProvider 实现数据的封装，而相对应的 ContentResolver 是访问数据的接口。一般情况下，ContentProvider 是单实例的，但是可以有多个 ContentResolver 在不同的应用项目和不同的进程之间与 ContentProvider 交互。

在 Activity 代码中通过 getContentResolver() 方法来获取 ContentResolver 对象。ContentResolver 提供的方法与 ContentProvider 需要实现的方法对应，同样使用 query()、insert()、delete()、update()、getType()等方法来操作数据。这些方法的名称和参数与 ContentProvider 中的一样，在此不做赘述。

3. URI

URI 是指向数据的一个资源标识符，在 ContentProvider 中代表了要操作的数据。通过 URI 使得 ContentResolver 知道与哪一个 ContentProvider 对应，并且来操作哪些数据，示例如下。

```
content://com.wirelessqa.content.provider/person/16
```

一个 URI 由以下三部分组成。

（1）content://——scheme，是 Android 定义的一种协议。ContentProvider(内容提供者)的 scheme 由 Android 规定为 content。

（2）com. wirelessqa. content. provider——authority，用于唯一标识这个 ContentProvider 的主机名。它相当于访问万维网时使用的 URI 中的 www. XXXX. com，代表的是 ContentProvider 所在的"域名"，而且这个"域名"在 Android 中是唯一的。一般情况下，建议采用完整的包名加类名来标识这个 ContentProvider 的 authority。

（3）/person/16——路径，用来表示要操作的数据。要操作的数据不一定来自数据库，也可以是文件、XML 或网络等其他存储方式。路径的构建应根据业务而定。比如以下例子。

要操作 person 表中 ID 为 16 的记录，可以构建路径为/person/16。

要操作 person 表中 ID 为 16 的记录的 name 字段，可以构建路径为/person/16/name。

要操作 person 表中的所有记录，可以构建路径为/person。

要操作 XML 文件中 person 节点下的 name 节点，可以构建路径为/person/name。

通常 URI 比较长，在编程时容易出错，且难以理解。所以在 Android 应用开发中通常会定义辅助类，在辅助类中定义一些常量用来代替这些长字符串的 URI。对于用字符串表示的 URI，可以使用 Uri 类中的 parse()方法将字符串转换成 URI，示例如下。

```
Uri uri = Uri.parse("content://com.wirelessqa.content.provider/person");
```

在 ee.example.activity_contentprovider 包内，如果在辅助类中定义了包名为常量 AUTHORITY，这时，在实现 ContentProvider 接口的代码文件中可以按如下方法定义 URI 常量。

```
public static final Uri CONTENT_URI = Uri.parse("content://" + AUTHORITY + "/person");
```

代码中的 CONTENT_URI 就相当于：content://ee.example.activity_contentprovider/person。

4. 设置权限

需要注意的是，无论对哪一个 ContentProvider 中的数据进行操作，进行哪一类操作，都需要在 AndroidManifest.xml 文件中添加相应的权限。例如要对手机的通讯录进行查询和修改操作，则需要在 AndroidManifest.xml 文件的< manifest >元素内添加下列权限设置。

```
uses - permission android:name = "android.permission.READ_CONTACTS" /
uses - permission android:name = "android.permission.WRITE_CONTACTS" /
```

Android 6.0(API 23)以后，系统在运行时权限上增加了对用户隐私权的保护功能。这里，READ_CONTACTS、WRITE_CONTACTS 就属于危险权限。如果在应用中需要使用这两个权限，不仅在 AndroidManifest.xml 文件中需要添加这两个权限，还必须在代码中申请用户授权。

9.5.2 ContentProvider 应用案例

本节应用案例将通过对手机通讯录中的联系人相关信息的操作来说明 ContentProvider 如何实现应用项目之间的数据共享。

在 Android 4.0(API 14)之后的版本，有关手机通讯录中所有联系人的信息都定义在 ContactsContract 类中。ContactsContract 类位于 android.provider.ContactsContract 包中。ContactsContract 可以看作是一个存储有联系人相关信息的可以扩展的数据库，该数据库的结构分为三层：Contacts、RowContacts、Data，在此称其为子类。

ContactsContract 主要包括：Contacts、RowContacts、Data、CommonDataKinds 等子类。

子类 ContactsContract.Contacts 是一张表，是一条或多条 RawContacts 的集合，用于描述一个联系人的概要信息。其中最重要的变量是_ID，它是关联其他子类的关键字。

子类 ContactsContract.RawContacts 包含通讯录中用户的基本信息，是原始数据，每一行描述了一个人的相关信息，并与一个账户相关。

子类 ContactsContract.Data 包含各种具体的数据表，可以存储任何数据，里面的每一项都指向一条 rawcontact，通常是用来存放一个联系人的详细信息，比如一个 Phone 或者 E-mail 的元数据等。Data 是一个可以存放任何类型数据的通用表，每一行的数据类型由该行的 MIMETYPE 字段决定，并决定了对应的从 DATA1 到 DATA15 各字段的意义。比如，如果一行的类型为 Phone. CONTENT_ITEM_TYPE，那么该行的 DATA1 字段存储的就是电话号码。

ContactsContract. Data 中的类型是可以扩展的。目前这些自带的类型都定义在子类 ConstractContact. CommonDataKind 中。ContactsContract. CommonDataKinds 还定义了包含于 ContactsContract. Data 表中的各数据类型的常量，如 Phone、E-mail 的 URI 等。

【案例 9.3】 自定义一个应用，实现对手机通讯录中的信息进行添加和查看功能。

说明：本案例是自定义一个应用来对手机通讯录中的数据进行操作。所以需要使用 ContentProvider 来对两个应用进行数据共享。编程中使用 ContentResolver 接口，从指定的 URI 中获取 Cursor 数据集对象，然后对这个 Cursor 对象进行插入或浏览操作。

本案例将使用简单列表对话框来显示通讯录中联系人的姓名、电话和邮箱地址信息。如果来电人有多个电话号码，则显示多个电话号码。

本案例需要对手机通讯录进行读写，所以在 AndroidManifest. xml 文件中需要添加读、写通讯录权限，并且在代码中增加运行时动态请求授权的代码。

开发步骤及解析：过程如下。

1）创建项目

在 Android Studio 中创建一个名为 Activity_ContentProvider 的项目。其包名为 ee. example. activity_contentprovider。

2）添加权限

在 AndroidManifest. xml 中添加对手机通讯录的读写权限，代码片段如下。

```
<?xml version = "1.0" encoding = "utf - 8"?>
< manifest xmlns:android = "http://schemas.android.com/apk/res/android"
    package = "com.example.ee.activity_contentprovider">
    <!-- 联系人 -->
    < uses - permission android:name = "android.permission.READ_CONTACTS" />
    < uses - permission android:name = "android.permission.WRITE_CONTACTS" />
    < application
        ...
    </application >
</manifest >
```

3）设计布局

本项目有三个页面布局需要设计。应用的主界面由 activity_main. xml 设计，activity_result. xml 设计一个 ListView 控件的布局，listitem. xml 设计 ListView 中每一项的布局。

activity_main. xm 是主 Activity 的布局。在其中设计三个 EditText 控件，用于输入联系人的简单信息。其中 ID 为 ed_nm 的输入联系人姓名，ID 为 ed_ph 的输入联系人电话，ID 为 ed_em 的输入联系人 E-mail。这三个 EditText 都设置 android:hint 属性以示提示用户的输入行为。在三个 EditText 控件的下面添加两个按钮，ID 为 bt_add 的按钮触发向手机通讯录添加联系人信息的操作，ID 为 bt_brw 的按钮触发查看手机通讯录联系人的操作。

activity_result. xml 是查看联系人信息的布局。其中，只添加一个 ListView 列表控件，其 ID 为 lv,用于显示联系人信息。

listitem. xml 是 ListView 的每一项的显示布局。其中添加了三个 TextView 控件,ID 为 contactsname 的显示联系人姓名,ID 为 contactsphone 的显示联系人电话,ID 为 contactsemail 的输入联系人 E-mail。

4) 开发逻辑代码

打开 java/com. example. ee. activity_contentprovider 包下的 MainActivity. java 文件,并进行编辑,代码如下。

```
package ee.example.activity_contentprovider;

import androidx.appcompat.app.AppCompatActivity;
import androidx.core.app.ActivityCompat;
import androidx.core.content.ContextCompat;

import android.Manifest;
import android.app.AlertDialog;
import android.content.ContentUris;
import android.content.ContentValues;
import android.content.pm.PackageManager;
import android.database.Cursor;
import android.net.Uri;
import android.os.Bundle;
import android.provider.ContactsContract;
import android.view.View;
import android.widget.Button;
import android.widget.EditText;
import android.widget.ListView;
import android.widget.SimpleAdapter;
import android.widget.Toast;

import java.util.ArrayList;
import java.util.HashMap;
import java.util.List;
import java.util.Map;

import static android.text.TextUtils.isEmpty;

public class MainActivity extends AppCompatActivity {
    private Button bt1, bt2;
    private ListView Lview;
    private static final int PERMISSION_CONTACTS_REQUEST_CODE = 1; //申请通讯录权限请求码

    @Override
    protected void onCreate(Bundle savedInstanceState) {
        super.onCreate(savedInstanceState);
        setContentView(R.layout.activity_main);
        bt1 = (Button) findViewById(R.id.bt_add);
        bt1.setOnClickListener(new AddPersonClick());
        bt2 = (Button) findViewById(R.id.bt_brw);
        bt2.setOnClickListener(new LookPresonClick());

        checkPermissionAndCamera();
```

```
    }

    //检查权限,授权后调用手机通讯录
    private void checkPermissionAndCamera() {
        int hasCameraPermission = ContextCompat.checkSelfPermission(getApplication(),
                                        Manifest.permission.READ_CONTACTS);
        if (hasCameraPermission == PackageManager.PERMISSION_GRANTED) {
            //有权限,正常往下执行
        } else {
            //没有权限,申请权限
            ActivityCompat.requestPermissions(
                    this,
                    new String[]{
                            Manifest.permission.READ_CONTACTS,
                            Manifest.permission.WRITE_CONTACTS},
                    PERMISSION_CONTACTS_REQUEST_CODE);
        }
    }

    private class AddPersonClick implements View.OnClickListener {
        @Override
        public void onClick(View v) {
            //获取程序界面中的文本框内容
            String name = ((EditText) findViewById(R.id.ed_nm)).getText().toString();
            String phone = ((EditText) findViewById(R.id.ed_ph)).getText().toString();
            String email = ((EditText) findViewById(R.id.ed_em)).getText().toString();
            //创建一个空的 ContentValue
            ContentValues values = new ContentValues();
            //向 RawContacts.CONTNT_URI 执行一个空值插入
            //目的是获取系统返回的 rawContactId
            Uri rawContactsUri = getContentResolver().insert(
                ContactsContract.RawContacts.CONTENT_URI, values);
            long rawContactId = ContentUris.parseId(rawContactsUri);
            values.clear();
            values.put(ContactsContract.Data.RAW_CONTACT_ID, rawContactId);
            //设置内容类型
            values.put(ContactsContract.Data.MIMETYPE,
                ContactsContract.CommonDataKinds.StructuredName.CONTENT_ITEM_TYPE);
            //设置联系人名字
            values.put(ContactsContract.CommonDataKinds.StructuredName.GIVEN_NAME, name);
            //向联系人 Uri 添加联系人名字
            getContentResolver().insert(
                                android.provider.ContactsContract.Data.CONTENT_URI, values);
            values.clear();
            values.put(ContactsContract.Data.RAW_CONTACT_ID, rawContactId);
            values.put(ContactsContract.Data.MIMETYPE,
                    ContactsContract.CommonDataKinds.Phone.CONTENT_ITEM_TYPE);
            //设置联系人的电话
            values.put(ContactsContract.CommonDataKinds.Phone.NUMBER, phone);
            //设置电话类型
            values.put(ContactsContract.CommonDataKinds.Phone.TYPE,
                    ContactsContract.CommonDataKinds.Phone.TYPE_MOBILE);
            //向联系人电话 Uri 添加电话号码
            getContentResolver().insert(
```

```
                        android.provider.ContactsContract.Data.CONTENT_URI, values);
            values.clear();
            values.put(ContactsContract.Data.RAW_CONTACT_ID, rawContactId);
            values.put(ContactsContract.Data.MIMETYPE,
                        ContactsContract.CommonDataKinds.Email.CONTENT_ITEM_TYPE);
            //设置联系人的 E-mail 地址
            values.put(ContactsContract.CommonDataKinds.Email.DATA, email);
            //设置 E-mail 的类型
            values.put(ContactsContract.CommonDataKinds.Email.TYPE,
                        ContactsContract.CommonDataKinds.Email.TYPE_WORK);
            getContentResolver().insert(
                        android.provider.ContactsContract.Data.CONTENT_URI, values);
            Toast.makeText(MainActivity.this, "添加联系人信息成功", Toast.LENGTH_LONG)
                  .show();
        }
    }

    class LookPresonClick implements View.OnClickListener {
        @Override
        public void onClick(View v) {
            //加载 result.xml 界面布局代表的视图
            View resultDialog = getLayoutInflater().inflate(R.layout.activity_result, null);
            Lview = resultDialog.findViewById(R.id.lv);
            //创建适配器对象
            SimpleAdapter listItemAdapter = new SimpleAdapter(
                  resultDialog.getContext(),
                  getData(),
                  R.layout.listitem,
                  new String[]{"lvname","lvphone","lvemail"},
                  new int[]{R.id.contactsname,R.id.contactsphone,R.id.contactsemail});
            Lview.setAdapter(listItemAdapter);

            new AlertDialog.Builder(MainActivity.this).setView(resultDialog)
                    .setPositiveButton("确定", null).show();
        }
    }

    private List<Map<String, String>> getData() {
        ArrayList<Map<String, String>> listitem = new ArrayList<Map<String, String>>();
        Map<String, String> map = new HashMap<>();
        //使用 ContentResolver 查找联系人数据
        final Cursor cursor = getContentResolver().query(
              ContactsContract.Contacts.CONTENT_URI, null, null, null, null);
        //遍历结果,获取系统所有联系人信息
        while (cursor.moveToNext()) {
            //获取联系人 ID
            String contactid = cursor.getString(
                    cursor.getColumnIndex(ContactsContract.Contacts._ID));
            //获取联系人的名字
            String csname = cursor.getString(
                    cursor.getColumnIndex(ContactsContract.Contacts.DISPLAY_NAME));
            map.put("lvname", csname);

            //使用 ContentResolver 查找联系人的电话号码
```

```
128             Cursor phones = getContentResolver().query(
                    ContactsContract.CommonDataKinds.Phone.CONTENT_URI,
                    null,
                    ContactsContract.CommonDataKinds.Phone.CONTACT_ID + " = ?",
                    new String[] { contactid },
                    null);
129             String csphone = "";
130             //遍历查询结果,获取该联系人的多个电话
131             while (phones.moveToNext()) {
132                 //获取查询的结果中的电话号码列
133                 if (isEmpty(csphone))
134                     csphone = "电话号码:";
135                 else
136                     csphone = csphone + ", ";
137                 String phoneNumber = phones.getString(
                      phones.getColumnIndex(ContactsContract.CommonDataKinds.Phone.NUMBER));
138                 csphone = csphone + phoneNumber;
139             }
140             map.put("lvphone", csphone);
141             phones.close();
142
143             //使用 ContentResolver 查找联系人的 E-mail 地址
144             Cursor emails = getContentResolver().query(
                  ContactsContract.CommonDataKinds.Email.CONTENT_URI,
                  null,
                  ContactsContract.CommonDataKinds.Email.CONTACT_ID + " =?",
                  new String[] { contactid },
                  null);
145             //遍历查询结果,获取该联系人的多个 E-mail 地址
146             String csemail = "";
147             while (emails.moveToNext()) {
148                 //获取查询的结果中 E-mail 地址中列的数据
149                 if (isEmpty(csemail))
150                     csemail = "email 地址:";
151                 else
152                     csemail = csemail + ", ";
153                 String emailAddress = emails.getString(
                        emails.getColumnIndex(ContactsContract.CommonDataKinds.Email.DATA));
154                 csemail = csemail + emailAddress;
155             }
156             map.put("lvemail", csemail);
157             emails.close();
158
159             listitem.add(map);
160             map = new HashMap<>();
161         };
162         cursor.close();
163         return listitem;
164     }
165 }
```

(1) 第 44 行调用自定义方法 checkPermissionAndCamera()检查读、写通讯录权限是否授权。该方法在第 48～56 行实现。第 49 行定义一个整型变量 hasCameraPermission，其

值为是否拥有读取通讯录授权的返回值。如果没有授权则申请权限，申请授权代码见第54行。

(2) 第58～98行，是实现“添加联系人信息”按钮的单击事件 AddPersonClick()的监听接口。

① 第66行，创建一个空的 ContentValue 对象 values。这个 ContentValue 对象是用来存储一系列的键-值对数据，类似于一个字典对象。

② 第69行，向 RawContacts. CONTENT_URI 插入一个 ContentValue 对象 values，从而获取 ContactsContract. RawContacts 的新插入行的 URI。第70行将字符串值转换为可识别的 URI 格式。

③ 第71～76行，向对象 values 分别设置 ID 号、姓名属性的内容类型、联系人姓名等键-值对信息。第78行，向 ContactsContract. Data 子类的联系人名字 URI 中添加载有联系人姓名的相关键-值对信息。

④ 第79～85行，向对象 values 分别设置 ID 号、电话属性的内容类型、联系人电话及电话类别等键-值对信息。第87行，向 ContactsContract. Data 子类的联系人电话 URI 中添加载有联系人电话的相关键-值对信息。

⑤ 第88～94行，向对象 values 分别设置 ID 号、邮箱属性的内容类型、联系人邮箱及邮箱类别等键-值对信息。第95行，向 ContactsContract. Data 子类的联系人邮箱 URI 中添加载有联系人邮箱的相关键-值对信息。

(3) 第100～112行，是实现“查看联系人”按钮的单击事件 LookPresonClick()的监听接口。

① 第104行，创建一个视图对象 resultDialog，该视图由 activity_result. xml 布局文件实例化。

② 第107行，创建一个简单的适配器对象。其中第一个参数 resultDialog. getContext()指出这个 ListView 控件将置于 resultDialog 视图中；第二个参数指出这个 ListView 控件中的数据 ArrayList 由 getData()方法提供；第三个参数指出 ListView 的视图资源文件是 listitem. xml；第四个参数指出 Map 对象中的名称映射的列名分别是 lvname、lvphone 和 lvemail；第五个参数指出第四个参数对应的值将映射到视图中的控件分别是 contactsname、contactsphone 和 contactsemail。

③ 第110行，动态创建并显示一个 AlertDialog 对话框，该对话框呈现的内容由列表对象 resultDialog 提供，该对话框只有一个“确定”按钮。

(4) 第114～164行，定义 getData()方法的实现。通过第118行，getContentResolver(). query()对其手机通讯录进行查询，获取游标对象 cursor。再通过遍历数据集，将从 cursor 中获取的数据一一添加到 ListView 的列表项中。

① 第122行中，getColumnIndex(ContactsContract. Contacts. _ID)从 cursor 中获取当前联系人的 ID 号。

② 第124～125行，实现向 Map 对象传入当前联系人姓名。在第124行中，getColumnIndex(ContactsContract. Contacts. DISPLAY_NAME)从 cursor 中获取当前联系人的姓名。第125行，将联系人的姓名置入 Map 对象的 lvname 中。

③ 第128～141行，实现向 Map 对象传入当前联系人电话号码。由于有些联系人可能

有多个电话号码，所以另外创建一个 phones 游标对象，查询当前联系人的所有电话号码。第 129～139 行，遍历 phones 游标，通过第 137 行代码 getColumnIndex(ContactsContract.CommonDataKinds.Phone.NUMBER)从 phones 中获取电话号码，组合成一个字符串赋值给 csphone 变量。第 140 行，将联系人的电话串置入 Map 对象的 lvphone 中。

④ 第 144～157 行，实现向 Map 对象传入当前联系人邮箱地址。由于一个人可能有多个邮箱地址，所以处理代码同上面第 128～141 行。

⑤ 第 159 行，将 Map 对象获得的值添加到 ListView 的项中。

运行结果：本案例是用一个自定义的应用实现与手机通讯录中的数据进行操作控制，这是自定义联系人通讯录管理与系统通讯录管理两个应用之间的数据共享应用。

在运行应用之前，首先运行手机系统通讯录应用，完成向通讯录添加一个联系人操作。在 Android Studio 支持的模拟器上，找到桌面中的 Contacts 应用项目，即通讯录，完成添加联系人操作，如图 9-7 所示。

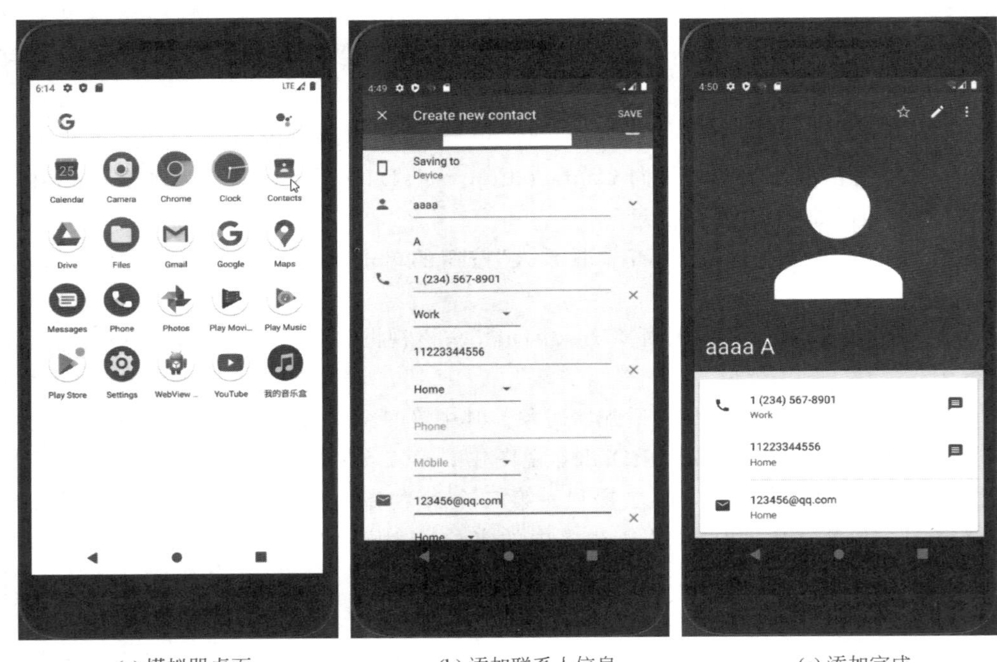

(a) 模拟器桌面　　(b) 添加联系人信息　　(c) 添加完成

图 9-7　在模拟器的通讯录中添加联系人运行效果

然后再运行本案例，完成向通讯录添加一个联系人的操作。运行 Activity_ContentProvider 项目，运行结果如图 9-8 所示。

最后在手机通讯录中可以看到通过两种方式都成功添加了联系人，如图 9-9 所示。本案例说明，无论是使用手机内置应用，还是自定义应用，都可以共享通讯录数据，可以共同对通讯录进行管理操作。

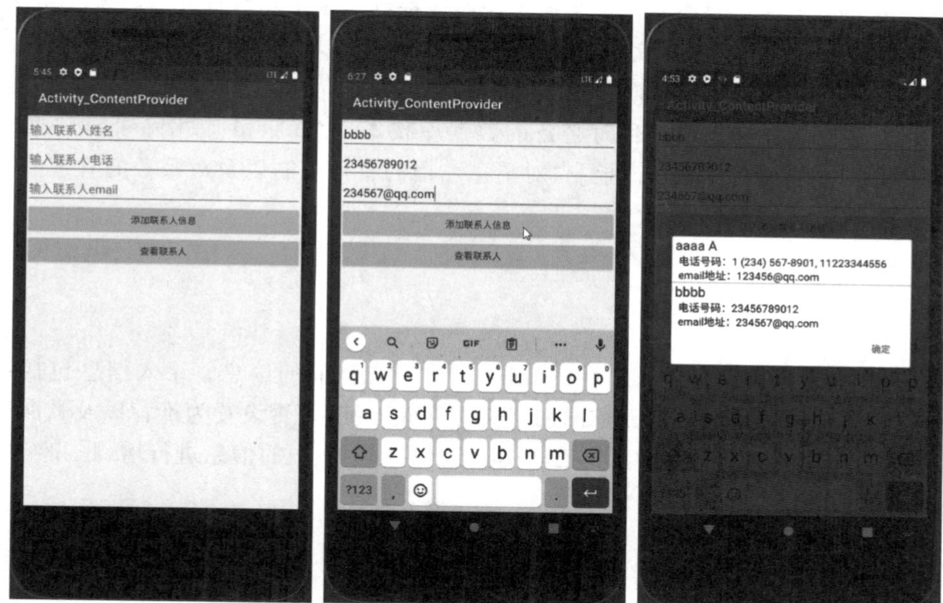

(a) 初始运行界面　　(b) 添加一个联系人信息　　(c) 查看联系人信息

图 9-8　本案例添加联系人与浏览通讯录运行效果

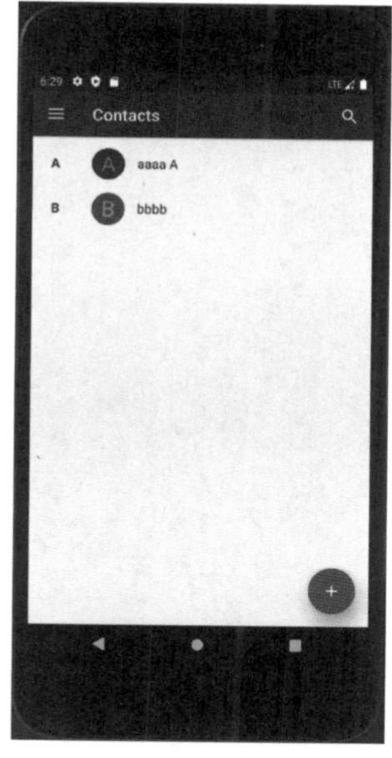

图 9-9　模拟器的通讯录中联系人列表

小结

本章重点介绍了应用项目的数据存储与数据共享技术，包括 SharedPreferences 存取、SQLite 数据库操作的方法技术、外部存储设备 SD 卡读写文件技术，以及使用 ContentProvider

在应用项目之间的数据共享等技术。掌握了这些数据存储技术和数据共享技术的用法，加之Java的文件读写处理技术，要开发一个Android的本地应用就会毫无问题。

在介绍SQLite数据库操作时涉及数据库的数据描述语言DDL和数据操纵语言DML相关内容，这些也是开发应用项目时必备的数据库技术。Android应用除了对数据的存储管理外，更重要的是对线程的管理控制、对消息广播的管理控制以及对服务的管理控制。在第10章将进一步学习这些内容。

练习

创建一个本地数据库表，存放“我的音乐盒”中一首歌曲的信息。录入信息包括歌曲名称、歌曲原唱歌手姓名、歌曲时长、歌曲网络文件存储地址、是否设置为推荐以及歌曲简介等内容。要求：设计录入页面和浏览页面，并且可以对数据表中的信息进行增、删、改、查。

第 10 章 后台处理

在 Android 的应用中，有一些应用是没有界面的，不需要用户去显式启动的，并且可以在运行其他应用的同时，仍然运行着的这些操作，被称为 Android 的后台处理。例如，消息提示，接收、发送广播消息，线程内的数据控制，播放音乐等，这些都是在 Android 后台中运行着的应用。

10.1 消息通知 Notification

Android 系统提供一套友好的消息通知机制，不会打断用户当前的操作。常用的方式是使用 Notification。下面介绍这种消息通知的用法。

10.1.1 Notification 简介

Notification 类在 android. app. Notification 包中。它是一种提示简短、即时消息的通知方式。Notification 无须 Activity，将消息内容以图标的形式显示在手机状态栏中。状态栏位于手机屏幕的顶端，如图 10-1(a)所示，通常推送 App 即时提醒用户的消息，以及手机电池电量、信息强度、时间等状态信息。在 Android 手机中，用手指按住状态栏往下拉，以抽屉式操作方式拉开通知栏，在通知栏中查看系统推送的详细消息，如图 10-1(b)所示。

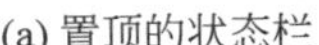
(a) 置顶的状态栏

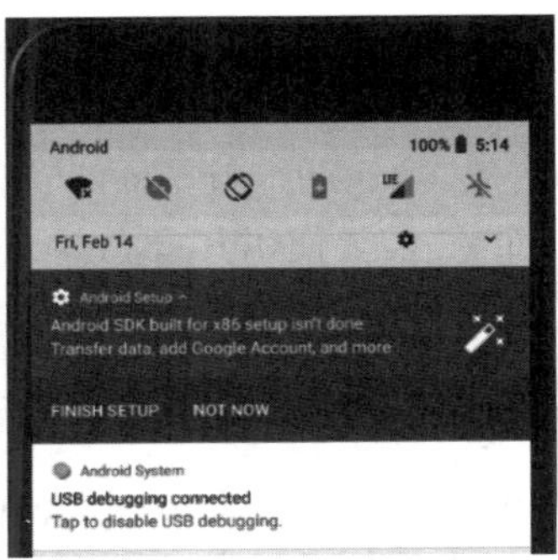

(b) 拉开的通知栏

图 10-1 通知信息的两种显示效果

在Android中，Notification包含如下功能：创建新的通知栏图标，在扩展的通知栏窗口还可以推送额外的信息（也可以发起一个Intent），闪烁/LED，让手机振动，发出声音（铃声、媒体库歌曲）等。从Android 5.0（API 21）开始，增加了浮动通知窗口。当手机设备接收到一些重要级别的通知，会在浮动通知窗口中短暂显示，以提醒用户即时了解重要通知。在Android 8.0（API 26）及更高版本的设备上，支持用彩色图标表示通知标志。

1. 创建通知

每个通知都是一个Notification对象，通知的内容可以由某个UI信息或某个操作指定。从Android 8.0（API 26）开始，所有通知都必须分配一个渠道，否则通知将不会显示。要创建Notification对象，可以使用Notification. Builder创建，也可以使用NotificationCompat. Builder创建。

1）Notification. Builder类

Notification. Builder位于android. app. Notification. Builder包中，该类可以十分方便地创建通知栏中的各个域，并生成各个域中的视图内容。但是在Android 8.0之后，Notification. Builder创建了通知对象后必须添加通知渠道。NotificationChannel是通知渠道类，位于android. app. NotificationChannel包中。可使用setChannelId(String channelId)方法为Notification. Builder对象添加渠道。channelId是渠道ID，由开发人员自己定义，用于区分其他渠道的通知。

例如创建一个名为not1的通知对象，渠道号为channelId，代码片段如下。

```
Notification.Builder builder = new Notification.Builder(ctx);
builder.setContentIntent(contentIntent)
        .setContent(notify_music)
        .setTicker(song)
        .setSmallIcon(R.drawable.ic_notif_s);
if (Build.VERSION.SDK_INT >= Build.VERSION_CODES.O) { //如果在Android 8.0以上版本
    builder.setChannelId(channelId);
}
Notification not1 = builder.build();
```

2）NotificationCompat. Builder类

从Android 8.0之后，越来越多的开发者使用AndroidX支持库中的NotificationCompat API来创建通知对象。NotificationCompat类位于androidx. core. app. NotificationCompat包中。创建Notification对象，需要使用其子类NotificationCompat. Builder来创建通知对象并设置通知内容和渠道。NotificationCompat. Builder类的构造函数如下。

```
NotificationCompat.Builder(Context context, String channelId)
```

其中，参数context为上下文；参数channelId为渠道ID。

3）设置通知栏内容的常用方法

使用NotificationCompat. Builder或Notification. Builder创建Notification对象，二者的用法基本相似，用来设置通知栏内容和显示视图的方法也大致相同。常用的设置方法见表10-1。

表 10-1 构建 Notification 对象的常用方法及说明

方 法	说 明
setAutoCancel(boolean autoCancel)	设置该通知是否自动清除。若为 true,则单击该通知后,通知会自动消失;若为 false,则单击该通知后通知不消失
setContentInfo(CharSequence info)	设置通知栏右下方的文本。若调用该方法,则 setNumber()的设置失效
setContentIntent(PendingIntent intent)	设置内容的延迟意图 PendingIntent,单击该通知时触发该意图。通常调用 PendingIntent 的 getActivity()方法获得延迟意图对象
setContentText(CharSequence text)	设置通知栏里的内容文本
setContentTitle(CharSequence title)	设置通知栏里的标题文本
setDeleteIntent(PendingIntent intent)	设置删除延迟意图 PendingIntent,滑掉该通知时触发该动作
setLargeIcon(Bitmap b/Icon icon)	设置通知栏里的大图标
setNumber(int number)	设置通知栏右下方的数字,可与 setProgress()联合使用,表示当前的进度数值
setProgress(int max, int progress, boolean indeterminate)	设置进度条与当前进度。进度条位于标题文本与内容文本中间
setSmallIcon(int icon/Icon icon)	设置状态栏里的小图标
setSubText(CharSequence text)	设置通知栏里的附加说明文本,位于内容文本下方。若调用该方法,则 setProgress()的设置失效
setTicker(CharSequence tickerText)	设置状态栏里的提示文本
setUsesChronometer(boolean b)	设置是否显示计数器。若为 true,不显示推送时间,动态显示从通知被推送到当前的时间间隔,以"分:秒"格式显示
setWhen(long when)	设置推送时间,格式为"时:分"。推送时间在通知栏右方显示
setShowWhen(boolean show)	设置是否显示推送时间
build()	构建 Notification 对象。在以上参数都设置完毕后,调用该方法返回 Notification 对象

在设置 Notification 对象时,至少包含以下内容。

(1) 调用 setSmallIcon()方法,否则在状态栏就不会显示通知消息。

(2) 调用 setContentTitle()方法,否则在通知栏就不会显示标题。

(3) 调用 setContentText()方法,否则在通知栏就不会显示详细文本。

其他的参数都是可选项,在应用中根据需求设置。例如使用 NotificationCompat.Builder 来创建一个 Notification 对象 noti,代码片段如下。

```
Notification noti = new NotificationCompat.Builder(mContext,mChannelId)
                        .setContentTitle("My notification")
                        .setContentText("Hello World!")
                        .setSmallIcon(R.drawable.notification_icon)
                        .setLargeIcon(aBitmap)
                        .build();
```

注意: *在通知栏中有些设置方法是互斥的,在同一时间不能同时调用。*

例如:

setWhen()与 setUsesChronometer()同一时间只能调用一个,因为推送时间与计数器都位于通知栏的右边,无法同时显示。

setSubText()与 setProgress()同一时间只能调用一个，因为附加说明与进度条都位于标题文本的下方，无法同时显示。

setNumber()与 setContentInfo()同一时间只能调用一个，因为计数值与提示都位于通知栏的右下方，无法同时显示。

还可以为通知添加一些操作，通知操作允许用户直接从通知转到应用中的 Activity。一个通知可以提供多个操作。例如，暂停闹铃或立即答复短信等。在 Notification 内部，操作本身由 PendingIntent 定义，如果要在用户单击抽屉式通知栏中的通知文本时启动 Activity，则可通过调用 setContentIntent()来添加 PendingIntent。

释疑：

什么是 PendingIntent？ 它与 Intent 有什么不同？

PendingIntent 类是一个 Intent 的描述，它位于 android. app 包下。Pending 一词的含义是即将发生或来临的事情。PendingIntent 这个类用于处理即将发生的事情。例如在通知 Notification 中用于跳转页面，但不是马上跳转。

Intent 是即时启动，随所在的 Activity 消失而消失。而 PendingIntent 可以看作是对 Intent 的包装，通常通过 getActivity()、getBroadcast()、getService()得到 PendingIntent 的实例，当前 Activity 并不能马上启动它所包含的 Intent，而是在外部执行 PendingIntent 时调用 Intent 的。Intent 一般是用于在 Activity、Service、BroadcastReceiver 之间传递数据，而 PendingIntent 一般用在 Notification 上或其他的方法参数中。

2. 管理通知

所有的 Notification 对象由 NotificationManager（通知管理器）来管理，NotificationManager 类也位于 android. app. NotificationManager 包中，是用来处理 Notification 的系统服务。通过使用 NotificationManager，可以触发新的 Notification，修改现有的 Notification 或者删除那些不再需要的 Notification。

NotificationManager 类主要负责将 Notification 在状态栏显示出来和取消，常用的方法见表 10-2。

表 10-2　NotificationManager 的常用方法及说明

方　　法	说　　明
cancel(int id)	取消指定 ID 的 Notification。如果是一个短暂的 Notification，试图将其隐藏；如果是一个持久的 Notification，将从状态栏中移走
cancelAll()	取消以前显示的所有 Notification
notify(int id, Notification notification)	把 Notification 持久地推送到状态栏上，id 是指该 Notification 的 ID 号

在创建 NotificationManager 类对象时，必须使用 getSystemService(String)方法，其参数是 Context. NOTIFICATION_SERVICE，用于初始化 NotificationManager 对象。

在 Android 8.0 及更高版本上使用通知，首先必须向系统中注册应用的通知渠道，传递通知渠道所需要的信息。在创建 NotificationChannel 实例时，NotificationChannel 构造函数需要一个描述重要级别参数 importance，它由 NotificationManager 类中的其中一个常量提供。NotificationManager 类中与重要级别相关的常量见表 10-3。

表 10-3 NotificationManager 类关于重要级别的常量及说明

常 量 名	值	说 明
IMPORTANCE_HIGH	4	开启通知,发出提示音,状态栏中显示,会在屏幕上方弹出悬浮弹框
IMPORTANCE_DEFAULT	3	开启通知,发出提示音,状态栏中显示,但是不会弹出
IMPORTANCE_LOW	2	开启通知,状态栏中显示,但是不发出提示音,不会弹出
IMPORTANCE_MIN	1	开启通知,但没有提示音,状态栏中无显示,不会弹出
IMPORTANCE_NONE	0	关闭通知

在 Android 8.0(API 26)以后,要求每一个通知对象必须有一个对应的通知渠道号,但是较低版本不要求。研发人员开发出的应用将会在不同的 Android 版本上运行,为了考虑到应用的兼容性,保证开发出的代码能在不同的 SDK_INT 版本上正常运行,因此,在创建通知渠道实例时通常需要判断 SDK 的版本,代码片段如下。

```
private void createNotificationChannel() {
    if (Build.VERSION.SDK_INT >= Build.VERSION_CODES.O) {
        String channelId = getString(R.string.channel_id);
        String channelName = getString(R.string.channel_name);
        int importance = NotificationManager.IMPORTANCE_DEFAULT;
        NotificationChannel channel = new NotificationChannel(channelId, channelName, importance);
        NotificationManager notificationManager = getSystemService(NotificationManager.class);
        notificationManager.createNotificationChannel(channel);
    }
}
```

10.1.2 简单通知应用

Android 应用项目中的通知内容及显示视图由系统模板规划,通知栏从创建到显示,以及更新、移除等,是由 Notification、NotificationChannel 和 NotificationManager 等类联合使用的结果。下面以基本通知的应用为例,讲述创建和显示通知的基本使用方法。

应用项目中使用通知,其基本步骤如下。

(1) 创建 NotificationManager 实例。

(2) 如果是 Android 8.0 以上的版本,创建 NotificationChannel 实例。

(3) 创建 Notification 对象。

(4) 通过 NotificationManager 实例把通知推送到状态栏上。

【案例 10.1】 输入通知的标题和内容,并为通知设置小图标和大图标,通过单击按钮发送到状态栏中。

说明: 设计两个 Activity 页面,第一个页面设计一个按钮,单击该按钮跳转到第二个页面。在第二个页面中设计两个输入框和一个按钮,输入通知的标题和内容,单击按钮发送简单通知到状态栏,并且在展开通知信息列表中,单击该条通知信息可跳转回到第一个页面。

开发步骤及解析: 过程如下。

1) 创建项目

在 Android Studio 中创建一个名为 Activity_SimpleNotify 的项目。其包名为

ee. example. activity_simplenotify。

2）准备颜色资源

编写 res/values 目录下的 colors. xml 文件，分别声明 black、white、darkgrey 和 blue 等颜色。在前面案例中已经多次展示过 colors. xml 的代码，在此就不赘述了。

3）准备图片和背景资源

将 ic_smlp. png、ic_largp. jpeg 分别作为通知的小图和大图资源复制到本项目的 res/drawable 目录中。在该目录中，edittext_selector. xml 文件声明了文本输入框的选择样式，shape_edit_focus. xml 声明输入框获取焦点时的图形背景，shape_edit_normal. xml 声明文本输入框失去焦点时（即通常状态）的图形背景。例如 shape_edit_normal. xml 文件，声明一个圆角矩形块，其代码如下。

```
<shape xmlns:android = "http://schemas.android.com/apk/res/android" >
    <solid android:color = "@color/white" />
    <stroke
        android:width = "1dp"
        android:color = "@color/darkgrey" />
    <corners
        android:bottomLeftRadius = "5dp"
        android:bottomRightRadius = "5dp"
        android:topLeftRadius = "5dp"
        android:topRightRadius = "5dp" />
     <padding
        android:bottom = "2dp"
        android:left = "2dp"
        android:right = "2dp"
        android:top = "2dp" />
</shape>
```

edittext_selector. xml 文件的代码如下。

```
<?xml version = "1.0" encoding = "utf-8"?>
<selector xmlns:android = "http://schemas.android.com/apk/res/android">
    <item android:state_focused = "true" android:drawable = "@drawable/shape_edit_focus"/>
    <item android:drawable = "@drawable/shape_edit_normal"/>
</selector>
```

4）设计布局

在 res/layout 目录下有两个布局文件，activity_main. xml 是第一个页面的布局文件，activity_simplenotify. xml 是第二个页面的布局文件。

在 activity_main. xml 代码中，设计一个按钮，其 ID 为 btn_notify_simple。

在 activity_simplenotify. xml 代码中，设计两个输入框和一个按钮，输入框的背景由 drawable 目录中的 edittext_selector. xml 定义。其中，ID 为 et_title 的文本输入框输入通知标题，ID 为 et_message 的文本输入框输入通知的内容；按钮的 ID 为 btn_send_simple。

5）开发逻辑代码

在 java/ee. example. activity_simplenotify 包下有两个 Activity 类的实现代码文件，分别用于控制第一个页面和第二个页面的操作逻辑。

MainActivity. java 文件是第一个类实现代码，代码如下。

```
package ee.example.activity_simplenotify;

import androidx.appcompat.app.AppCompatActivity;

import android.content.Intent;
import android.os.Bundle;
import android.view.View;

public class MainActivity extends AppCompatActivity {

    @Override
    protected void onCreate(Bundle savedInstanceState) {
        super.onCreate(savedInstanceState);
        setContentView(R.layout.activity_main);
        findViewById(R.id.btn_notify_simple).setOnClickListener(new View.OnClickListener() {
            @Override
            public void onClick(View v) {
                Intent intent = new Intent(MainActivity.this, NotifySimpleActivity.class);
                startActivity(intent);
            }
        });
    }

}
```

NotifySimpleActivity.java 文件,是创建通知类实现代码,代码如下。

```
package ee.example.activity_simplenotify;

import android.annotation.TargetApi;
import androidx.annotation.RequiresApi;
import androidx.appcompat.app.AppCompatActivity;
//import androidx.core.app.NotificationCompat;
//import androidx.core.content.ContextCompat;
import android.app.Notification;
import android.app.NotificationChannel;
import android.app.NotificationManager;
import android.app.PendingIntent;
import android.content.Intent;
import android.graphics.BitmapFactory;
import android.os.Build;
import android.os.Bundle;
import android.view.View;
import android.view.View.OnClickListener;
import android.widget.EditText;

@TargetApi(Build.VERSION_CODES.JELLY_BEAN)
public class NotifySimpleActivity extends AppCompatActivity implements OnClickListener {

    private EditText et_title;
    private EditText et_message;
    private NotificationManager mManager;
    private String channelId = "msg";
    private String channelName = "消息";
```

```

    @Override
    protected void onCreate(Bundle savedInstanceState) {
        super.onCreate(savedInstanceState);
        setContentView(R.layout.activity_simplenotify);
        et_title = (EditText) findViewById(R.id.et_title);
        et_message = (EditText) findViewById(R.id.et_message);
        mManager = (NotificationManager) getSystemService(NOTIFICATION_SERVICE);

        if (Build.VERSION.SDK_INT >= Build.VERSION_CODES.O) {
            int importance = NotificationManager.IMPORTANCE_DEFAULT;
            createNotificationChannel(channelId, channelName, importance);
        }

        findViewById(R.id.btn_send_simple).setOnClickListener(this);
    }

    @Override
    public void onClick(View v) {
        if (v.getId() == R.id.btn_send_simple) {
            String title = et_title.getText().toString();
            String message = et_message.getText().toString();
            sendSimpleNotify(title, message);
        }
    }

    @RequiresApi(api = Build.VERSION_CODES.O)
    private void createNotificationChannel(String channelId, String channelName, int importance) {
        NotificationChannel channel = new NotificationChannel(channelId, channelName, importance);
        mManager.createNotificationChannel(channel);
    }

    private void sendSimpleNotify(String title, String message) {
        Intent clickIntent = new Intent(this, MainActivity.class);
        PendingIntent contentIntent = PendingIntent.getActivity(this,
            R.string.app_name, clickIntent, PendingIntent.FLAG_UPDATE_CURRENT);

        //使用 Notification.Builder 创建 notification
        Notification.Builder notification_builder = new Notification.Builder(this)
                .setContentIntent(contentIntent)
                .setContentTitle(title)
                .setContentText(message)
                .setWhen(System.currentTimeMillis())
                .setSmallIcon(R.drawable.ic_smlp)
                .setLargeIcon(BitmapFactory.decodeResource(getResources(), R.drawable.ic_largp))
                .setAutoCancel(true);
        if (Build.VERSION.SDK_INT >= Build.VERSION_CODES.O) {
            notification_builder.setColor(getResources().getColor(R.color.colorPrimary,null));
            notification_builder.setChannelId(channelId);
        }
        Notification notification = notification_builder.build();

/*      //使用 NotificationCompat.Builder 创建 notification
        Notification notification = new NotificationCompat.Builder(this, "msg")
```

```
82                    .setContentIntent(contentIntent)
83                    .setContentTitle(title)
84                    .setContentText(message)
85                    .setWhen(System.currentTimeMillis())
86                    .setColor(ContextCompat.getColor(this, R.color.colorPrimary))
87                    .setSmallIcon(R.drawable.ic_smlp)
88                    .setLargeIcon(BitmapFactory.decodeResource(getResources(), R.drawable.ic_largp))
89                    .setAutoCancel(true)
90                    .build();
91    */
92            mManager.notify(1, notification);
93        }
94
95    }
```

(1) 第 20、54 行，是 Annotation 的常用注解，这些注解的作用只是去除 Lint 的错误警告，并不会影响任何的代码逻辑。第 20 行表明这段代码只能在 JELLY_BEAN 及以上的版本上正常运行，Build. VERSION_CODES. JELLY_BEAN 指定的是 Android 4.1(API 16)，如果代码在 API 16 以下的系统上运行，则会报错。第 54 行注解的作用与第 20 行相同，表明代码必须在指定的版本及以上的 API 运行，Build. VERSION_CODES. O 即 Android 8.0(API 26)。建议在编程时，在编写方法代码时在内部对版本进行判断，以确保运行不会出问题。

Build. VERSION_CODES 是一个类，该类封装了已经存在的 SDK 框架及 Android 版本。VERSION_CODES 表示 SDK 版本，其成员就是从最早版本开始到当前运行的系统的版本号别名常量。

(2) 第 37～40 行，对版本进行判断。Build. VERSION 可获取 Android 系统的版本信息，Build. VERSION. SDK_INT 表示 SDK 的版本号。

(3) 第 55～58 行，这段代码是在 Android 8.0 及以上的版本需要运行的代码。所以在这个方法的前一行添加@RequiresApi(api = Build. VERSION_CODES. O)注解。

(4) 第 61～63 行，创建一个 PendingIntent 对象 contentIntent，它是一个延迟的 intent 对象，用于设置后面创建的 Notification 对象的单击跳转属性。即在运行中，显示该通知信息时，单击该通知会触发页面跳转操作。

(5) 第 66～78 行，使用 Notification. Builder 创建一个 Notification 对象 notification。首先是创建 Notification. Builder 对象 notification_builder，调用 set. XXX()方法设置 notification 所需要的内容属性；其中第 74～77 行设置 Android 8.0 以上版本需要设计的属性；最后才调用 built()方法创建 notification。

(6) 第 80～91 行，被注释了，该段代码是使用 NotificationCompat. Builder 创建一个 Notification 对象 notification。

(7) 第 92 行，使用 NotificationManager 对象 mManager，调用 notify()方法，将通知对象 notification 推送到状态栏上。

6) 配置清单文件

因为有两个 Activity 类，所以在 AndroidManifest. xml 文件的<application>元素节点内，需要添加一个<activity>子节点，添加代码如下。

```
<activity android:name = ".NotifySimpleActivity" />
```

运行结果：在 Android Studio 支持的模拟器上，运行 Activity_SimpleNotify 项目，运行结果如图 10-2 所示。

图 10-2(a)是初始页面，单击“转发送简单消息界面”按钮，进入通知信息设置页面，如图 10-2(b)所示。输入通知标题和内容，单击“发送简单消息”按钮，即刻在状态栏新增该通知的小图标，如图 10-2(c)所示，从屏幕顶部往下滑动，拉开通知栏，可看到通知的详细信息，如图 10-2(d)所示。如果单击该通知信息所在位置，即可回到如图 10-2(a)所示的初始页面。

(a) 初始运行页面

(b) 输入通知标题

(c) 单击“发送简单消息”按钮

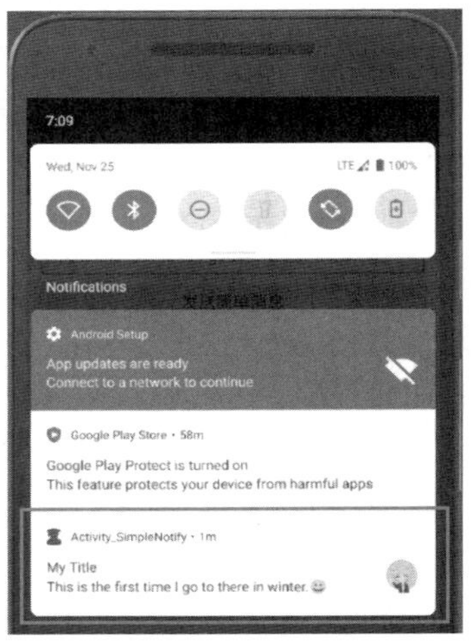

(d) 向下拉开通知栏

图 10-2　发送简单通知消息的运行效果

10.1.3　自定义通知栏

在 Notification.Builder 类和 NotificationCompat.Builder 类中，有一个共同的方法 setContent(RemoteViews views)，其功能是为通知栏设置一个用户定制的通知栏视图，取代系统默认的视图模板。RemoteViews 被称为远程视图。

1. RemoteViews 简介

RemoteViews 类位于 android.widget.RemoteViews 包中，它就是一个小型的简化的

页面，与 Activity 页面相比，在功能上有所减少，受限制较多。但是和 Activity 一样，它拥有自己的布局文件。

注意：RemoteViews 不支持所有的控件，也不支持自定义控件。它能支持的布局类型有 FrameLayout、LinearLayout、RelativeLayout、GridLayout；支持的控件有 Button、ImageView、ImageButton、ProgressBar、TextView、ListView、GridView、Chronometer（计时器）、AnalogClock（模拟时钟）等。在设计布局文件时，不能使用它不支持的 View。例如，不能使用 EditText，否则会发生异常。

通常创建 RemoteViews 对象使用的构造方法如下。

```
public RemoteViews(String packageName, int layoutId){
    this(getApplicationInfo(packageName, UserHandle.myUserId()), layoutId);
}
```

其中，第一个参数是应用所在包名，第二个参数是布局文件的 ID。

RemoteViews 不能使用 findViewById()方法，从而就不能获取 RemoteViews 中的子 View。它通过一系列的 set 方法来设置控件。RemoteViews 通过使用 PendingIntent，并调用 setOnClickPendingIntent()方法来实现对 View 的单击事件。RemoteViews 中常用的方法见表 10-4。

表 10-4 RemoteViews 类常用方法及说明

方法	说明
setChronometer(int viewId, long base, String format, boolean started)	设置计时器信息
setImageViewBitmap(int viewId, Bitmap bitmap)	设置 ImagView 或 ImageButton 控件的位图对象
setImageViewResource(int viewId, int srcId)	设置 ImagView 或 ImageButton 控件的图像资源 ID
setImageViewUri(int viewId, Uri uri)	设置 ImagView 或 ImageButton 控件的图像的 URI
setProgressBar(int viewId, int max, int progress, boolean indeterminate)	设置进度条信息，包括最大值和当前进度
setTextColor(int viewId, int color)	设置指定 TextView 或 Button 控件的文字颜色
setTextViewCompoundDrawables(int viewId, int left, int top, int right, int bottom)	设置指定 TextView 或 Button 控件的文字周围图标
setTextViewText(int viewId, CharSequence text)	设置指定 TextView 或 Button 控件的文字内容
setTextViewTextSize (int viewId, int units, float size)	设置指定 TextView 或 Button 控件的文字大小
setViewPadding(int viewId, int left, int top, int right, int bottom)	设置指定控件的间距
setViewVisibility(int viewId, int visibility)	设置指定控件是否可见
setOnClickPendingIntent(int viewId, PendingIntent pendingIntent)	设置指定控件的单击响应事件，需要先定义 PendingIntent

为什么称 RemoteViews 为远程视图呢？是因为它可以在其他进程中显示，可以跨进程更新它的界面。在 Android 中，RemoteViews 主要用于通知栏和桌面小部件。下面介绍使用它来定制通知栏。

2. RemoteViews 在通知栏中的应用

自定义通知栏视图将使用 RemoteViews。其基本步骤如下。

(1) 定义一个布局文件。

(2) 创建一个 RemoteViews 对象。

(3) 调用一系列 set 方法实例化 RemoteViews 对象。

(4) 通过 Notification. setContent()方法加载到通知对象中。

【案例 10.2】 在播放音乐时或暂停音乐时发送通知，要求定制通知栏视图，设置通知栏的信息包括：大图标，进度，进度条，通知内容，计时与按钮。

说明：设计两个 Activity 页面，第一个布局是页面的界面，设计两个按钮，一个是播放时发送通知，另一个是暂停时发送通知。第二个布局是通知栏界面，设计一个进度条，用于呈现发送时刻音乐播放的进度。在本例中使用 RemoteViews 加载自定义界面到通知对象。单击通知栏返回主页面。

开发步骤及解析：过程如下。

1) 创建项目

在 Android Studio 中创建一个名为 Activity_Notification 的项目。其包名为 ee. example. activity_notification。

2) 准备颜色资源

编写 res/values 目录下的 colors. xml 文件，分别声明 black、red、white、darkgrey 和 blue 等颜色。

3) 准备图片和背景资源

复制 ic_notif_s. png、ic_notif. png 到本项目的 res/drawable 目录中分别作为通知的小图和大图资源。在该目录中，edittext_selector. xml 文件声明了文本输入框的选择样式，shape_edit_focus. xml 声明输入框获取焦点时的图形背景，shape_edit_normal. xml 声明文本输入框失去焦点时（即通常状态）的图形背景。

4) 设计布局

在 res/layout 目录下有两个布局文件，activity_main. xml 是 Activity 页面的布局文件，notify_remoteview. xml 是通知栏的布局文件。

在 activity_main. xml 代码中，设计一个输入框用于输入音乐的播放进度数据，其 ID 为 et_song。设计两个并排的按钮，第一个按钮的 ID 为 btn_start，单击它表示音乐正在播放，推送播放通知；第二个按钮的 ID 为 btn_stop，单击它表示音乐暂停播放，推送暂停通知。

notify_remoteview. xml 文件，定制通知栏的布局。在该布局中将屏幕横向划分为三个区域，左侧是通知大图标；中段是通知的主要内容，分三行显示，上行为音乐播放进度数，中间行是进度条，下行是通知内容；右侧上部是一个计时器，下部是一个按钮，控制音乐的播放和暂停。其中，声明进度条的代码片段如下。

```
<ProgressBar
    android:id = "@ + id/pb_play"
    style = "?android:attr/progressBarStyleHorizontal"
    android:layout_width = "match_parent"
    android:layout_height = "0dp"
```

```
6          android:layout_weight = "1"
7          android:max = "100"
8          android:progress = "10" />
```

第 3 行是系统定义的进度条形状样式。?android:attr/progressBarStyleHorizontal 表示水平形状，?android:attr/progressBarStyle 表示圆圈形状。

声明计时器的代码片段如下。

```
<Chronometer
    android:id = "@ + id/chr_play"
    android:layout_width = "match_parent"
    android:layout_height = "0dp"
    android:layout_weight = "1" />
```

5）开发逻辑代码

在 java/ee.example.activity_notification 包下打开 MainActivity.java 文件，代码如下。

```
package ee.example.activity_notification;

import androidx.appcompat.app.AppCompatActivity;

import android.os.Build;
import android.os.Bundle;
import android.app.Notification;
import android.app.NotificationManager;
import android.app.NotificationChannel;
import android.app.PendingIntent;
import android.content.Context;
import android.content.Intent;
import android.os.SystemClock;
import android.view.View;
import android.view.View.OnClickListener;
import android.widget.EditText;
import android.widget.RemoteViews;

public class MainActivity extends AppCompatActivity implements OnClickListener{
    private EditText et_song;
    private String PAUSE_EVENT = "ee.example.musicplayer.pause";
    private String PLAY_EVENT = "ee.example.musicplayer.play";
    private String channelId = "1";
    private String channName = "channel_name";

    @Override
    protected void onCreate(Bundle savedInstanceState) {
        super.onCreate(savedInstanceState);
        setContentView(R.layout.activity_main);
        et_song = (EditText) findViewById(R.id.et_song);
        findViewById(R.id.btn_start).setOnClickListener(this);
        findViewById(R.id.btn_pause).setOnClickListener(this);
    }

    @Override
```

```
    public void onClick(View v) {

        Notification notify = new Notification();
        if (v.getId() == R.id.btn_start) {
            notify = getNotify(this, PAUSE_EVENT, "荷塘月色", true,
                        Integer.parseInt(et_song.getText().toString()), SystemClock.
                        elapsedRealtime());
        }else if (v.getId() == R.id.btn_pause) {
            notify = getNotify(this, PLAY_EVENT, "荷塘月色", false,
                    Integer.parseInt(et_song.getText().toString()), SystemClock.
                    elapsedRealtime());
        }

        NotificationManager notifyMgr =
                (NotificationManager) getSystemService(Context.NOTIFICATION_SERVICE);
        if(Build.VERSION.SDK_INT >= Build.VERSION_CODES.O) {
            NotificationChannel channel = new NotificationChannel(channelId, channName,
                                        NotificationManager.IMPORTANCE_HIGH);
            notifyMgr.createNotificationChannel(channel);
        }
        notifyMgr.notify(R.string.app_name, notify);
    }

private Notification getNotify(Context ctx, String event, String song, boolean isPlay,
                        int progress, long time) {
    Intent pIntent = new Intent(event);
    PendingIntent nIntent = PendingIntent.getBroadcast(
                ctx, R.string.app_name, pIntent, PendingIntent.FLAG_UPDATE_CURRENT);
    RemoteViews notify_music = new RemoteViews(ctx.getPackageName(),
                                        R.layout.notify_remoteview);
     if (isPlay == true) {
        notify_music.setTextViewText(R.id.btn_play, "暂停");
        notify_music.setTextViewText(R.id.tv_progress, "播放进度:" + String.valueOf
             (progress) + "%");
        notify_music.setTextViewText(R.id.tv_play, song + "正在播放");
        notify_music.setChronometer(R.id.chr_play, time, "%s", true);
    } else {
        notify_music.setTextViewText(R.id.btn_play, "继续");
        notify_music.setTextViewText(R.id.tv_progress, "播放进度:" + String.valueOf
             (progress) + "%");
        notify_music.setTextColor(R.id.tv_progress,getResources().getColor(R.color.red));
        notify_music.setTextViewText(R.id.tv_play, song + "暂停播放");
        notify_music.setTextColor(R.id.tv_play,getResources().getColor(R.color.red));
        notify_music.setChronometer(R.id.chr_play, time, "%s", false);
     }
     notify_music.setProgressBar(R.id.pb_play, 100, progress, false);
     notify_music.setOnClickPendingIntent(R.id.btn_play, nIntent);

     Intent intent = new Intent(ctx, MainActivity.class);
     PendingIntent contentIntent = PendingIntent.getActivity(
                ctx, R.string.app_name, intent, PendingIntent.FLAG_UPDATE_CURRENT);

```

```
78          Notification.Builder builder = new Notification.Builder(ctx);
79          builder.setContentIntent(contentIntent)
80                      .setContent(notify_music)
81                      .setTicker(song)
82                      .setSmallIcon(R.drawable.ic_notif_s);
83          if (Build.VERSION.SDK_INT >= Build.VERSION_CODES.O) {
84                  builder.setChannelId(channelId);
85          }
86          Notification notify = builder.build();
87          return notify;
88      }
89
90 }
```

(1) 第21、22行，声明两个常量，分别表示另一个应用项目包中的服务。在本案例中使用字符串代替。

(2) 第36～51行，实现OnClickListener接口的onClick()事件。第40、42行，根据不同的按钮被单击，调用getNotify()方法实现通知对象实例化，传递不同的参数，获取通知对象notify实例。第45行创建NotificationManager对象notifyMgr，第46～49行，在Android 8.0及以上的版本中运行，实现创建渠道对象并在notifyMgr对象中注册。第50行，notifyMgr对象调用notify()方法，将通知对象notification推送到状态栏上。

(3) 第53～88行，实现方法getNotify()。其中第57行创建RemoteViews对象notify_music。第58～71行使用一系列的set方法设置notify_music对象的视图内容及属性。第72行，设置notify_music对象内的单击事件，其中的PendingIntent对象nIntent定义于第55、56行。第75、76行定义的PendingIntent对象contentIntent，用于实例化通知对象。在第80行，调用setContent(notify_music)，将RemoteViews对象notify_music加载到通知对象中。

运行结果：在Android Studio支持的模拟器上，运行Activity_Notification项目，运行结果如图10-3所示。

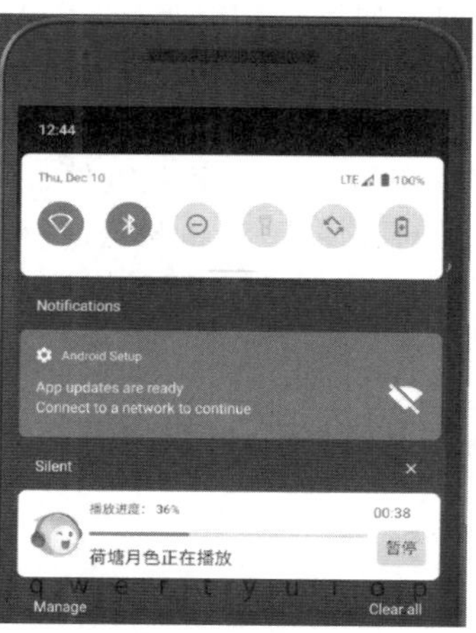

(a) 初始运行页面

(b) 正在播放通知消息

图10-3 定制通知栏视图的运行效果

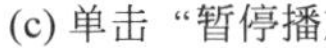
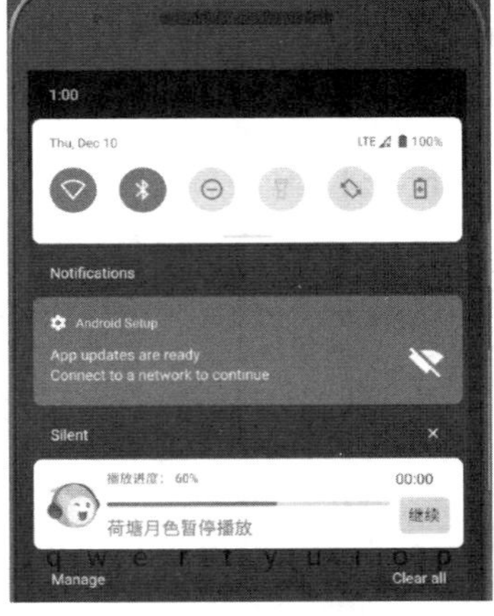

(c) 单击“暂停播放”按钮　　　　(d) 暂停播放通知消息

图 10-3 （续）

图 10-3(a)是初始页面，当输入 36 并单击“开始播放”按钮后，发送通知信息。下滑屏幕拉开通知栏，可见通知内容，如图 10-3(b)所示。在本案例的通知栏上单击，可以回到初始页面。这时输入 60，然后单击“暂停播放”按钮，如图 10-3(c)所示。这时也发送了通知，滑开通知栏，可以见到不一样的通知信息，如图 10-3(d)所示。同样，单击该通知信息所在位置，即可回到初始页面。

开发 Notification 应用，实现的是在同一应用项目中进行的消息传递。如果想做到跨应用和普通用户流之外进行消息传递，就需要用到广播。

10.2　广播接收器 BroadcastReceiver

Broadcast(广播)是一种广泛运用的应用项目之间传输信息的机制，与 Activity 不同。Activity 只能一对一通信，而 Broadcast 可以一对多地通信。

BroadcastReceiver(广播接收器)是用于接收来自系统和应用的广播，对广播进行过滤并响应的组件，是 Android 的四大组件之一，其所在类位于 android. content. BroadcastReceiver 包中。BroadcastReceiver 是一种后台组件，当所关注的事件发生时发送广播。例如，系统的时区改变、系统时间改变、电池电量低时会发送广播；某个应用在特定的事件触发时，也发送广播。但是，在应用中不要滥用后台响应广播，否则会导致系统变慢。

10.2.1　广播的内容及分类

1. 广播的内容

要发送的广播内容封装在一个 Intent 对象中，这个 Intent 可以携带要传送的数据。广播的内容可以来自系统，也可以来自其他应用项目。

Android 内置了很多系统级别的广播。只要涉及手机的基本操作(如开机、网络状态变化、拍照等)，都会发出相应的广播。每个广播都有特定的Intent-Filter(包括具体的action)，通过Intent-Filter对象的action来确定系统操作。常用于系统广播的intent的action常量见表10-5。

表10-5　常用的系统操作对应action常量

常　量	action值	系统操作
ACTION_AIRPLANE_MODE_CHANGED	android. intent. action. AIRPLANE_MODE	关闭或打开飞行模式
ACTION_BATTERY_CHANGED	android. intent. action. BATTERY_CHANGED	充电时或电量发生变化
ACTION_BATTERY_LOW	android. intent. action. BATTERY_LOW	电池电量低
ACTION_BATTERY_OKAY	android. intent. action. BATTERY_OKAY	电池电量充足(即从电量低变化到饱满时会发出广播)
ACTION_BOOT_COMPLETED	android. intent. action. BOOT_COMPLETED	系统启动完成后(仅广播一次)
ACTION_CAMERA_BUTTON	android. intent. action. CAMERA_BUTTON	按下照相时的拍照按键(硬件按键)时
ACTION_CLOSE_SYSTEM_DIALOG	android. intent. action. CLOSE_SYSTEM_DIALOGS	屏幕锁屏
ACTION_DATE_CHANGED	android. intent. action. DATE_CHANGED	日期改变
ACTION_MEDIA_EJECT	android. intent. action. MEDIA_EJECT	插入或拔出外部媒体
ACTION_PACKAGE_ADDED	android. intent. action. PACKAGE_ADDED	成功安装APK
ACTION_PACKAGE_REMOVED	android. intent. action. PACKAGE_REMOVED	成功删除APK
ACTION_REBOOT	android. intent. action. REBOOT	重启设备
ACTION_SHUTDOWN	android. intent. action. ACTION_SHUTDOWN	关闭系统时
ACTION_TIME_CHANGED	android. intent. action. TIME_SET	时间改变
ACTION_TIMEZONE_CHANGED	android. intent. action. TIMEZONE_CHANGED	时区改变

Android其他应用的广播开发，可以通过继承BroadcastReceiver基类来自定义广播，接收某些我们感兴趣的广播来实现业务需求。

2. 广播的分类

Android广播可分为普通广播、有序广播、黏性广播和本地广播。其中黏性广播在Android 5.0(API 21)以后已经失效，在此不做介绍。

普通广播(Normal Broadcast)，是一种完全异步执行的广播。当广播发出后，所有广播接收器几乎会在同一时刻接收到这条广播。这类广播效率比较高，但所有接收器的执行顺序不确定。缺点是接收者不能将处理结果传递给下一个接收者，并且无法截断广播的传播。

有序广播(Ordered Broadcast)，是一种按照优先级依次执行的广播。数值越大优先级越高，广播发出之后同一时刻只有一个广播接收器能收到这条广播消息。当广播接收器接收到广播后，只有这个广播接收器中的逻辑执行完毕后，广播才继续传递到下一个广播接收器，并且可以把执行结果传递给下一个接收器，或截断正在传递的广播。

本地广播(Local Broadcast)，是局部广播，只在进程内传播，可避开其他程序的广播，比系统全局广播更高效，因此无须担心安全漏洞隐患。

前两种广播都是全局广播，所有应用都可以接收到，这样就带来安全隐患，而本地广播只在进程内传播，可以起到保护数据安全的作用。

10.2.2 注册广播接收器

Android应用可以通过两种方式来注册广播：一种方式是静态的，即使用清单文件声明广播接收器。另一种是动态的，在逻辑代码中通过上下文注册广播接收器。

1. 静态注册

静态注册是在AndroidManifest.xml的< application >元素里添加< receiver >元素，并设置要接收的action。使用清单文件声明广播接收器，系统软件包管理器会在应用安装时注册接收器。当系统收到广播消息时会发送广播到Android应用上，如果该应用还没有启动，系统也可以启动该应用并发送广播。

注意：在Android 8.0（API 26）版本以后，限制了大多数系统级别的隐式广播在清单文件中声明，只有少数的常用的隐式广播可以在清单文件中声明。例如，系统启动、时区更改、时间更改、APK安装等。

在清单文件中声明广播接收器，使用< receiver > 元素。以注册电量变化广播为例，清单文件代码片段如下。

```
<receiver android:name = ".BatteryChangedReceiver" android:exported = "true">
    <intent - filter>
        <action android:name = "android.intent.action.BATTERY_CHANGED"/>
        <category android:name = "android.intent.category.DEFAULT" />
    </intent - filter>
</receiver>
```

如果是注册有序广播，则需要添加priority属性，设置优先级别，优先级的范围为从－1000到1000，代码片段如下。

```
<receiver android:name = ".BatteryChangedReceiver" android:exported = "true">
    <intent - filter android:priority = "1000">
        <action android:name = "android.intent.action.BATTERY_CHANGED"/>
        <category android:name = "android.intent.category.DEFAULT" />
    </intent - filter>
</receiver>
```

在逻辑代码文件中，有相应的广播接收器子类实现，代码片段如下。

```
public class BatteryChangedReceiver extends BroadcastReceiver {
    private static final String TAG = "BatteryChangedReceiver ";
    @Override
    public void onReceive(Context context, Intent intent) {
        //在这里定义事件响应处理
        int currLevel = intent.getIntExtra(BatteryManager.EXTRA_LEVEL, 0);  //当前电量
        int total = intent.getIntExtra(BatteryManager.EXTRA_SCALE, 1);  //总电量
        int percent = currLevel * 100 / total;
```

```
            StringBuilder sb = new StringBuilder();
            sb.append("battery: " + percent + "%");        //生成当前电量百分比字符串
            String log = sb.toString();
            Log.d(TAG, log);
            Toast.makeText(context, log, Toast.LENGTH_LONG).show();
        }
    }
```

注意： 在我们的应用首次启动的时候，系统会自动实例化静态注册的 BroadcastReceiver，然后将这个 BroadcastReceiver 注册到系统中，无论该应用项目是否处于活动状态，BroadcastReceiver 始终处于被监听状态。当系统接收到广播之后，就会做出相应的判断，调用 onReceive()方法。通过这种方式注册的广播，即使我们的应用被销毁，依然能收到广播。

由此可知，静态注册的广播，耗电、占内存、不受程序生命周期影响，所以在 Android 8.0 以后禁止了大部分广播的静态注册。本地广播也无法使用静态注册方式。

2. 动态注册

动态注册是在类的实现代码中使用上下文来注册广播接收器的。实现的步骤是先创建 BroadcastReceiver 的对象，然后创建 IntentFilter 并调用 registerReceiver(BroadcastReceiver, IntentFilter)来注册接收器，代码片段如下。

```
BroadcastReceiver br = new MyBroadcastReceiver();
IntentFilter filter = new IntentFilter(ConnectivityManager.CONNECTIVITY_ACTION);
filter.addAction(Intent.ACTION_AIRPLANE_MODE_CHANGED);      //添加 Intent 的 action
this.registerReceiver(br, filter);
```

registerReceiver()方法有两个参数，第一个参数是指定接收器对象，第二个参数是 IntentFilter 对象，注意，IntentFilter 对象内要设置接收的 action 属性。

通过动态的方式注册的广播接收器，如果不再需要该接收器或上下文不再有效时，需要调用 unregisterReceiver(android.content.BroadcastReceiver)方法来注销接收器。

动态注册的广播接收器，只要在该上下文有效时，注册的接收器就会接收广播。例如，如果在名为 Activity_1 这个上下文中注册，那么只要 Activity_1 没有被销毁，就可以收到广播；如果在一个应用的上下文中注册，那么只要这个应用在运行，就可以收到广播。由此可见，通过动态注册的广播接收器，在内存占用、信息安全方面都比静态注册的好。因此，从 Android 8.0 版本开始，推荐使用动态注册方式。

10.2.3 广播接收器的生命周期

BroadcastReceiver 作为 Android 的组件，也存在生命周期。当接收到广播的时候开始创建 BroadcastReceiver 对象，到 onReceiver()方法执行完成之后，BroadcastReceiver 就不再活跃。

使用静态注册接收器，当系统监听到有广播消息发出时，会创建新的 BroadcastReceiver 对象。此对象调用 onReceive(Context, Intent)方法，在此期间 BroadcastReceiver 对象有效。一旦从 onReceiver()方法返回代码，系统便会认为该组件不再活跃，BroadcastReceiver 的生命周期结束。

使用动态注册接收器，如果在 Activity 的上下文 onCreate(Bundle)中注册接收器，则应在 onDestroy()中注销；如果在 onResume()中注册接收器，则应在 onPause() 中注销。

注意：BroadcastReceiver 的生命周期很短暂，onReceiver()方法的执行在 10 秒之内，否则 Android 系统会弹出一个超时异常。正因为 BroadcastReceiver 的生命周期很短暂，所以不要在 onReceiver()方法中执行耗时的操作，因为 BroadcastReceiver 被销毁后，这个子进程就会成为空进程，很容易被杀死。如果需要在 BroadcastReceiver 中执行耗时的操作，可以通过 Intent 启动 Service 去完成，但不能绑定 Service。

10.2.4 发送广播

BroadcastReceiver 注册完之后，这个 BroadcastReceiver 就能够接收响应的广播了。在开发中通过 Context 类下的方法发送广播。对于不同类别的广播，发送广播的方法也不同。下面分别介绍。

1. sendBroadcast()方法

使用 context. sendBroadcast()方法发送的广播是普通广播，所有接收器的执行顺序不确定。这种广播传递效率比较高，但接收器不能将处理结果传递给下一个接收器，并且无法终止广播的传播。

方法格式如下。

```
sendBroadcast(Intent intent)
```

其中，参数 intent 是发送广播的意图对象。

2. sendOrderedBroadcast()方法

使用 context. sendOrderedBroadcast()方法发送的广播是有序广播，发送有序广播类似于上级下发文件，是一级一级地往下发，如果中间级接收失败，发送就会终止。所有的广播接收器都有一个优先级参数，范围为－1000 到 1000，通过 receiver 的 intent-filter 中的 android:priority 属性来设置，数值越大优先级越高。当广播接收器接收到广播后，可以使用 setResult()方法把结果传递给下一个接收者，通过 getResult()方法获取上一个接收者传递过来的结果；也可以通过 abortBroadcast()方法丢弃该广播，截断该广播传递下去。

sendOrderedBroadcast()方法格式如下。

```
sendOrderedBroadcast(Intent intent, String receiverPermission,
                     BroadcastReceiver resultReceiver, Handler scheduler,
                     int initialCode, String initialData, Bundle initialExtras)
```

方法中有七个参数，分别如下。

intent：是一个 Intent，是发送广播的意图对象，可以携带广播信息。

receiverPermission：是一个字符串，用于指定接收权限，如果为空，则不需要权限。

resultReceiver：是一个广播接收器，用于指出有序广播的最终广播接收者，该最终接收器可以不在清单文件中配置，仍会接收到广播。

scheduler：是一个线程，若传 null，则默认是在主线程中。

initialCode：是一个整数，用于初始化的一个值。默认值为 Activity. RESULT_OK。

initialData：是一个字符串，用于发送广播的初始化数据(相当于一条广播数据)，可为 null。

initialExtras：是一个 Bundle，用于绑定数据传递(例如 Intent 对象 extras 数据)，通常为 null。

3. LocalBroadcastManager. sendBroadcast()方法

LocalBroadcastManager. sendBroadcast()方法会将广播发送给与发送器位于同一应用中的接收器，即本地广播。如果不需要跨应用发送广播，请使用本地广播。这种实现方法的效率更高(无须进行进程间通信)，而且无须担心安全漏洞隐患。

在 Android 4.0 之后，将本地广播封装到 LocalBroadcastManager 类中。在使用方式上，本地广播与全局广播几乎相同，只是注册、发送、注册广播接收器时将主调 context 实例变成了 localBroadcastManager 实例。例如 localBroadcastManager. sendBroadcast(intent)。

注意：从 Android 8.0 版本开始，对于静态注册的自定义广播，在调用 sendBroadcast() 或 sendOrderedBroadcast()方法之前，一定要先指定 intent 参数的所在包名，即调用 intent. setPackage(getPackageName())方法添加 intent 所在包名，否则广播无法发送给接收器。动态注册的广播接收器不需要调用此方法。

10.2.5 BroadcastReceiver 的应用案例

BroadcastReceiver 组件没有提供可视化的界面来显示广播信息，但可以使用 Notification 来实现广播信息的显示，也可以用即时消息提示来短暂显示。下面通过一个用户自定义的 BroadcastReceiver 发送广播案例，介绍如何使用静态、动态注册的方式来实现 BroadcastReceiver 组件的编程开发。

【案例 10.3】 通过两个按钮，一个按钮发出一个无序广播，另一个按钮发出有序广播。要求分别将广播信息发送到状态栏和页面的 Toast 提示消息中；并且有序广播接收器要传递信息到下一个接收器。

说明：本案例自定义三个 BroadcastReceiver，其中第一个接收器使用静态方式注册，接收信息后分别向状态栏和 Toast 提示消息区发送广播；第二个接收器使用动态方式注册，只向 Toast 提示消息区发送广播，并向下一个接收器传递信息；第三个接收器是在发送有序广播时作为最终接收者。

本案例应用需要定义 4 个类实现代码，一个类是 Activity，在该类中定义用户界面的交互、用于动态注册和注销广播接收器，在按钮的 onClick()方法中使用 sendBroadcast()、sendOrderedBroadcast()方法发送广播。另外 3 个类是 BroadcastReceiver 的子类，在子类中重写 onReceive()方法，用于定义接收器接收到广播 Intent 传递的信息之后的各种操作。

开发步骤及解析：过程如下。

1）创建项目

在 Android Studio 中创建一个名为 Activity_BroadcastReceiver 的项目。其包名为 ee.example.activity_broadcastreceiver。

2）准备图片

将用于通知的图标的图片资源复制到本项目的 res/drawable 目录中。

3）设计布局

在 res/layout 目录下编写布局文件 activity_main.xml。在布局中设计两个按钮，ID 为 btn_send 的按钮，单击后发送无序广播；ID 为 btn_orderedsend 的按钮，单击后发送有序广播。

4）开发逻辑代码

在 java/ee.example.activity_broadcastreceiver 包下有四个 Java 代码文件，一个是 MainActivity 的类实现代码，三个是 BroadcastReceiver 子类的实现代码。

MainActivity.java 文件，是实现 MainActivity 的类定义，代码如下。

```
package ee.example.activity_broadcastreceiver;

import androidx.appcompat.app.AppCompatActivity;
import android.content.Intent;
import android.content.IntentFilter;
import android.os.Bundle;
import android.view.View;

public class MainActivity extends AppCompatActivity {
    private IntentFilter myIntentfilter = null;
    private MyBcReceiver2 myReceiver2 = null;          //第 2 个接收器
    private MyBcReceiver3 myReceiver3 = null;          //第 3 个接收器
    private Intent sendintent = null;                  //发送广播的 Intent

    @Override
    protected void onCreate(Bundle savedInstanceState) {
        super.onCreate(savedInstanceState);
        setContentView(R.layout.activity_main);
        //设置广播消息 Intent 的属性值:
        sendintent = new Intent();
        sendintent.setAction("myaction");
        sendintent.putExtra("msg", "广播消息:教育部下发 2020 新规定.");
        sendintent.setPackage(getPackageName());  //用于添加静态注册的接收器所在包名

        findViewById(R.id.btn_send).setOnClickListener(new BtnOnClick());
        findViewById(R.id.btn_orderedsend).setOnClickListener(new BtnOnClick());
    }

    @Override
    protected void onResume() {
        super.onResume();
        myReceiver2 = new MyBcReceiver2();
        myIntentfilter = new IntentFilter("android.intent.action.MyBcReceiver2");
        myIntentfilter.addAction("myaction");
        myIntentfilter.setPriority(99);
```

```
36             registerReceiver(myReceiver2, myIntentfilter);
37         }
38 
39         @Override
40         protected void onPause() {
41             super.onPause();
42              if(myReceiver2!= null) {                    //注销接收器
43                 unregisterReceiver(myReceiver2);
44             };
45             if(myReceiver3 != null) {                    //注销接收器
46                 unregisterReceiver(myReceiver3);
47             };
48         }
49 
50         private class BtnOnClick implements View.OnClickListener {
51             @Override
52             public void onClick(View v) {
53 
54                 switch (v.getId()) {
55                     case R.id.btn_send: {
56                         //发送无序广播
57                         sendBroadcast(sendintent);
58                         break;
59                 }
60                 case R.id.btn_orderedsend: {
61                     myReceiver3 = new MyBcReceiver3();
62                     //发送有序广播
63                     sendOrderedBroadcast(sendintent, null,myReceiver3,
64                             null, MainActivity.this.RESULT_OK, null, null);
65                     break;
66                 }
67             }
68         }
69     }
70 
71 }
```

(1) 第 20～22 行，在 onCreate()方法中设置携带广播内容的 Intent 实例 sendintent 的 Action 和 Extrant 属性值，所有注册的接收器，只要其 Intent 的 action 值与 sendintent 的 action 值相同，都可以接收到该广播消息。

(2) 第 23 行，指定接收器所在包的包名。用于对静态注册的广播接收器的 Intent 添加包名。对于 Android 8.0 以上的版本，必须调用该方法，否则静态注册的广播接收器不能接收广播。

(3) 第 30～37 行，重写 onResume()方法。第 32～36 行动态注册第二个广播接收器，第 35 行为 Intent 实例设置一个优先级别，为发送有序广播设定接收广播的优先级别。

(4) 第 57 行，发送一个无序广播。执行该方法，静态注册的接收器 1 和动态注册的接收器 2 都会收到广播，由于接收器 3 没有注册，所以不会接收到广播。

(5) 第 63 行，发送一个有序广播。在该方法中的第三个参数指定了最终接收者是第三个接收器，不用注册就可以接收广播。

下面是三个实现 BroadcastReceiver 子类的 Java 文件，分别重写三个广播接收器的

onReceive()方法。

MyBcReceiver1.java 文件，实现第一个广播接收器，代码如下。

```
package ee.example.activity_broadcastreceiver;

import androidx.core.app.NotificationCompat;
import android.app.Notification;
import android.app.NotificationChannel;
import android.app.NotificationManager;
import android.content.BroadcastReceiver;
import android.content.Context;
import android.content.Intent;
import android.os.Build;
import android.util.Log;
import android.view.Gravity;
import android.widget.Toast;
import androidx.core.content.ContextCompat;
import static android.content.ContentValues.TAG;
import static android.content.Context.NOTIFICATION_SERVICE;

public class MyBcReceiver1 extends BroadcastReceiver {

    private NotificationManager mManager;

    @Override
    public void onReceive(Context context, Intent intent) {

        mManager = (NotificationManager) context.getSystemService(NOTIFICATION_SERVICE);

        if (Build.VERSION.SDK_INT >= Build.VERSION_CODES.O) {
            String channelId = "brd";
            String channelName = "广播";
            int importance = NotificationManager.IMPORTANCE_DEFAULT;
            NotificationChannel channel = new NotificationChannel(channelId, channelName,
                                                   importance);
            mManager.createNotificationChannel(channel);
        }

        String title = "第 1 个接收器,接收到广播消息";
        String message = intent.getStringExtra("msg");

        Notification notification = new NotificationCompat.Builder(context, "brd")
                .setContentTitle(title)
                .setContentText(message)
                .setWhen(System.currentTimeMillis())
                .setColor(ContextCompat.getColor(context, R.color.colorPrimary))
                .setSmallIcon(R.drawable.bc1)
                .setAutoCancel(true)
                .build();
        mManager.notify(1, notification);

        Toast tt1 = Toast.makeText(context,
                    "这是第 1 个接收器\n" + intent.getStringExtra("msg"),
                    Toast.LENGTH_LONG);
```

```
51              tt1.setGravity(Gravity.TOP,0,650);
52              tt1.show();
53              Log.e(TAG, "onReceive: " + "第 1 个广播接收器" );
54
55          }
56      }
```

(1) 第 27～33 行,对于 Android 8.0 以上的版本,创建通知的渠道实例。如果不是 Android 8.0 以上的版本,该段代码不须执行。

(2) 第 36 行,从 intent 的 Extra 的 msg 中获取广播内容,并存入 message 变量中。

(3) 第 38～45 行,创建一个 Notification 实例,通知内容显示 message 广播内容。

(4) 第 46 行,将创建的通知实例发送到状态栏中。

(5) 第 48～50 行,创建一个 Toast 消息实例 tt1。其中第 49 行是从 intent 中获取 Extra 的 msg 中的信息,其内容就是接收到的广播内容。

(6) 第 51 行,设置该 Toast 消息实例的显示位置,该设置表示即时消息框会出现在屏幕的顶部向下偏移 650 的纵向位置。注意,在 Android 11(API 30)版本的模拟器上,Toast.setGravity()失效,在低于 API 30 的模拟器上可以看到 Toast 消息框的位置变化。

(7) 第 52 行,显示该 Toast 消息。

MyBcReceiver2.java 文件,代码片段如下。

```
1   public class MyBcReceiver2 extends BroadcastReceiver {
2
3       @Override
4       public void onReceive (Context context, Intent intent) {
5           Toast tt2 = Toast.makeText(context,
6                       "这是第 2 个接收器\n" + intent.getStringExtra("msg") + "\n 并传递消息",
7                       Toast.LENGTH_LONG);
8           tt2.setGravity(Gravity.CENTER,0,200);
9           tt2.show();
10
11          Log.e(TAG, "onReceive: " + "第 2 个广播接收器,并传递消息" );
12
13          Bundle bundle  =  new Bundle();
14          bundle.putString("first", "第 2 个接收器传递的消息啦!");
15          setResultExtras(bundle);
16      }
17
18  }
```

(1) 第 8 行,设置 Toast 消息实例的显示位置,该设置使得 Toast 消息框会出现在屏幕的中央向下偏移 200 的纵向位置。同样,在低于 API 30 的模拟器上运行才有效。

(2) 第 13～15 行,创建一个 Bundle 实例,绑定第二个接收器自身的一条信息,用于发送有序广播时,传递给下一个接收器。

MyBcReceiver3.java 文件,代码片段如下。

```
public class MyBcReceiver3 extends BroadcastReceiver {

    @Override
    public void onReceive(Context context, Intent intent) {
```

```
5
6         String message =  intent.getExtras().getString("msg");
7         Bundle bundle = getResultExtras(true);
8         String first = bundle.getString("first");
9
10        Toast tt3 = Toast.makeText(context,
11            "这是第 3 个接收器\n"  + message + "\n\n"
12            + "同时接收到第 2 个接收器传来的消息:" + first,Toast.LENGTH_LONG);
13        tt3.setGravity(Gravity.BOTTOM,0,120);
14        tt3.show();
15
16        String ss = "onReceive: " + "第 3 个广播接收器,并接收第 2 个接收器传出的消息" + first;
17        Log.e(TAG, ss);
18
19        }
20
21    }
```

(1) 第 7、8 行，创建一个 Bundle 实例，获取从第二个接收器绑定的名-值对 first 中的信息。

(2) 第 13 行，设置 Toast 消息实例的显示位置，该设置使得 Toast 消息框会出现在屏幕的底部向上偏移 120 的纵向位置。

5）静态注册设置

在 AndroidManifest.xml 中注册第一个广播接收器，其注册代码添加在< application >元素下，代码片段如下。

```
1    < receiver android:name = ".MyBcReceiver1">
2        < intent - filter android:priority = "100">
3            < action android:name = "myaction" />
4        </intent - filter >
5    </receiver >
```

第 3 行添加了 intent-filter 的属性 android:priority＝"100"，用于在执行发送有序广播时的优先级设置。

运行结果：在 Android Studio 支持的模拟器上，运行 Activity_BroadcastReceiver 项目。由于在 API 30 以上的模拟器中，Toast 的 setGravity()方法失效，本例在 API 29 模拟器中运行，运行结果如图 10-4 所示。

在项目的初始页面中，单击“发送无序广播”按钮，即可看到如图 10-4(a)所示的效果。从屏幕下部显示的 Toast 消息来看，第一个接收器和第二个接收器的 Toast 消息内容同时出现，两个接收器几乎同时收到广播；此时可在状态栏上看到新增了一个通知图标，向下拉开通知栏，可以看到第一个接收器发送的通知信息，是接收到的广播内容，如图 10-4(b)所示。

单击“发送有序广播”按钮，即可看到如图 10-4(c)所示的效果。如果仔细观察可以发现，显示在屏幕下部的即时消息框，是按第一个、第二个和第三个接收器依次出现的，几秒后消息框消失。

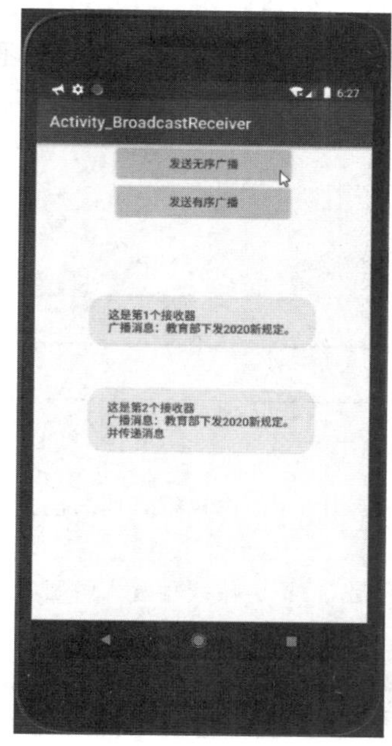

(a) 单击“发送无序广播”按钮

(b) 向下拉开通知栏

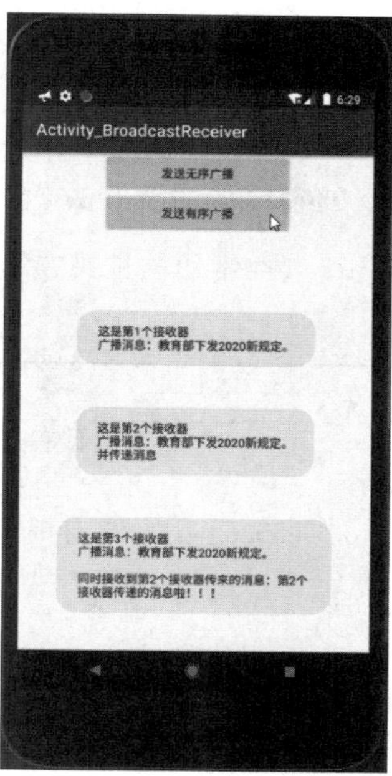

(c) 单击“发送有序广播”按钮

图 10-4 发送广播的运行效果

10.3 Android 后台线程

当一个程序第一次启动时，Android 会启动一个 Linux 进程和一个主线程（Main Thread）。线程是比进程更小的执行单位。

主线程主要负责处理与 UI 相关的事件，如用户的按键事件、用户接触屏幕的事件以及屏幕绘图事件等，并把相关的事件分发到对应的组件进行处理。所以主线程通常又被叫作 UI 线程。Android 的主线程就是 UI 线程，用户对 UI 的操作必须在 UI 线程中执行。

在开发 Android 应用时必须遵守单线程（Single-threaded）模型的原则。Android 希望 UI 线程能根据用户的要求做出快速响应，如果 UI 线程花太多时间处理后台的工作，当 UI 事件发生时，让用户等待时间超过 5 秒而未处理，Android 系统就会给用户显示 ANR 提示信息。主线程除了处理 UI 事件之外，还要处理 Broadcast 消息。所以在 BroadcastReceiver 的 onReceive() 函数中，也不宜占用太长的时间，否则导致主线程无法处理其他的 Broadcast 消息或 UI 事件。如果占用时间超过 10 秒，Android 系统就会给用户显示 ANR 提示信息。解决办法自然还是解放 UI 主线程，将耗时操作交给子线程，避免阻塞。

10.3.1 线程 Thread

所有非主线程就是子线程，子线程一般都是后台线程，它是由用户在程序中定义的。由于 Android 的 UI 是单线程的，为了保证应用项目的响应效率，一般使用后台线程，把所有运行慢的、耗时的操作移出主线程，放到子线程中。例如，Activity 在 5 秒内不响应任何输入事件，或者 BroadcastReceiver 在 10 秒后仍未完成 onReceive 的处理操作，等等。

Thread 类位于 java. lang. Thread 包中，是主要用于处理耗时操作的 Java 线程管理机制。在 Java 虚拟设备上允许一个应用可以并发处理多个线程的执行操作。

1. Thread 的主要方法

Thread 类常用的主要方法见表 10-6。

表 10-6　Thread 类常用方法及说明

方　法	说　明
run()	包含线程运行时所执行的代码
start()	用于启动线程
sleep(long millis)	线程休眠，以毫秒为单位。线程休眠时交出 CPU，让 CPU 去执行其他的任务，然后线程进入阻塞状态，sleep 方法不会释放锁
yield()	使当前线程交出 CPU，让 CPU 去执行其他的任务，但不会使线程进入阻塞状态，而是重置为就绪状态，yield 方法不会释放锁
currentThread()	获取当前线程。是静态方法，返回值为当前线程
interrupt()	中断线程。注意只能中断已经处于阻塞的线程

Thread 线程主要有 5 种状态：新建(New)、可运行(Runnable)、运行(Running)、阻塞(Blocked)和死亡(Dead)。当 Thread 对象被实例化之后就处于 New 状态；调用了 start()之后就处于 Runnable 状态；线程被 CPU 执行，调用 run()之后就处于 Running 状态；调用 join()、sleep()之后，线程处于 Blocked 状态；在线程的 run()方法运行完毕或被中断或被异常退出，线程将会到达 Dead 状态。

2. 开启 Thread

在 Android 中开启一个子线程有两种方法：一是定义一个继承 Thread 类的子类；二是实现 Runnable 接口。

(1) 使用继承 Thread 类方式开启子线程，代码片段如下。

```
private class SyncThread extends Thread {
    SyncThread(String name) {              //构造方法
        super(name);
    }
    @Override
    public void run() {                    //重写 run()
        //执行耗时操作
    }
}
```

然后在 Activity 的实现代码中创建该线程对象并启动，代码片段如下。

```
SyncThread syncThread1 = new SyncThread("线程一");
SyncThread syncThread2 = new SyncThread("线程二");

syncThread1.start();
syncThread2.start();
```

(2) 使用实现 Runnable 接口方式开启子线程,代码片段如下。

```
private class SyncRunnable implements Runnable {
    @Override
    public void run() {
        //执行耗时操作
    }
}
```

然后在 Activity 的实现代码中创建该线程对象并启动,代码片段如下。

```
SyncRunnable syncRunnable = new SyncRunnable();
Thread syncThread1 = new Thread(syncRunnable, "线程一");
Thread syncThread2 = new Thread(syncRunnable, "线程二");

syncThread1.start();
syncThread2.start();
```

注意:在子线程中操作 UI 对象是不安全的。如果后台的子线程执行了 UI 对象,Android 就会发出错误信息 CalledFromWrongThreadException。为了解决后台子线程与主线程(UI 线程)间的信息交互问题,Android 设计了一种 Handler 消息传递机制或 AsyncTask 后台运行事务,来处理多线程之间的数据传递问题。

10.3.2 Handler 消息传递机制

1. Handler 类

Handler 类位于 android. o. Handlers 包中。它负责消息(Message)的发送和消息内容的执行处理,主要完成 Android 的控件(Widget)与应用项目中子线程之间的交互。自定义的后台线程可与 Handler 通信,一个 Handler 对应一个 Activity,Handler 在主线程中运行。

Handler 的作用有两个:发送消息和处理消息。在传递消息时与下列几个类一起工作。

Message(消息)类,位于 android. os. Message 包中。Message 可以理解为线程间进行交流的消息。

MessageQueue(消息队列)类,位于 android. os. MessageQueue 包中。它采用先进先出的方式来管理 Message。

Looper 类,位于 android. os. Looper 包中。每个 Looper 对应一个 Message Queue,Looper 的 loop()方法负责读取 Message Queue 中的消息,读到信息后就把消息交给 Handler 进行处理。所有线程有且只有一个 Looper,应用在启动的时候,系统就自己创建了一个 Looper 并与主线程绑定。

在创建 Message 对象时,构造方法中的主要参数的含义说明见表 10-7。

表 10-7 Message 类构造方法主要参数说明

参数	参数类型	说明
what	int	是消息的标识,用于标识本次消息的唯一编号
arg1	int	存放消息的处理结果

续表

参　数	参数类型	说　明
arg2	int	存放消息的处理代码
obj	Object	存放返回消息的数据结构
repkyTo	Messager	回应信使,在跨进程通信中使用,多线程通信无须用

通过 Handler 对象处理线程之间的消息传递,常用方法见表 10-8。

表 10-8　Handler 类消息传递常用方法及说明

方　法	返回类型	说　明
handleMessage(Message msg)	void	子类对象通过该方法接收消息。该方法处于主线程(UI 线程)中,主线程通过执行该方法处理子线程发出的消息
hasMessages(int what)	boolean	监测消息队列中是否还有 what 值(指定标识)的消息
obtainMessage (int what)	Messager	获取指定 what 值的消息对象
post(Runnable r)	boolean	将线程 r 添加到消息队列中
sendEmptyMessage(int what)	boolean	发送一个只含有 what 值的空消息
sendEmptyMessageDelayed(int what)	boolean	延迟一段时间后发送只含有 what 值的空消息
sendMessage(Message msg)	boolean	发送消息
sendMessageAtTime(Message msg, long uptimeMillis)	boolean	在指定时间点发送消息,时间单位为毫秒
sendMessageDelayed (Message msg, long delayMillis)	boolean	延迟一段时间后发送消息
removeMessages (int what)	void	从消息队列中移除指定 what 值的消息

2. Handler 类的使用

主线程处理子线程发出的消息需要实现 Handler 对象的 handleMessage()方法,根据 Message 消息的具体内容分别进行相应处理。所以在应用开发中,Handler 与 Message 要结合一起使用。带有 Handler 类的应用项目,开发步骤如下。

(1) 在 Activity 或在子线程中创建 Handler 类的对象,重写 handlerMessage()方法。

Handler 对象用于发送和处理消息,所以,当创建 Handler 对象时,必须要指定一个 Looper 对象。Handler 的构造方法为: Handler(Looper looper),不带参数的构造方法 Handler()已被摒弃。对于每个 Activity,系统自动为 UI 线程(即主线程)初始化了一个 Looper 对象。

在主线程中创建一个 Handler 对象,可使用如下代码。

```
public Handler mHandler = new Handler(Looper.getMainLooper ());     //创建对象 mHandler
```

子线程默认是没有消息循环和消息队列的,如果想让该线程具有消息队列和消息循环,需要在线程中首先调用 Looper. prepare()来创建消息队列,然后调用 Looper. loop()进入消息循环。在 Looper. prepare()和 Looper. loop()之间创建 Handler 对象实例。

在子线程中创建 Handler 对象,可使用如下代码片段。

```
//创建子线程
new Thread(new Runnable() {
    @Override
    public void run() {
        Looper.prepare();  //在子线程中初始化一个 Looper 对象,即为当前线程创建消息队列
        handler = new Handler(Looper.myLooper()){      //实例化 Handler 对象
            @Override
            public void handleMessage(Message msg) {
                super.handleMessage(msg);
                //把 UI 线程发送来的消息显示到屏幕上
                Log.i("main", "what = " + msg.what + "," + msg.obj);
                Toast.makeText(WorkThreadActivity.this, "what = " + msg.what + "," + msg.obj,
                                                 Toast.LENGTH_SHORT).show();
            }
        };
        Looper.loop(); //运行刚才初始化的 Looper 对象,循环消息队列的消息
    }
}).start();
```

(2) 在新启动的线程中调用 sendEmptyMessage()或 sendMessage()方法向 Handler 发送消息。

(3) Handler 类的对象用 handlerMessage()方法接收消息,然后根据消息执行相应的操作。

下面通过多线程机制实现一个幻灯片多线程滚动轮播的案例,介绍 Handler 和 Message 结合使用的用法。

【案例 10.4】 使用 Handler 机制以幻灯片的形式显示 IT 界的名人照片,每张图片停留 5 秒。

说明:本案例使用两个类来定义代码,一个是 MainActivity 类,在其中构建 Handler 对象,并重写 handlerMessage()方法,在该方法内根据消息的值确定 UI 的显示内容。另一个是继承 Thread 的子类,并重写 run()方法,在该方法内使用 sendEmptyMessage()方法向 Handler 发送消息。

开发步骤及解析:过程如下。

1) 创建项目

在 Android Studio 中创建一个名为 Activity_Thread 的项目。其包名为 ee.example.activity_thread。

2) 准备图片

将要显示的图片资源复制到本项目的 res/drawable 目录中。

3) 准备字符串资源

编写 res/values 目录下的 strings.xml 文件,分别定义下列字符串信息,包括标题 title 和 andy、bill、edgar、torvalds、turing 等人的一行简介信息。

4) 设计布局

在 res/layout 目录下编写布局文件 activity_main.xml,设计布局内容包括一个标题控件,ID 为 picTitle; 一个图片控件,ID 为 myPic; 一个简介控件,ID 为 picName。

5）开发逻辑代码

在 java/ee.example.activity_thread 包下有两个代码文件，一个是 MainActivity 的子类代码实现，另一个是 Thread 子类的代码实现。

MainActivity.java 是实现 Activity 的子类定义，代码如下。

```
package ee.example.activity_thread;

import androidx.appcompat.app.AppCompatActivity;

import android.os.Bundle;
import android.os.Handler;
import android.os.Looper;
import android.os.Message;
import android.widget.ImageView;
import android.widget.TextView;

public class MainActivity extends AppCompatActivity {
    ImageView myPicture;
    TextView myPicname;
    //创建一个 Handler 对象
    Handler myHandler = new Handler(Looper.getMainLooper()){
        @Override
        public void handleMessage(Message msg) {
            super.handleMessage(msg);
            switch(msg.what){           //判断 what 的值
                case 0:
                    myPicture.setImageResource(R.drawable.andy);
                    myPicname.setText(R.string.andy);
                    break;
                case 1:
                    myPicture.setImageResource(R.drawable.bill);
                    myPicname.setText(R.string.bill);
                    break;
                case 2:
                    myPicture.setImageResource(R.drawable.edgar);
                    myPicname.setText(R.string.edgar);
                    break;
                case 3:
                    myPicture.setImageResource(R.drawable.torvalds);
                    myPicname.setText(R.string.torvalds);
                    break;
                case 4:
                    myPicture.setImageResource(R.drawable.turing);
                    myPicname.setText(R.string.turing);
                    break;
            };
        }
    };

    @Override
    protected void onCreate(Bundle savedInstanceState) {
        super.onCreate(savedInstanceState);
        setContentView(R.layout.activity_main);
```

```
49
50            myPicture = (ImageView) findViewById(R.id.myPic);
51            myPicname = (TextView) findViewById(R.id.picName);
52            MyThread myThread = new MyThread(this); //初始化 MyThread 线程
53            myThread.start();                        //启动线程
54        }
55    }
```

(1) 第 16～43 行创建一个 Handler 对象 myHandler，并重写了 handleMessage()方法，在该方法中根据传入的 Message 对象 msg 的 what 值，确定 ImageView 和 TextView 中的显示内容。

(2) 第 52 行，初始化一个线程对象，该线程类由 MyThread.java 实现。

(3) 第 53 行，调用 start()方法启动线程。

MyThread.java 是实现 Thread 子类的文件，代码如下。

```
1    package ee.example.activity_thread;
2
3    public class MyThread extends Thread{
4        MainActivity activity;                    //activity 的引用
5        int what = 0;                             //发送消息的 what 值
6        public MyThread(MainActivity activity){
7            this.activity = activity;             //得到 activity 的引用
8        }
9        @Override
10       public void run() {                       //重写的 run 方法
11           while(true){
12               activity.myHandler.sendEmptyMessage((what++) % 5);  //发送消息
13               try{
14                   Thread.sleep(5000);           //睡眠 5 秒
15               }
16               catch(Exception e){
17                   e.printStackTrace();
18               }
19           }
20       }
21   }
```

(1) 第 4 行引用 MainActivity 类的对象 activity，用于在本类中使用 MainActivity 中的 Handler 对象 myHandler。

(2) 第 10～20 行重写 run()方法，在该方法中执行一个永真循环，每隔 5 秒，调用 sendEmptyMessage()方法向 MainActivity 中的 myHandler 对象发送一个消息值，该值由表达式(what++)%5 计算，%为整除运算符，其计算结果分别是 0、1、2、3、4。

运行结果：在 Android Studio 支持的模拟器上，运行 Activity_Thread 项目，运行结果是以 5 秒的时间间隔逐一播放资源中提供的人物照片，并且循环往复。其显示效果如图 10-5 所示。

使用 Handler+Message 还可以实现多线程的通信处理，本书在后续章节会陆续给出有关线程应用的案例。对于多线程的应用，在子线程中要更新 UI 还必须要引入 Activity

(a) 显示当前一张照片

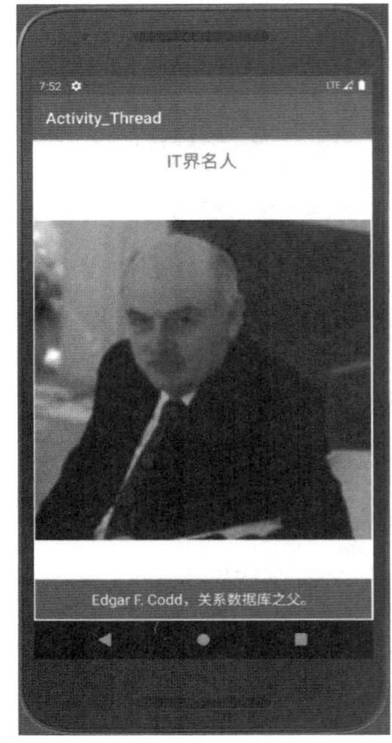

(b) 5秒后显示下一张照片

图 10-5 幻灯片滚动轮播的运行效果

中的 Handler，这样处理容易与子线程自己的 Handler 混在一起，不宜维护。为了解决这一问题，Android 提供了 AsyncTask 这个轻量级的异步任务工具。

10.3.3 异步任务 AsyncTask

AsyncTask 是一种简单的实现后台运行事务的方式，内部封装了 Handler＋Message 的线程通信机制。使用 AsyncTask 可以将耗时的操作放在后台来处理，而不需要另外编写代码来处理线程或 Handler 实现通信。在开发中只需按照业务流程编写代码，无须关心线程通信的复杂流程。AsyncTask 通常用于网络访问操作，例如 HTTP 接口调用、文件上传与下载等。尽管 AsyncTask 类已在 Android 11(API 30)中被摒弃，但是使用它能够简化线程处理，所以仍然介绍给读者。

1. AsyncTask 类

AsyncTask 是抽象类，位于 android.os.AsyncTask 包中。AsyncTask 使用异步任务实现后台运行事务。异步任务，是 Android 1.5 引入的一个新特性，主线程可以把某些操作留给后台运行的线程去运行，Android 自动新建和删除后台线程，无须开发者编写代码实现。所以，AsyncTask 的特点是任务在主线程之外运行，而回调方法是在主线程中执行，这就有效地避免了使用 Handler 带来的麻烦。

AsyncTask 是一个模板抽象类(AsyncTask < Params，Progress，Result >)，从它派生出来的子类需要指定模板的参数类型。这里的三个参数是一种范式类型，定义这种范式参数的格式为“<数据类型>...<参数名>”。它们被用于一个异步任务。在一个异步任务里，不是所有的类型总被使用。假如一个类型不被使用，可以简单地使用 Void 类型来替代，如 doInBackground(Void...Params)。下面就 AsyncTask < Params,Progress,Result >的三个参数进行说明。

Params：String 类型或自定义的数据结构。当执行时，向后台任务的执行方法传递参数，例如 HTTP 请求的 URL。

Progress：int 类型。是后台任务执行过程中的进度，例如在后台计算期间，后台任务执行的百分比。

Result：String 类型或自定义的数据结构。后台任务执行完返回的结果参数，例如 HTTP 调用的执行结果、返回报文等。

对于抽象类，必须通过继承该类的子类来实现应用。在编写 AsyncTask 子类时，需要实现下列方法，见表 10-9。

表 10-9　AsyncTask 子类需要实现的方法及说明

方　法	返回类型	说　明
onPreExecute()	void	准备执行任务时触发。该方法在 doInBackground()方法执行之前调用
doInBackground(Params...)	Result	在后台执行，完成任务的主要工作，通常需要较长的时间。输入参数对应 execute()方法的输入参数，输出参数对应 onPreExecute()方法的输入参数。注意，该方法运行于子线程，不能操作界面，其他方法都能操作界面。一般网络请求等异步处理操作都放在该方法中
onProgressUpdate(Progress...)	void	用于刷新任务执行的进度条。在 doInBackground()方法中调用 publicProgress()方法时触发
onPostExecute(Result)	void	返回任务执行的结果。在 doInBackground()方法执行完毕后调用，输入参数对应 doInBackground()方法的输出参数
onCancelled(Result)	void	调用任务对象的 cancel()方法时触发。表示取消任务并返回

AsyncTask 类除了上述需要重写的方法外，还有一些常用的可直接调用的公共方法，见表 10-10。

表 10-10　AsyncTask 子类可直接调用的常用方法及说明

方　法	返回类型	说　明
cancel (boolean mayInterruptIfRunning)	boolean	取消任务。该方法调用后，doInBackground()方法中的处理可能不会马上停止。如果要立即停止处理，则可在 doInBackground()方法中加入 isCancelled 的判断
execute (Params... params)	AsyncTask	开始执行异步处理任务。这个方法将会返回此任务本身。此方法必须在主线程中调用
publicProgress(Progress...)	void	更新任务的进度。该方法只能在 doInBackground()方法中调用，调用后会触发 onProgressUpdate()方法
get (long timeout, TimeUnit unit)	Result	等待计算结束并返回结果，最长等待时间为：timeout(超时时间)
getStatus()	AsyncTask.Status	获得任务的当前状态。状态有三种：PENDING(还未执行)、RUNNING(正在执行)、FINISHED(执行完毕)
isCancelled()	boolean	判断该任务是否取消。true 表示取消，false 表示未取消

2. AsyncTask 执行的步骤

AsyncTask 的执行分为若干步骤，每一步都对应一个回调方法，这些方法需要重写。注意：在任务的执行过程中，这些方法不由应用项目调用，而是被自动调用的。

一个 AsyncTask 运行的过程中，经历以下 4 个步骤。

1）onPreExecute()

在 execute 调用后立即在主线程中执行。这步通常被用于设置任务，例如在用户界面显示一个进度条。

2）doInBackground(Params...)

当 onPreExecute()执行完成后，立即在后台线程中运行。这步被用于执行较长时间的后台计算。异步任务的参数也被传到这步。计算的结果必须在这步返回，将传回到上一步。在执行过程中可以调用 publishProgress(Progress...)来更新任务的进度。

3）onProgressUpdate(Progress...)

在调用 publishProgress(Progress...)后，在主线程中运行。执行的时机是不确定的。当后台计算还在进行时，这个方法用于在用户界面显示进度。例如：这个方法可以被用于一个进度条动画或在文本域显示日志。

4）onPostExecute(Result)

当后台计算结束时，调用主线程。后台计算结果作为一个参数传递到前台。

由于 AsyncTask 的特殊性，很多开发工作都被省略了，比如下面的工作是不需要开发者来完成的：产生自己的后台线程；在适当的时候终止后台线程等。AsyncTask 在线程产生后不用用户干涉，全部自发完成。因此，在编程中必须遵守如下线程规则。

（1）AsyncTask 任务实例必须创建在主线程中。

（2）execute(Params...)必须在主线程上调用。

（3）不要手动调用 onPreExecute()、onPostExecute(Result)、doInBackground(Params...)和 onProgressUpdate(Progress...)方法。

（4）每个 AsyncTask 只能有 1 个实例被执行，同时运行 2 个以上的 AsyncTask，将会抛出异常。

下面通过一个简单的案例来介绍如何利用 AsyncTask 实现后台操作任务的用法。

【案例 10.5】 使用 AsyncTask 来模拟一个页面的加载进度。使用进度条对话框的形式展示进度更新操作。

说明：本案例使用两个类来定义代码，一个是 MainActivity 类，用于构建用户的交互界面，在该界面上（即 UI 线程）会有一些显示信息、按钮和进度条等内容，同时监听 AsyncTask 任务。另一个是实现继承 AsyncTask 的子类，并重写相应的方法，在该子类的 doInBackground()方法中模拟网络通信的时间消耗。

开发步骤及解析：过程如下。

1）创建项目

在 Android Studio 中创建一个名为 Activity_ProgressAsyncTask 的项目。其包名为 ee.example.activity_progressasynctask。

2）准备颜色资源

编写 res/values 目录下的 colors.xml 文件，分别声明 black 等颜色。

3）准备图片和背景资源

将要显示的图片资源复制到本项目的 res/drawable 目录中。在该目录中，还有 edittext_selector.xml、shape_edit_focus.xml、shape_edit_normal.xml 文件分别声明文本输入框的相关操作样式。在前面案例中有类似的样式声明，在此就不赘述了。

4）设计布局

在 res/layout 目录下编写布局文件 activity_main.xml。在界面上设计一个输入框，ID 为 et_booknm，设置背景为 android:background="@drawable/edittext_selector"，用于输入需要加载的信息；一个按钮，ID 为 but_load，单击它执行开始加载操作；下面是一个文本框，ID 为 tv_async，显示加载的进度和结果。

5）开发逻辑代码

在 java/ee.example.activity_progressasynctask 包下有两个代码文件，一个是 MainActivity 类代码实现，另一个是 AsyncTask 子类的代码实现。

MainActivity.java 是实现 MainActivity 的子类定义，代码如下。

```
package ee.example.activity_progressasynctask;

import androidx.appcompat.app.AppCompatActivity;

import android.os.Bundle;
import android.app.ProgressDialog;
import android.view.View;
import android.widget.Button;
import android.widget.EditText;
import android.widget.TextView;

public class MainActivity extends AppCompatActivity implements
                                        ProgressAsyncTask.OnProgressListener {

    private EditText et_async;
    private TextView tv_async;
    private ProgressDialog mDialog;

    @Override
    protected void onCreate(Bundle savedInstanceState) {
        super.onCreate(savedInstanceState);
        setContentView(R.layout.activity_main);
        et_async = (EditText) findViewById(R.id.et_booknm);
        tv_async = (TextView) findViewById(R.id.tv_async);

        Button bok = (Button)this.findViewById(R.id.but_load);
            bok.setOnClickListener(
                new View.OnClickListener(){
                    @Override
            public void onClick(View v) {
                 //创建后台任务的对象，并通过 execute()启动后台线程
                String msg = et_async.getText().toString();
                ProgressAsyncTask asyncTask = new ProgressAsyncTask(msg);
                asyncTask.setOnProgressListener(MainActivity.this);
                asyncTask.execute(msg);
```

```
35                  }
36              }
37          );
38      }
39
40      private void closeDialog() {
41          if (mDialog != null && mDialog.isShowing() == true) {
42              mDialog.dismiss();
43          }
44      }
45
46          @Override
47          public void onFinish(String result) {
48              String desc = String.format("您要阅读的«%s»已经加载完毕", result);
49              tv_async.setText(desc);
50              closeDialog();
51          }
52
53          @Override
54          public void onCancel(String result) {
55              String desc = String.format("您要阅读的«%s»已经取消加载", result);
56              tv_async.setText(desc);
57              closeDialog();
58          }
59
60          @Override
61          public void onUpdate(String request, int progress, int sub_progress) {
62              String desc = String.format("%s当前加载进度为%d%%", request, progress);
63              tv_async.setText(desc);
64              mDialog.setProgress(progress);
65              mDialog.setSecondaryProgress(sub_progress);
66          }
67
68          @Override
69          public void onBegin(String request) {
70              tv_async.setText(request + "开始加载");
71              if (mDialog == null || mDialog.isShowing() != true) {
72                  mDialog = new ProgressDialog(this);
73                  mDialog.setTitle("请稍等");
74                  mDialog.setMessage(request + "页面加载中……");
75                  mDialog.setIcon(R.drawable.loading_sml);
76                  mDialog.setProgressStyle(ProgressDialog.STYLE_HORIZONTAL);
77                  mDialog.show();
78              }
79          }
80
81      }
```

(1) 第12行，声明实现ProgressAsyncTask类中的一个监听接口OnProgressListener。

(2) 第32～34行，创建一个ProgressAsyncTask子类的实例，并将文本框输入的内容作为参数传到该实例，通过execute()方法来启动后台线程。

(3) 第48、55、62行，在format()方法中，使用后面的字符串变量值来替换前面字符串中的"%s"，使用后面的整型变量值来替换前面字符串中的"%d"。

(4) 第 64、65 行，更新当前进度条对话框实例的进度数据。

(5) 第 72～77 行，创建进度对话框实例，并设置标题、提示内容、图标和进度条样式等信息。最后显示该进度对话框。

(6) 在该类中，定义了 onBegin()、onUpdate()、onFinish()、onCancel()四个方法的具体实现。描述了在主线程中，该类实例在整个生命周期中的操作内容。

ProgressAsyncTask.java 是继承 AsyncTask 类的实现，代码如下。

```
package ee.example.activity_progressasynctask;

import android.os.AsyncTask;

public class ProgressAsyncTask extends AsyncTask<String, Integer, String> {
    private String mArtical;
    public ProgressAsyncTask(String title) {
        super();
        mArtical = title;
    }

    @Override
    protected String doInBackground(String... params) {
        int ratio = 0;
        for (; ratio <= 100; ratio += 2) {
            //睡眠 0.2 秒模拟网络通信处理
            try {
                Thread.sleep(200);
            } catch (InterruptedException e) {
                e.printStackTrace();
            }
            publishProgress(ratio);              //更新进度条
        }
        return params[0];
    }

    @Override
    protected void onPreExecute() {
        mListener.onBegin(mArtical);
    }

    @Override
    protected void onProgressUpdate(Integer... values) {
        mListener.onUpdate(mArtical, values[0], 0);
    }

    @Override
    protected void onPostExecute(String result) {
        mListener.onFinish(result);
    }

    @Override
    protected void onCancelled(String result) {
        mListener.onCancel(result);
    }
```

```
46
47          private OnProgressListener mListener;
48          public void setOnProgressListener(OnProgressListener listener) {
49              mListener = listener;
50          }
51
52          public static interface OnProgressListener {
53              public abstract void onBegin(String request);
54              public abstract void onUpdate(String request, int progress, int sub_progress);
55              public abstract void onFinish(String result);
56              public abstract void onCancel(String result);
57          }
58
59      }
```

（1）在该子类中，定义后台线程的操作内容。重写了 onPreExecute()、doInBackground()、onProgressUpdate()、onPostExecute()、onCancelled()方法，除了 doInBackground()方法外，其他的方法都分别调用了主线程中相应的 onBegin()、onUpdate()、onFinish()、onCancel()方法。

（2）第 15～23 行，调用 Thread. sleep(200)方法，模拟网络数据传输，每隔 0.2 秒发送一次进度数据。

（3）第 52～57 行，定义一个 OnProgressListener 接口，其中声明了四个方法：onBegin()、onUpdate()、onFinish()和 onCancel()。这四个方法在主线程中被实现。

运行结果：在 Android Studio 支持的模拟器上，运行 Activity_ProgressAsyncTask 项目。在输入框输入要阅读文章的标题，如图 10-6(a)所示。单击“开始加载”按钮，即可弹出加载该文章的进度条对话框，显示加载的进度比例，如图 10-6(b)所示。加载完毕会给出完成信息，如图 10-6(c)所示。

(a) 输入文章名称

(b) 加载页面进度

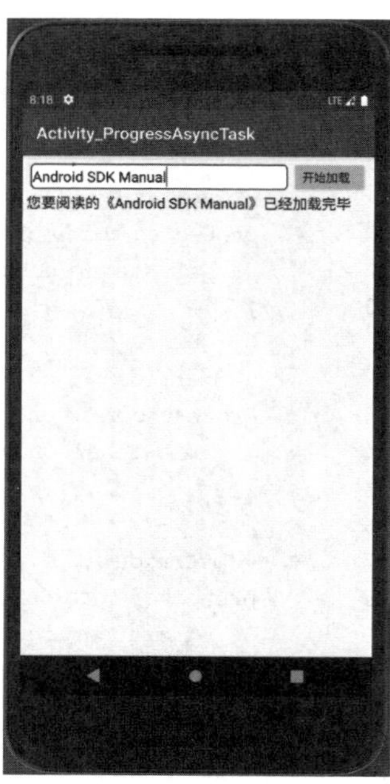

(c) 加载完毕

图 10-6　加载要阅读的文章页面的运行效果

10.3.4 Android 线程池简介

在 Android 中会经常用非 UI 线程来处理耗时的逻辑，即使用线程处理异步任务，但每个线程的创建和销毁都需要一定的开销。假设每次执行一个任务都需要开一个新的线程去执行，则这些线程的创建和销毁将消耗大量的资源，并且这些线程都是“各自为政”，很难对其进行控制，更别说一堆线程了。为了解决这些问题就需要线程池来对线程进行管理。

线程池是在队列中并行运行任务的托管线程集合。当现有线程变为空闲状态时，新任务会在这些线程上执行。

1. 线程池的优势

在应用开发中使用线程池，有如下优势。

(1) 通过重用已存在的线程，可以降低线程创建和销毁造成的系统资源消耗。

(2) 当有任务到达时，通过复用已存在的线程，无须等待新线程的创建便能立即执行，从而提高系统响应速度。

(3) 方便线程并发数的管控。如果线程无限制地创建，可能会导致内存占用过多而产生内存溢出，以及线程过度切换而造成 CPU 付出更多的时间成本。

(4) 提供更强大的功能，延时定时线程池。

2. ThreadPoolExecutor 类的构造方法

Android 中线程池的概念来源于 Java 中的 Executor，线程池真正的实现类是 ThreadPoolExecutor，它间接实现了 Executor 接口。ThreadPoolExecutor 提供了一系列参数来配置线程池，通过不同的参数配置实现不同功能特性的线程池。ThreadPoolExecutor 类一共有 4 个构造方法。其中，拥有最多参数的构造方法如下。

```
public ThreadPoolExecutor(int corePoolSize,
                          int maximumPoolSize,
                          long keepAliveTime,
                          TimeUnit unit,
                          BlockingQueue<Runnable> workQueue,
                          ThreadFactory threadFactory,
                          RejectedExecutionHandler handler){
...
}
```

参数说明如下。

corePoolSize：线程池的核心线程数，默认情况下线程池是空的，只在任务提交时才会创建线程。如果当前运行的线程数小于 corePoolSize，则会创建新线程来处理任务；如果大于或等于 corePoolSize，则不再创建。

maximumPoolSize：最大的线程数，包括核心线程，也包括非核心线程，在线程数达到这个值后，新来的任务将会被阻塞。

keepAliveTime：超时的时间，闲置的非核心线程超过这个时长，将会被销毁回收；当

allowCoreThreadTimeOut 为 true 时，这个值也作用于核心线程。

unit：超时时间的时间单位。可选的单位有天 Days、小时 Hours、分钟 Minutes、秒 Seconds、毫秒 Milliseconds 等。

workQueue：线程池的任务队列，通过 execute()方法提交的 runnable 对象会存储在这个队列中。

threadFactory：线程工厂，为线程池提供创建新线程的功能。

handler：任务无法执行时，回调 handler 的 rejectedExecution()方法来通知调用者。

3. Android 中线程池分类及用法

Android 中的线程池都是直接或间接通过配置 ThreadPoolExecutor 来实现的。在 Android 中的 Executors 类提供了 4 个工厂方法用于创建 4 种不同特性的线程池给开发者使用。如需将任务发送到线程池，请使用 ExecutorService 接口。

1) FixedThreadPool

FixedThreadPool 被称为定长线程池，其特点是：只有核心线程数，并且没有超时机制，因此核心线程即使闲置时，也不会被回收。这种线程池能更快地响应外界的请求。适合控制线程最大并发数的应用场景。

下面是定长线程池的编程步骤示例，代码片段如下。

```
//1.通过构造方法获得定长线程池
public static ExecutorService newFixedThreadPool(int nThreads) {
    return new ThreadPoolExecutor(nThreads, nThreads, 0L, TimeUnit.MILLISECONDS,
                                   new LinkedBlockingQueue<Runnable>());
}
//2.创建定长线程池对象,设置线程池线程数量固定为 3
ExecutorService fixedThreadPool = Executors.newFixedThreadPool(3);
//3.创建 Runnable 类线程对象,定义需执行的任务
Runnable task = new Runnable(){
    public void run(){
        System.out.println("执行任务啦");
    }
};
//4.向线程池提交任务:execute()
fixedThreadPool.execute(runnable 对象);
//5.关闭线程池
fixedThreadPool.shutdown();
```

2) CachedThreadPool 线程池

CachedThreadPool 被称为可缓存线程池，其特点是：没有核心线程，非核心线程数量没有限制，超时为 60 秒。这种线程池适用于执行大量耗时较少的任务，当线程闲置超过 60 秒时就会被系统回收掉，当所有线程都被系统回收后，它几乎不占用任何系统资源。

下面是可缓存线程池的编程步骤示例，代码片段如下。

```
//1.通过构造方法获得定长线程池
public static ExecutorService newCachedThreadPool() {
    return new ThreadPoolExecutor(0, Integer.MAX_VALUE,
                                   60L, TimeUnit.SECONDS,
```

```
                                        new SynchronousQueue<Runnable>());
    }
    //2.创建可缓存线程池对象
    ExecutorService cachedThreadPool  =  Executors.newCachedThreadPool();
    //3.创建 Runnable 类线程对象,定义需执行的任务
    Runnable task  = new Runnable(){
        public void run(){
            System.out.println("执行任务啦");
        }
    };
    //4.向线程池提交任务:execute()
    cachedThreadPool.execute(runnable 对象);
    //5.关闭线程池
    cachedThreadPool.shutdown();
```

3）ScheduledThreadPool

ScheduledThreadPool 被称为定时线程池，其特点是：核心线程数是固定的，非核心线程数量没有限制，没有超时机制。这种线程池主要用于执行定时任务和具有固定周期的重复任务。

下面是定时线程池的编程步骤示例，代码片段如下。

```
//1.通过构造方法获得定时线程池
public ScheduledThreadPoolExecutor(int corePoolSize) {
    super(corePoolSize, Integer.MAX_VALUE, 0, NANOSECONDS,
        new DelayedWorkQueue());
}
//2.创建定时线程池对象,设置线程池线程数量固定为 5
ScheduledExecutorService scheduledThreadPool  =  Executors.newScheduledThreadPool(5);
//3.创建 Runnable 类线程对象,定义需执行的任务
Runnable task  = new Runnable(){
    public void run(){
        System.out.println("执行任务啦");
    }
};
//4.向线程池提交任务:schedule()
scheduledThreadPool.schedule(task, 2000, TimeUnit.MILLISECONDS);
                                                //延迟 2000 毫秒后执行任务
scheduledThreadPool.scheduleAtFixedRate(task,1,10,TimeUnit.SECONDS);
                                                //延迟 1 秒后每隔 10 秒执行任务
//5.关闭线程池
scheduledThreadPool.shutdown();
```

4）SingleThreadExecutor

SingleThreadExecutor 被称为单线程化线程池，其特点是：只有一个核心线程，并没有超时机制。用意在于统一所有的外界任务到一个线程中，这使得在这些任务之间不需要处理线程同步的问题。这种线程池用于不适合并发但可能引起 I/O 阻塞性及影响 UI 线程响应的操作，例如数据库操作、文件操作等。

下面是单线程化线程池的编程步骤示例，代码片段如下。

```
//1.通过构造方法获得定时线程池
public static ExecutorService newSingleThreadExecutor() {
```

```
    return new FinalizableDelegatedExecutorService
        (new ThreadPoolExecutor(1, 1,
                                0L, TimeUnit.MILLISECONDS,
                                new LinkedBlockingQueue<Runnable>()));
}
//2.创建单线程化线程池
ExecutorService singleThreadExecutor = Executors.newSingleThreadExecutor();
//3.创建 Runnable 类线程对象,定义需执行的任务
Runnable task = new Runnable(){
    public void run(){
        System.out.println("执行任务啦");
            }
    };
//4.向线程池提交任务:execute()
singleThreadExecutor.execute(task);
//5.关闭线程池
singleThreadExecutor.shutdown();
```

10.4 服务 Service

服务(Service)是 Android 的四大组件之一，是一种可以在 Android 系统后台长时间运行而不需要界面的服务程序。这种服务程序可以由其他应用组件启动，也可以由系统内部的某一个触发条件启动，而且当用户切换到其他应用后，服务程序仍将在后台继续运行。例如：播放音乐，后台数据计算(如记录用户的地理信息位置的改变)，发出通知和广播，检测 SD 卡上文件的变化，接受短信或电话等后台服务。

Service 的地位和级别与 Activity 组件差不多，只不过没有 Activity 使用的频率高。与 Activity 组件一样，Service 也存在生命周期。

10.4.1 Service 的生命周期

Service 有自己的生命周期。要想用好 Service，必须清楚 Service 的生命周期以及在其生命周期中经常需要调用的方法。

1. Service 生命周期方法

Service 生命周期的方法比 Activity 的要少一些，有些是 Service 自己独有的，其相关方法见表 10-11。

表 10-11 Service 生命周期方法及说明

方法	返回类型	说明
onCreate()	void	创建服务。在 Service 第一次被创建的时候调用
onStart(Intent intent, int startId)	void	启动服务。从 Android 2.0 以后，这个方法已废弃。由 onStartCommand()方法代替
onStartCommand(Intent intent, int flags, int startId)	int	启动服务。在该方法被调用后，会告诉系统如何重启服务；在生命周期内，如果 Service 被系统杀死，在重启 Service 时会重新调用 onStartCommand()方法

续表

方 法	返回类型	说 明
onDestroy()	void	销毁服务。在 Service 销毁前调用。这是 Service 接收的最后一个调用。当 Service 不再使用时,由系统调用
onBind(Intent intent)	IBinder	绑定服务。返回进程间通信渠道给其绑定的服务
onRebind(Intent intent)	void	重新绑定。该方法只有当上次 onUnbind()返回 true 的时候才会被调用
onUnbind(Intent intent)	boolean	解除绑定。返回值为 true 表示允许再次绑定,再绑定时调用 onRebind()方法;返回值为 false 表示只能绑定一次,不能再次绑定,默认为 false

启动服务时依次执行 onCreate()和 onStartCommand()方法。如果在系统显示调用 stopService()和 stopSelf()之前终止服务,Service 可再次重启,onStartCommand()方法会重新被调用。调用 onStartCommand()方法的返回值是一个整型,有四个值,通常用系统常量表示,即 START_STICKY、START_NOT_STICKY、START_REDELIVER_INTENT、START_STICKY_COMPATIBILITY。它们的含义分别如下。

START_STICKY:如果 Service 进程被杀死,保留 Service 的状态为开始状态,但不保留递送的 Intent 对象。随后系统会尝试重新创建 Service,由于服务状态为开始状态,所以创建服务后一定会调用 onStartCommand(Intent,int,int)方法。如果在此期间没有任何启动命令被传递到 Service,那么参数 Intent 将为 null。

START_NOT_STICKY:"非黏性的"。使用这个返回值时,如果在执行完 onStartCommand()后,服务被异常杀死,系统不会自动重启该服务。

START_REDELIVER_INTENT:重传 Intent。使用这个返回值时,如果在执行完 onStartCommand()后,服务被异常杀死,系统会自动重启该服务,并将 Intent 的值传入。

START_STICKY_COMPATIBILITY:START_STICKY 的兼容版本,但不保证服务被杀死后一定能重启。

2. Service 的启动方式

Service 没有可交互的用户界面,它不能自己启动,必须要通过某一个 Activity 或其他的 Context 对象来启动之。启动一个 Service 有两种方式,一种是通过调用 startService()启动一个 Service,另一种是使用 bindService()方法来绑定一个存在的 Service。不同的启动方式对 Service 生命周期的影响是不一样的。

1) 通过 startService()启动

通过 startService()启动 Service,在启动时会依次执行 onCreate()→onStartCommand()方法。注意:如果 Service 已经启动了,当再次启动 Service 时,不会再执行 onCreate()方法,而是直接执行 onStartCommand()方法。

当 Service 被自己或用户停止,但没有调用 stopService()方法时,Service 会一直在后台运行,直到下次调用者再次启动起来。当调用 stopService()方法后,Service 会停止并直接进入 onDestroy()方法。Service 才会真正结束。

2) 通过 bindService()启动

通过 bindService()启动 Service,在启动时只会运行 onCreate()方法,这时将调用者和

Service 绑定在一起。

如果调用者退出了，Service 也会依次调用 onUnbind()→onDestroy()方法，与调用者同时退出。这就是所谓绑定的含义。

使用这两种方式启动 Service，其生命周期可通过图 10-7 来描述，其中左侧为 startService()启动，右侧为 bindService()启动。

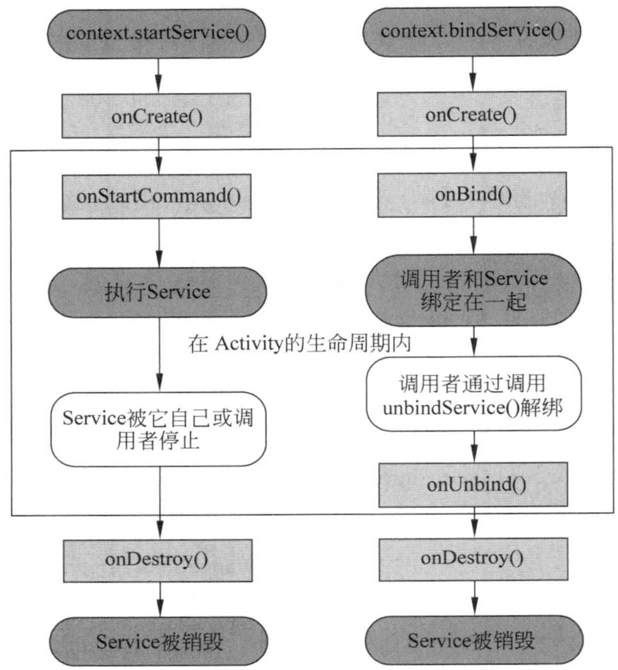

图 10-7　两种启动 Service 的生命周期表现

Service 拥有较高的优先级。一般在下列几种情况下都不会被系统杀死。

如果 Service 正在调用 onCreate()、onStart()或 onDestroy()方法，那么用于当前 Service 的进程则变为前台进程，不会被杀死。

如果 Service 已经被启动，拥有它的进程仅次于可见的进程，而比不可见的进程重要，这时 Service 一般不会被杀死。

如果客户端已经连接到 Service，那么拥有它的进程则拥有最高的优先级，可以认为该 Service 是可见的，不会被杀死。

如果 Service 可以使用 startForeground(int, Notification)方法来将 Service 设置为前台状态，那么系统就认为是用户可见的服务，所以在内存不足时不会被杀死。

10.4.2　使用 Service

Service 位于 android. app. Service 包中。Service 没有自己的页面，它一般由 Activity 启动，但不依赖于 Activity。也可以由其他的 Service 或者 BroadcastReceiver 启动。在应用中使用 Service 编程，需要做好以下几方面的工作。

1. 创建 Service 子类

使用 Service 进行编程，首先要创建一个 Service 子类，创建方法比较简单，只要定义一个类继承于 Service，并且要重写该类中相应的方法即可。这些方法如下。

onCreate()方法：当 Service 第一次被创建时，由系统调用。

onStartCommand(Intent intent, int flag, int startId)方法：当 startService()方法启动 Service 时，该方法被调用。

onBind(Intent intent)方法：另一种启动 Service 的实现方法，返回一个绑定的接口给 Service。

onDestroy()方法：当 Service 不再使用时，由系统调用。

2. 启动和停止 Service

一旦创建好一个 Service，就可以在其他组件中启动该 Service 来使用它了。启动一个 Service，由 Context 类的实例调用方法实现。例如 Context. startService(Intent intent)方法，这与启动一个 Activity 非常相似，也是传递一个 Intent。或使用 Context. bindService (Intent service, ServiceConnection conn, int flags)来绑定 Service，其中第一个参数是 Intent；第二个参数是绑定 Service 的对象；第三个参数是创建 Service 的方式，一般指定为系统常量 BIND_AUTO_CREATE，即绑定时自动创建。

启动一个 Service 可通过类名称来显式启动。以调用 startService()方法为例，如果已经创建了一个名为 MyService 的服务类，则启动代码如下。

```
Intent myIntent = new Intent(this, MyService.class);
myIntent.putExtra("TOPPING","Margherita");
startService(myIntent);
```

或使用隐式启动，用一条代码代替上面的三条代码，代码如下。

```
startService(new Intent(this, MyService.class));
```

停止一个 Service，可调用 stopSelf()方法，由 Service 自己停止，也可调用 stopService()方法，代码示例如下。

```
stopService(new Intent(this, MyService.class));
```

3. 推送 Service 到前台

前面介绍的启动和停止服务都是由该服务以外的实例完成的操作，对于服务本身，也可以实现类似于 Service 启动和停止的操作，准确地说，是 Service 把自己推送到 Activity 前台。把 Service 推送到前台操作能够引起用户的关注。每一个前台服务都会在状态栏上显示一个带有优先级的通知，该通知消息一直会保持在状态栏，直到前台服务被终止或被移出前台。

从 Android 9 (API 28) 及以上的版本，使用前台服务需要在 AndroidManifest. xml 清单文件中声明前台服务权限，这是一个普通级权限，所以系统会自动授权给 App 的权限请求。在清单文件中添加的权限代码如下。

```
<uses-permission android:name="android.permission.FOREGROUND_SERVICE"/>
```

在 Service 类中，下面两个方法与服务推送有关。

startForeground (int id, Notification notification)，把当前服务推送到前台运行。其中第一个参数表示通知的编号，第二个参数表示通知的对象。这个方法实现将该服务切换到前台的通知栏中展示。

stopForeground(boolean)，停止前台服务的运行。当参数为 true 时，清除通知；当参数为 false 时，表示不清除通知。

4. 注册 Service 组件

在应用项目中使用 Service，需要在清单文件中注册，否则服务无法正常启动。这点很容易被初学者忽略。在 AndroidManifest.xml 文件中注册<service>元素，一定要显式地指定服务的名称，示例如下。

```
<service
    android:name=".MyService">
</service>
```

10.4.3 Service 的应用案例

由于 Service 应用时是没有交互界面的，所以通常要与 Activity、BroadcastReceiver 等组件联合使用。下面通过两个案例来说明 Service 的编程应用。

这是一个 Service 与一个 BroadcastReceiver 和一个 Thread 线程结合使用的案例。

【案例 10.6】 用两个按钮分别启动、停止服务。当单击“启动 Service”按钮时，Service 运行，并显示运行的时长；如果不单击“停止 Service”按钮，则让服务运行一段时间后自动停止。当单击“停止 Service”按钮时，立即中止 Service。

说明：本案例通过 Service 与 BroadcastReceiver 联合使用，通过 BroadcastReceiver 广播将服务运行的时间发送到界面上，设置当本服务启动 10 秒后自动停止。程序需要设计两个类，一个是 Activity 类，在其中布局按钮对象和文本框对象，文本框用来显示广播消息；另一个是继承 Service 的子类，重写其相关的方法，并且在该服务子类中定义 Thread 子类来创建后台线程，用于控制发送广播的间隔。

开发步骤及解析：过程如下。

1）创建项目

在 Android Studio 中创建一个名为 Activity_ServiceBroadcast 的项目，其包名为 ee.example.activity_servicebroadcast。

2）设计布局

在 res/layout 目录下编写布局文件 activity_main.xml。在界面上设计两个按钮控件和一个文本框控件，ID 为 myButton1 的按钮，单击它启动服务；ID 为 myButton2 的按钮，单击它停止服务；下面的文本框，ID 为 myTextView，用于显示服务持续的时间。

3）开发逻辑代码

在 java/ee.example.activity_servicebroadcast 包下有两个代码文件，一个是 Activity 的子类代码实现，另一个是 Service 的子类代码实现。

MainActivity.java 是实现 Activity 的子类定义，而且在 MainActivity 子类里还定义了一个内部类（MyBroadcastReceiver）的实现，代码如下。

```java
package ee.example.activity_servicebroadcast;

import androidx.appcompat.app.AppCompatActivity;

import android.os.Bundle;
import android.content.BroadcastReceiver;
import android.content.Context;
import android.content.Intent;
import android.content.IntentFilter;
import android.view.Gravity;
import android.view.View;
import android.widget.Button;
import android.widget.TextView;
import android.widget.Toast;

import java.text.SimpleDateFormat;

public class MainActivity extends AppCompatActivity {

    Button button1;                         //声明按钮 button 1 的引用
    Button button2;                         //声明按钮 button 2 的引用
    TextView myTextView;

    @Override
    protected void onCreate(Bundle savedInstanceState) {
        super.onCreate(savedInstanceState);
        setContentView(R.layout.activity_main);
        myTextView = (TextView) this.findViewById(R.id.myTextView);
        button1 = (Button) this.findViewById(R.id.myButton1);
        button2 = (Button) this.findViewById(R.id.myButton2);

        //单击"启动 Service"按钮
        button1.setOnClickListener(new View.OnClickListener(){
            @Override
            public void onClick(View v) {
                Intent i = new Intent(MainActivity.this, MyService.class);
                startService(i);
                long currentTime = System.currentTimeMillis();
                String timeNow = new SimpleDateFormat("yyyy-MM-dd HH:mm:ss")
                                        .format(currentTime);
                String desc = String.format("\n%s,开始启动服务", timeNow);
                myTextView.append(desc);
                Toast.makeText(MainActivity.this,
                                "Service 启动成功", Toast.LENGTH_SHORT).show();
            }
        });
        //单击"停止 Service"按钮
```

```
46            button2.setOnClickListener(new View.OnClickListener(){
47                @Override
48                public void onClick(View v) {
49                    Intent i = new Intent(MainActivity.this, MyService.class);
50                    stopService(i);
51                    long currentTime = System.currentTimeMillis();
52                    String timeNow = new SimpleDateFormat("yyyy - MM - dd HH:mm:ss")
                                      .format(currentTime);
53                    String desc = String.format("\n%s,结束服务了", timeNow);
54                    myTextView.append(desc);
55                    Toast.makeText(MainActivity.this,
                              "Service 停止成功", Toast.LENGTH_SHORT).show();
56                }
57            });
58
59            //动态注册一个 BroadcastReceiver
60            //创建过滤器
61            IntentFilter intentFilter = new IntentFilter(".ServBroad.myThread");
62            MainActivity.MyBroadcastReceiver MyBroadcastReceiver = new
                                      MainActivity.MyBroadcastReceiver();
63            registerReceiver(MyBroadcastReceiver, intentFilter);  //注册 BroadcastReceiver 对象
64
65
66        //实现 MyBroadcastReceiver
67        public class MyBroadcastReceiver extends BroadcastReceiver {
68            @Override
69            public void onReceive(Context context, Intent intent) {
70                Bundle myBundle = intent.getExtras();
71                int myInt = myBundle.getInt("myThread");
72                if(myInt < 10){              //后台 Service 运行 10 秒
73                    myTextView.append("\n 后台 Service 运行了" + myInt + "秒");
74                }else{
75                    Intent i = new Intent(MainActivity.this, MyService.class);
76                    stopService(i);
77                    myTextView.setText("后台 Service 在" + myInt + "秒后停止运行");
78                }
79            }
80        }
81    }
```

（1）第 33～44 行，实现第一个按钮被单击后的监听操作。其中第 34 行调用 startService()方法来启动服务。第 38 行获取系统的当前时间，该时间值是一个长整型数，第 39 行将获得的系统时间置换为 yyyy-MM-dd HH:mm:ss 的字符串格式数据。第 41 行将合成的字符串添加到 TextView 对象中，该实例将显示在屏幕上。

（2）第 46～57 行，实现第二个按钮被单击后的监听操作。其中第 50 行调用 stopService()方法来停止服务。

（3）第 61～64 行，注册一个广播接收器。第 61 行创建该广播接收器的 IntentFilter，并设置其 action 值是“.ServBroad.myThread”；第 62 行创建一个内部自定义的广播接收器的对象，该自定义的子类为 MyBroadcastReceiver（见第 67～80 行）；第 63 行注册该广播接收器。

（4）第 67～80 行，定义 MyBroadcastReceiver 子类的实现。在实现中重写 onReceive() 方法，编写接收广播后的操作，即服务运行过程中的界面显示操作。其中第 70、71 行获取一个 Bundle 对象，并得到 Bundle 对象绑定的键-值对 myThread 中的整型数值信息，存入 myInt 变量中，myInt 的值就是服务的运行时间数。第 72～78 行，当 myInt 值< 10 时，则向 TextView 对象中添加服务运行的时间数信息；否则停止服务，即第 76 行调用 stopService() 方法。

MyService.java 是继承 Service 类的实现，并且在 MyService 子类里，还定义了一个内部线程 MyThread 子类，用来实现模拟时钟的间隔秒数，代码如下。

```
package ee.example.activity_servicebroadcast;

import android.app.Service;
import android.content.Intent;
import android.os.IBinder;

public class MyService extends Service{
    MyThread myThread;
    @Override
    public void onCreate() {
        super.onCreate();
    }

    @Override
    public IBinder onBind(Intent intent) {
        return null;
    }

    @Override
    public void onDestroy() {
        myThread.flag = false;                  //停止线程运行
        super.onDestroy();
    }
    @Override
    public int onStartCommand(Intent intent, int flags, int startId) {
        myThread = new MyThread() ;
        myThread.start();                       //启动线程
        super.onStartCommand(intent, flags, startId);
        return START_REDELIVER_INTENT;
    }

    //定义线程类
    class MyThread extends Thread{
        boolean flag = true;
        int ct = 1;
        @Override
        public void run() {
            while(flag){
                Intent i = new Intent(".ServBroad.myThread");      //创建 Intent
                i.putExtra("myThread", ct);     //放入数据
                sendBroadcast(i);               //发送广播
                ct++;
```

```
43                  try{
44                      Thread.sleep(1000);        //睡眠指定毫秒数
45                  }catch(Exception e){
46                      e.printStackTrace();
47                  }
48              }
49          }
50      };
51  }
```

(1) 在该服务子类中定义了一个内部的自定义后台线程 MyThread。第 8 行,声明这个后台线程的对象 myThread。

(2) 在该服务子类中,重写生命周期的方法：onCreate()、onBind(Intent intent)、onStartCommand(Intent intent, int flag, int startId)和 onDestroy()方法。

(3) 第 25～30 行,重写 onStartCommand(Intent intent, int flag, int startId)方法,这是服务每次启动时要执行的方法。第 26 行创建一个内部线程对象 myThread,第 27 行启动该线程。并将该方法的返回值设置为 START_REDELIVER_INTENT,即重启服务时会重传 Intent。

(4) 第 33～50 行,定义一个线程子类 MyThread。在该子类中重写 run()方法。在 run()方法中,创建一个 Intent 对象,设置 action 值为.ServBroad.myThread,并设置 Extra 的 myThread 值为睡眠的次数。第 44 行设置睡眠的时间间隔是 1 秒。因此,只要该服务在运行,这个线程就会一秒一秒地记录时间,第 41 行就会把线程累加的时间发送广播。

4) 注册 Service

在 AndroidManifest.xml 中的< application >元素节点内添加服务注册,代码片段如下。

```
service
    android:name = ".MyService">
</service >
```

运行结果：在 Android Studio 支持的模拟器上,运行 Activity_ServiceBroadcast 项目,运行结果如图 10-8 所示。

在项目的初始页面中,如图 10-8(a)所示,单击“启动 SERVICE”按钮,即可看到如图 10-8(b)所示的效果。每隔 1 秒显示一条后台服务运行的时间。如果不做任何干预操作,则在第 10 秒后自动停止服务,并会给出广播消息；如果启动服务后单击“停止SERVICE”按钮,则会停止服务,并给出停止服务的时间等广播消息,如图 10-8(c)所示。

接下来是一个音乐播放案例,由一个 Service 与两个 BroadcastReceiver 联合使用来实现应用。

【案例 10.7】 设计一个歌曲播放器,可以对一首固定歌曲实现播放、暂停和停止,并显示歌词内容。

说明：本案例在播放和暂停时显示歌词内容,当歌词过多时可以通过上、下滑动,展示全部内容。本案例还将模拟一个开关按钮,即同一个按钮,当歌曲没有播放时,执行播放功能；当歌曲正在播放时,执行暂停功能。为此,设计两个 BroadcastReceiver,一个 BroadcastReceiver 用于控制开关按钮的通信传递,另一个 BroadcastReceiver 用于歌曲播放

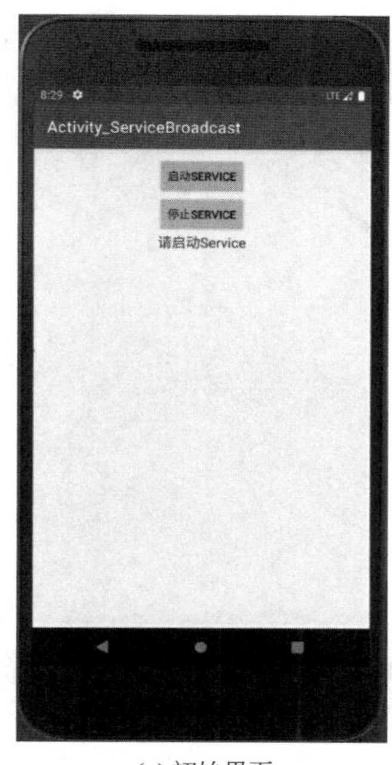

(a) 初始界面

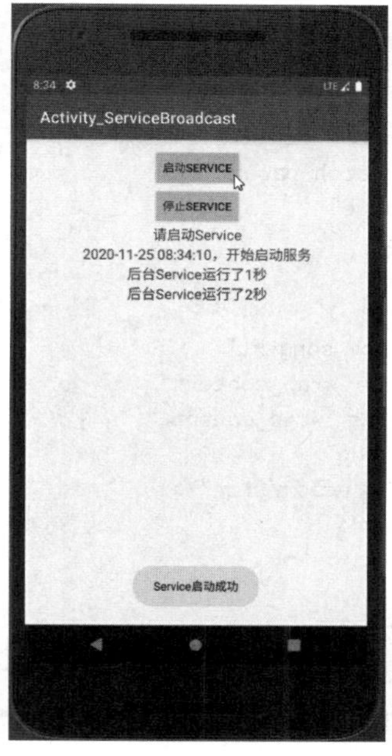

(b) 启动服务

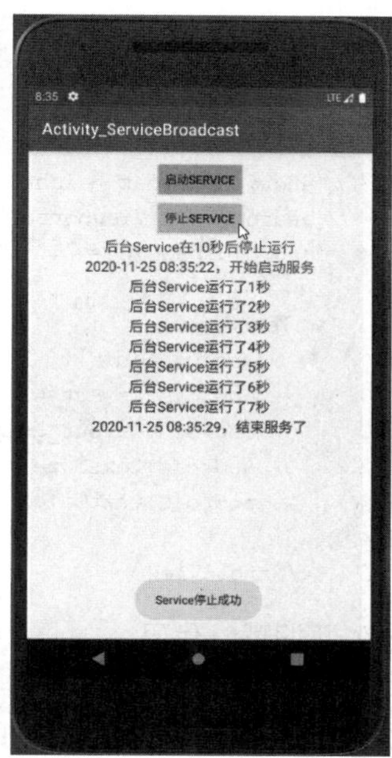

(c) 停止服务

图 10-8　启动和停止服务及广播服务的运行时间

服务的通信传递。

开发步骤及解析：过程如下。

1）创建项目

在 Android Studio 中创建一个名为 Activity_MusicPlayer 的项目。其包名为 ee.example.activity_musicplayer。

2）准备资源

在本案例中有比较多的资源需要准备，分别如下。

（1）图片资源，将背景图片、按钮图片复制到 res/drawable 目录下。

（2）媒体资源，在 res 下创建子目录 raw，复制歌曲文件 htys.mp3 到该子目录。

（3）颜色资源，在 res/values 目录下的 colors.xml 文件中声明需要设置的颜色。

（4）字符串资源，在 res/values 目录下的 strings.xml 文件中声明歌曲、歌手、作词作曲等信息字符串。

（5）数组资源，在 res/values 目录下的 arrs.xml 文件中声明歌曲歌词字符串数组。

3）设计布局

在 res/layout 目录下编写布局文件 activity_main.xml。该布局设计了 3 个嵌套的线性布局，将界面划分为三大纵横交错的区域，一是按钮区，二是歌曲作者区，三是歌词区。在按钮区设计两个 ImageButton 按钮，按钮的 ID 分别为 start_pause 和 stop；在歌曲作者区分别设计三个 TextView 控件，ID 分别为 textView1、textView2 和 textView3；歌词区占用屏幕的剩余部分。由于歌词比较长，一屏展示不全时，需要实现可上、下滑动以显示全部内容。因此在歌词区添加了< ScrollView >控件，把展示歌词的< TextView >控件包含其中，代码片段如下。

```
<!-- 歌词信息布局 -->
<ScrollView
android:layout_width = "match_parent"
android:layout_height = "match_parent"
android:fillViewport = "true"
android:padding = "10dp">

<TextView
    android:id = "@ + id/tv_songtxt"
    android:layout_width = "wrap_content"
    android:layout_height = "wrap_content"
    android:textSize = "20dp"
    android:textColor = "@color/stxt"/>

</ScrollView>
```

4）开发逻辑代码

在 java/ee.example.activity_ musicplayer 包下有两个代码文件，一个是 Activity 的子类代码实现，另一个是 Service 的子类代码实现。

MainActivity.java 是实现 Activity 的子类定义，在其内部，也定义了一个广播接收器的内部类实现，代码如下。

```
package ee.example.activity_musicplayer;

import androidx.appcompat.app.AppCompatActivity;

import android.os.Bundle;
import android.content.BroadcastReceiver;
import android.content.Context;
import android.content.Intent;
import android.content.IntentFilter;
import android.view.View;
import android.widget.ImageButton;
import android.widget.TextView;

public class MainActivity extends AppCompatActivity {
    ImageButton start_pause;                              //播放、暂停按钮
    ImageButton stop;                                     //停止按钮
    TextView tv_st;
    MainActivity.ActivityReceiver myReceiver;

    @Override
    protected void onCreate(Bundle savedInstanceState) {
        super.onCreate(savedInstanceState);
        setContentView(R.layout.activity_main);

        start_pause = (ImageButton) findViewById(R.id.start_pause);
        stop = (ImageButton) findViewById(R.id.stop);
        start_pause.setOnClickListener(new BtnOnClk());
        stop.setOnClickListener(new BtnOnClk());

```

```
        String[] songtxts = getResources().getStringArray(R.array.singtexts);
        tv_st = (TextView) findViewById(R.id.tv_songtxt);
        tv_st.setText(songtxts[0].toString());      //向 TextView 对象中加载歌词
        for(int i = 1;i < songtxts.length;i++){
            tv_st.append("\n" + songtxts[i].toString());
        }
        tv_st.setVisibility(View.INVISIBLE);

        myReceiver = new MainActivity.ActivityReceiver();  //创建广播接收器
        IntentFilter bfilter = new IntentFilter();
        bfilter.addAction("MusicPlayer.update");
        registerReceiver(myReceiver, bfilter);              //注册广播接收器

        Intent sintent = new Intent(this, MusicService.class); //创建服务的 Intent
        startService(sintent);                                 //启动后台 Service
    }

    /* 自定义的广播接收器 */
    public class ActivityReceiver extends BroadcastReceiver {
        @Override
        public void onReceive(Context context, Intent intent) {
            int mupdate = intent.getIntExtra("musicupdate", -1);        //获得 intent 传
                                                                        //的数据
            switch (mupdate) {
                case 1:                                                 //没有声音播放
                    start_pause.setImageResource(R.drawable.png2);      //显示播放图片
                    break;
                case 2:                                                 //正在播放声音
                    start_pause.setImageResource(R.drawable.png3);      //显示暂停图片
                    break;
                case 3:                                                 //暂停中
                    start_pause.setImageResource(R.drawable.png2);      //显示播放图片
                    break;
            }
        }
    }

    private class BtnOnClk implements View.OnClickListener{
        @Override
        public void onClick(View v) {
            Intent intent = new Intent("MusicPlayer.control");   //创建 Intent
            switch (v.getId()) {                                 //分支判断
                case R.id.start_pause:                           //按下播放、暂停按钮
                    tv_st.setVisibility(View.VISIBLE);
                    intent.putExtra("ACTION", 1);                //存放数据
                    sendBroadcast(intent);                       //发送广播
                    break;
                case R.id.stop:                                  //按下停止按钮
                    tv_st.setVisibility(View.INVISIBLE);
                    intent.putExtra("ACTION", 2);                //存放数据
                    sendBroadcast(intent);                       //发送广播
                    break;
            }
        }
```

```
83         }
84
85         @Override
86         protected void onDestroy() {                                  //释放时被调用
87             super.onDestroy();
88             Intent intent = new Intent(this, MusicService.class);     //创建 Intent
89             stopService(intent);                          //停止后台的 Service
90         }
91
92     }
```

（1）第 18 行，声明了一个内部类的对象 myReceiver，该内部类在第 48～64 行中定义。

（2）第 30～35 行，从数组元素资源文件中取出数组，然后逐个添加到用于显示歌词的 TextView 控件对象 tv_st 中。

（3）第 36 行，设置 tv_st 为不可见，以保证在项目运行的初始界面不能看到歌词，只有当播放按钮被单击后才显示歌词。

（4）第 38～41 行，注册一个广播接收器，该接收器的 IntentFilter 的 action 是 MusicPlayer. update。

（5）第 48～64 行，定义内部的广播接收器子类 ActivityReceiver 的实现。该类主要实现"广播/暂停"开关按钮的图片显示切换。该子类接收从服务子类中传入的 Intent 对象，因为对音乐的播放是在服务子类中进行的，这个 Intent 携带了音乐播放的当前状态，即接收的 Intent 的 Extra 传回的 musicupdate 值有三个，分别是 1、2、3，它们表示三种状态：1 表示没有播放；2 表示正在播放；3 表示暂停中。

（6）第 43、44 行，启动后台的 Service。

（7）第 86～90 行，重写 onDestroy()方法，在此方法中停止后台 Service 运行。

（8）第 66～83 行，实现两个按钮的本类接口 OnClickListener 中的回调方法 onClick()。在该 onClick()方法中定义发送广播，该广播的 Intent 对象的 action 是 MusicPlayer. control，并根据哪个按钮被单击发送不同的广播信息。注意，发送的广播不会被内部定义的广播接收器所接收，因为它们 Intent 的 action 不同。

MusicService. java 是继承 Service 类的实现，在该服务子类 MusicService 中，重写了 onBind()、onCreate()和 onDestroy()方法。并且在 MusicService 子类里，还定义了一个内部 ServiceReceiver 广播接收器子类的实现，代码如下。

```
package ee.example.activity_musicplayer;

import android.app.Service;
import android.content.BroadcastReceiver;
import android.content.Context;
import android.content.Intent;
import android.content.IntentFilter;
import android.media.MediaPlayer;
import android.os.IBinder;

public class MusicService extends Service{
    MediaPlayer mp;
    ServiceReceiver serviceReceiver;
```

```
    int status = 1;          //当前的状态,1 没有声音播放,2 正在播放声音,3 暂停

    @Override
    public IBinder onBind(Intent intent) {              //重写的 onBind()方法
        return null;
    }
    @Override
    public void onCreate() {
        serviceReceiver = new ServiceReceiver();        //创建 BroadcastReceiver
        IntentFilter filter = new IntentFilter();       //创建过滤器
        filter.addAction("MusicPlayer.control");        //添加 Action
        registerReceiver(serviceReceiver, filter);      //注册 BroadcastReceiver
        super.onCreate();
    }
    @Override
    public void onDestroy() {                           //重写的 onDestroy()方法
        unregisterReceiver(serviceReceiver);            //取消注册
        super.onDestroy();
    }

    public class ServiceReceiver extends BroadcastReceiver{      //自定义 BroadcastReceiver
        @Override
        public void onReceive(Context context, Intent intent) {      //重写的响应方法
            int action = intent.getIntExtra("ACTION", -1);           //得到需要的数据
            switch(action){
            case 1:                                                  //播放或暂停声音
                if(status == 1){                                     //当前没有声音播放
                    mp = MediaPlayer.create(context, R.raw.htys);
                    mp.start();
                    status = 2;
                    Intent sendIntent = new Intent("MusicPlayer.update");
                    sendIntent.putExtra("musicupdate", 2);
                    sendBroadcast(sendIntent);
                }
                else if(status == 2){                              //正在播放声音
                    mp.pause();                                    //停止
                    status = 3;                                    //改变状态
                    Intent sendIntent = new Intent("MusicPlayer.update");
                    sendIntent.putExtra("", 3);                    //存放数据
                    sendBroadcast(sendIntent);                     //发送 musicupdate 广播
                }
                else if(status == 3){                              //暂停中
                    mp.start();                                    //播放声音
                    status = 2;                                    //改变状态
                    Intent sendIntent = new Intent("MusicPlayer.update");
                    sendIntent.putExtra("musicupdate", 2);         //存放数据
                    sendBroadcast(sendIntent);                     //发送广播
                }
                break;
            case 2:                                                //停止声音
                if(status == 2 || status == 3){                    //播放中或暂停中
                    mp.stop();                                     //停止播放
                    status = 1;                                    //改变状态
                    Intent sendIntent = new Intent("MusicPlayer.update");
```

```
68                  sendIntent.putExtra("musicupdate", 1);     //存放数据
69                  sendBroadcast(sendIntent);                 //发送广播
70              }
71          }
72        }
73      }
74
75  }
```

(1) 第13行，声明一个内部广播接收器对象serviceReceiver。该内部广播接收器在第34～73行定义。

(2) 第21～27行，重写onCreate()方法，在该方法中注册了一个广播接收器对象，该接收器对象的Intent的action为MusicPlayer.control。

(3) 第29～32行，重写onDestroy()方法，在该方法中注销广播接收器。

(4) 第34～73行，定义内部广播接收器子类的实现。该子类主要实现音乐的播放服务，并发送音乐播放状态信息的广播。它既是一个广播的接收者，也是另一个广播的发送者。它接收的是携带MusicPlayer.control的广播消息，发送的是携带MusicPlayer.update的广播消息。MusicPlayer.control广播Intent的Extra传回ACTION值，其值由MainActivity类中按钮回调方法发送，值为1表示“播放/暂停”按钮被单击，值为2表示“停止”按钮被单击。

(5) 第40～62行，执行MusicPlayer.control广播传过来的值为1的操作，该广播在MainActivity中的第一个按钮“播放/暂停”被单击时发送。该按钮有三种状态：未播放、播放中、暂停。根据这三种状态，执行相应的操作，即当音乐处于未播放和暂停状态时，执行播放音乐操作，并向Intent中的musicupdate传送值为2；当音乐处于正在播放中状态时，执行暂停播放音乐操作，并向MusicPlayer.update广播的Intent中的musicupdate传送值为3；并发送MusicPlayer.update广播到MainActivity的接收器中。

(6) 第63～71行，执行MusicPlayer.control广播传过来的值为2的操作，该广播在MainActivity中的第二个按钮“停止”被单击时发送。在此执行停止播放音乐操作，并向Intent中的musicupdate传送值为1，发送MusicPlayer.update广播到MainActivity的接收器中。

(7) 第41行，实例化一个MediaPlayer对象mp，定义其音乐来源于raw目录下的htys.mp3文件。

5) 注册Service

在AndroidManifest.xml文件中声明一个Service组件即可。本案例的两个BroadcastReceiver都是在程序中动态注册的，所以不需要在该清单文件中再注册。

运行结果：在Android Studio支持的模拟器上，运行Activity_MusicPlayer项目，运行结果如图10-9所示。

在项目的初始页面，如图10-9(a)所示，为播放按钮，为停止按钮。单击开始播放歌曲，原来的播放按钮变为暂停按钮，并且显示歌词内容，如图10-9(b)所示。歌词内容较长，一屏显示不下，可以向上滑动，显示后面的歌词内容，如图10-9(c)所示。单击按钮，即可停止歌曲播放，返回初始页面状态。

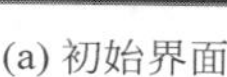
(a) 初始界面　　(b) 播放歌曲　　(c) 滑动歌词

图 10-9　歌曲播放器的运行效果

释疑：

案例 10.6 和案例 10.7 都是关于 Service 的应用。为什么案例 10.6 在 Service 子类只重写了 onStartCommand()方法，而案例 10.7 在 Service 子类中只重写了 onCreate()方法？

在案例 10.6 中，用户可以反复地执行对 Service 的启动和停止。因此在自定义 Service 子类中，需要重写 onStartCommand()方法，定义每次重新启动该 Service 时，应该执行的某些初始化操作和传入的相关信息。

在案例 10.7 中，用户只是通过启动一次服务，就可对所播放的音乐进行反复的播放、暂停与停止操作，退出服务后程序便结束了。因此在自定义的 Service 子类中，需要重写 onCreate()方法，只在 Service 子类首次创建时运行 onCreate()方法，执行与创建该 Service 有关的操作。

小结

本章介绍了 Android 应用项目编程的几种后台处理技术。消息提示 Notification 在同一个项目内传递消息，消息显示在状态栏中。BroadcastReceiver 在不同的应用项目中传递广播，传递广播要与其注册时 Intent 的 action 内容相匹配。后台线程与 UI 线程的信息交互可使用 Handler 和 AsyncTask 两种方式，使用中要掌握它们的消息传递机制、正确的编程步骤以及相关规则。Service 是 Android 中的重要组件之一，在应用上其使用频率仅次于 Activity。当 Service 提供的后台服务需要与用户进行交互时，这时通常借助 BroadcastReceiver 来与前台的 Activity 交换信息。熟练掌握这些后台编程技术，就为深层次的应用开发奠定了不可或缺的基础。

为了让用户界面更炫更精彩，为了让用户能更好地体验应用项目的操作使用，可以向应用项目中加入更多的表现形式，如在应用中添加一段音乐、插入一个视频等。第 11 章将介绍 Android 的多媒体应用开发。

练习

设计一个音乐播放应用，音乐文件存放在项目的 res/raw 目录下。要求在播放页面包括以下内容：一个图片动画，并且在播放歌曲时推送通知到状态栏，使用自定义通知栏。

第 11 章 多媒体应用

11.1 音频与视频播放

在应用中，适当地运用音频与视频的播放可以收到极好的应用效果。Android 系统提供了对常见格式媒体的编码、解码机制，可以非常容易地集成音频、视频和图片等多媒体文件到应用项目中，例如相册、播放器、录音和摄像等应用项目。当然，有些应用需要硬件的支持。

音频和视频的播放常用到 MediaPlayer 类。该类提供了播放、暂停、停止和重复播放等方法。下面分别对音频、视频的播放进行介绍。

11.1.1 音频播放

Android 平台中播放音频有两种方式：使用 MediaPlayer 类和 SoundPool 类进行播放。前一种方式适合比较长且对时间要求不高的情况(例如播放后台音乐、歌曲等)，后一种方式适合短促且对反应速度比较高的情况(例如播放游戏音效或按键声等)。Android 的音频文件存放在应用项目的 res/aw 目录下，这个 raw 目录需要开发者自己创建。

Android 支持的音频格式有 OGG、MP3、MID、WAV、AMR 等，其中 OGG 格式性能最佳。音频格式采样率有 11kHz、22kHz、44.1kHz，音频采样精度为 16 位的立体声。

1. 使用 MediaPlayer 类

MediaPlayer 是播放媒体文件最为广泛使用的类，位于 android.media.MediaPlayer 包中。MediaPlayer 不仅可以用来播放大容量的音频文件，也可以播放视频文件。在对流媒体播放控制方面支持播放操作(开始、暂停、停止等)，查找操作，媒体资源的加载操作以及与媒体操作相关的监听器。MediaPlayer 类常用的方法见表 11-1。

表 11-1 MediaPlayer 类常用的方法及说明

方　法	返回类型	说　明
start()	void	开始或恢复播放
pause()	void	暂停播放
stop()	void	停止播放
reset()	void	重置 MediaPlayer 对象为未初始化状态

续表

方　　法	返回类型	说　　明
prepare()	void	同步准备音、视频资源
prepareAsync()	void	异步准备音、视频资源。异步准备不会阻塞当前的UI线程，多适用于比较大的资源文件，如视频文件
seekTo(int msec)	void	搜索到指定时间的位置，参数 msec 以毫秒计
isLooping()	boolean	测试 MediaPlayer 对象是否重复播放
isPlaying()	boolean	测试 MediaPlayer 对象是否正在播放
create(Context context, int resid)	MediaPlayer	从 resid 资源 ID 对应的资源文件中加载音频文件，并返回新创建的 MediaPlayer 对象
create(Context context, Uri uri)	MediaPlayer	从指定 Uri 来加载音频文件，并返回新创建的 MediaPlayer 对象
setDataSource(String path)	void	指定加载 path 路径所代表的文件
setDataSource(FileDescriptor fd)	void	指定加载 fd 所代表的文件
setDataSource(Context context, Uri uri)	void	指定加载 Uri 所代表的文件
setOnCompletionListener(MediaPlayer.OnCompletionListener listener)	void	为 MediaPlayer 的播放完成事件绑定事件监听器
setOnPreparedListener(MediaPlayer.OnPreparedListener listener)	void	当 MediaPlayer 调用 prepare()方法时触发该监听器
setOnErrorListener(MediaPlayer.OnErrorListener listener)	void	为 MediaPlayer 的播放错误事件绑定事件监听器

要正确运用 MediaPlayer 对象来播放资源文件，要掌握好以下三个方面：一是 MediaPlayer 对象在不同的阶段所处的状态，二是 MediaPlayer 对象加载时媒体资源文件的来源，三是要会在特定的时候使用 MediaPlayer 事件监听器。

1）MediaPlayer 的状态

使用 MediaPlayer 来播放音频/视频文件或流的控制，是通过一个状态机来管理实现的。一个 MediaPlayer 对象刚被创建或被重新设置时，处于 Idle 状态，处于 Idle 状态的 MediaPlayer 还没有设置数据源；在向对象中加载了音频/视频资源后便处于 Initialized 状态；当对象调用了 prepare()方法后，对象便处于 Prepared 状态。对于处于 Prepared 状态的对象就可以调用 start()方法来播放音、视频了。如果调用 pause()方法，对象则处于 Paused 状态，再次调用 start()则恢复播放；如果调用 stop()方法，对象则处于 Stopped 状态，如果要再次播放，则必须重新执行 prepare()方法；如果调用 reset()方法，对象则回到 Idle 状态。在 MediaPlayer 对象的生命周期中，其各个阶段的状态与常用方法调用的关系如图 11-1 所示。

由图 11-1 可知，运用 MediaPlayer 对象播放音、视频文件，一般的编程步骤如下。

(1) 准备音、视频资源文件。这些资源文件可以存放在本地，也可以存放在外部存储器，或来自网络。

(2) 创建 MediaPlayer 对象。如果资源文件在本地，调用 create(Context context, int resid)方法通过资源 ID 直接加载音、视频文件，此时 MediaPlayer 对象状态为 Started，则第(3)、(4)步骤可以省略。

(3) 加载音、视频资源文件。可调用 setDataSource()方法。注意，资源文件来源不同，

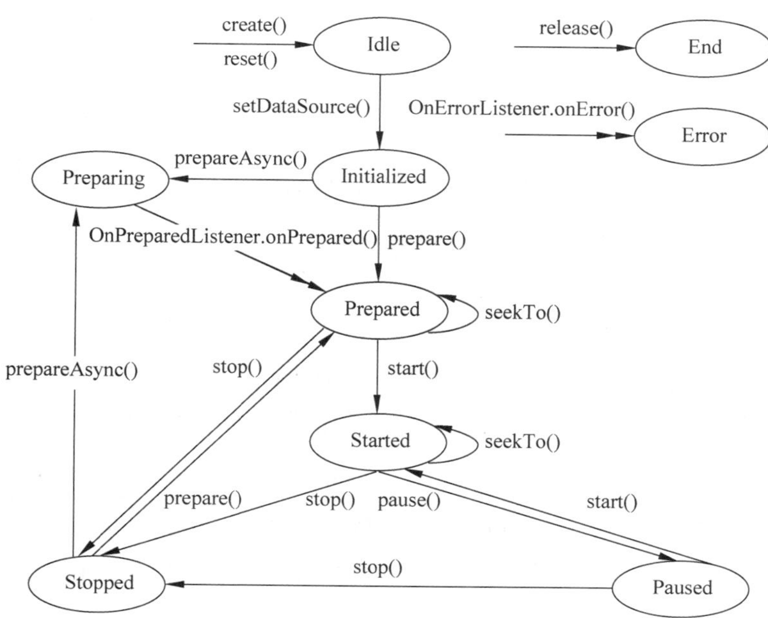

图 11-1　MediaPlayer 对象常用方法调用与状态转换之间的关系

调用 setDataSource()的参数不同。

(4) 准备音、视频，可调用 prepare()或 prepareAsync()方法。

(5) 播放、控制音、视频，可调用 start()、pause()、stop()等方法。

创建 MediaPlayer 对象可以使用两种方式：一种是使用 new MediaPlayer()，另一种是使用 MediaPlayer. create(...)。这两种方法创建的 MediaPlayer 对象所处的状态是不同的，下面分别说明之。

使用 new MediaPlayer()创建对象后，MediaPlayer 对象处于 Idle 状态，需要调用 setDataSource()方法为对象加载音频资源，setDataSource()支持从 path、Uri 和 FileDescriptor 三种途径获取音、视频资源；然后还需要调用 prepare()方法才能进入 Prepared 状态。

使用 create()方法创建 MediaPlayer 对象后，MediaPlayer 对象随即处于 Prepared 状态，无须调用 prepare()方法，否则系统会报错。这里，create()方法支持从 int(resid)和 Uri 两种途径获取音频或视频资源。

2) 加载资源文件

MediaPlayer 是用于播放音频、视频文件或流，而这些音、视频文件资源可能来自本地原生资源，注意本地的资源文件一般放在 Android 应用项目的 res/raw 中；可能来自外部存储设备，也可能来自网络。下面分别给出播放不同来源资源文件的代码片段。

(1) 播放本地音频文件。

例如，在应用项目 res/raw 中有一个音频文件 thesamesong. mp3。播放该音频的代码片段如下。

```
MediaPlayer mymediaplayer = MediaPlayer.create(this,R.raw.thesamesong);
mymediaplayer.start();
```

(2) 播放外部存储器上的音频文件。

例如,在外部存储器 SD 卡中存放有一个音频文件 thesamesong.mp3,该文件的具体存储路径是/mnt/sdcard/thesamesong.mp3。要播放该音频文件,代码片段如下。

```
MediaPlayer mymediaplayer = new MediaPlayer();
mymediaplayer.setDataSource("/mnt/sdcard/thesamesong.mp3");
mymediaplayer.prepare();
mymediaplayer.start();
```

为了保证对外部存储器上的文件的播放,还需要在 AndroidManifest.xml 文件中声明对外部存储设备的访问权限,在第 9 章的 9.3.2 节已做介绍。即需要添加声明如下。

```
<uses-permission android:name="android.permission.READ_EXTERNAL_STORAGE"/>
```

在 Android 6.0(API 23)及更高版本上,除了需要在 AndroidManifest.xml 中添加相应的访问权限外,还需要在 App 运行时向用户进行检查和请求读操作授权。例如对 SD 卡上的音频文件进行加载播放操作前,需要加入如下代码。

```
if(ActivityCompat.checkSelfPermission(MainActivity.this,
    Manifest.permission.READ_EXTERNAL_STORAGE) == PackageManager.PERMISSION_GRANTED)
{//如果已经授权
    …//开始加载音频文件并执行准备和播放操作
}else{//如果未授权,则请求外部存储器文件的读权限授权
ActivityCompat.requestPermissions(MainActivity.this,
        new String[]{Manifest.permission.READ_EXTERNAL_STORAGE},
        REQUEST_EXTERNAL_STORAGE);
}
```

(3) 播放来自网络的音频文件。

播放来自网络的音频文件有两种方式,一是直接使用 create(Context context, Uri uri)方法创建并加载音频文件,其编程步骤与播放本地音频文件相似,在此不做赘述。二是使用 new MediaPlayer()方式,调用 setDataSource()加载。例如,音频文件存放在网络上,其 URL 地址为 http://www.9ku.com/mp3/361140.mp3。要播放该音频文件,代码片段如下。

```
Uri uri = Uri.parse("http://www.9ku.com/mp3/361140.mp3");
MediaPlayer mymediaplayer = new MediaPlayer();
mymediaplayer.setDataSource(this, uri);
mymediaplayer.prepareAsync();
mymediaplayer.start();
```

因为从网络上获取资源,可能会遇到不可控的因素影响加载速度,建议调用异步准备方法 prepareAsync(),可以很好地解决不阻塞当前 UI 线程的问题。

3) MediaPlayer 监听器

对 MediaPlayer 对象可以定义如下监听器：OnCompletionListener、OnPrepareListener、OnErrorListener、OnBufferingUpdateListener、OnInfoListener, OnVideoSizeChangedListener 和

OnSeekCompleteListener 等。

如果要在播放过程中到达媒体源末端时做某些处理，如从列表中播放下一首歌曲或释放媒体播放器对象等，可以在 OnCompletionListener 监听器里的 onCompletion(MediaPlayer mp)事件中编码实现。如果在准备播放媒体源时需要做某些处理，可在 OnPrepareListener 监听器的 onPrepared(MediaPlayer mp)事件中进行编码实现。如果在异步操作过程中出现错误时做某些处理(其他错误将在调用方法时抛出异常)，可在 OnErrorListener boolean onError(MediaPlayer mp, int what, int extra)事件中进行编码实现，这里，参数 what 指明了已发生错误的类型，它可能为 MEDIA_ERROR_UNKNOWN 或 MEDIA_ERROR_SERVER_DIED。参数 extra 指明了与错误相关的附加信息。

MediaPlayer 类可以很好地完成音频文件的播放、暂停、进度和停止的控制，但是对音量的管理控制还是有所欠缺。为此，Android 提供了一个 AudioManager 类。

2. 使用 AudioManager 类管理音量

AudioManager 类位于 android. media. AudioManager 包中，该类是音频管理类，提供了大量的 API，这些 API 包括控制音量、铃声模式、蓝牙、耳机等管理操作。

AudioManager 实例对象是通过 getSystemService(Context. AUDIO_SERVICE)方法获得。例如实例化对象 audiomanager，代码如下。

```
AudioManager audiomanager = (AudioManager)Context.getSystemService(Context.AUDIO_SERVICE);
```

AudioManager 中常用的调节音量的方法见表 11-2。

表 11-2　AudioManager 的常用方法及说明

方　法	说　明
adjustVolume(int direction, int flags)	控制手机音量，调大或者调小一个单位
adjustStreamVolume(int streamType, int direction, int flags)	对指定声音类型调节音量大小
setStreamVolume(int streamType, int index, int flags)	直接设置音量大小
setStreamMute(int streamType,boolean state)	将手机某个声音类型设置为静音。参数 state 若为 true，表示静音；若为 false，取消静音
getStreamVolume(int streamType)	获得手机的当前音量，最大值为 7，最小值为 0，当设置为 0 的时候，手机自动将模式调整为“振动模式”
getMode()	返回当前音频模式，例如：NORMAL(普通)，RINGTONE(铃声)，IN_CALL(通话)
setMode()	设置声音模式，可取值 NORMAL(普通)，RINGTONE(铃声)，IN_CALL(通话)
setRingerMode(int ringerMode)	设置铃声模式，ringerMode 为铃声模式，如： RINGER_MODE_NORMAL(普通) RINGER_MODE_SILENT(静音) RINGER_MODE_VIBRATE(振动)
getStreamMaxVolume(int streamType)	获得当前手机最大铃声

在上面的方法中，对于出现在多个方法的参数说明如下。

streamType 参数，指定声音类型。有下述几种声音类型：STREAM_ALARM 表示手机闹铃，STREAM_MUSIC 表示手机音乐，STREAM_RING 表示电话铃声，STREAM_SYSTEAM 表示手机系统，STREAM_DTMF 表示音调，STREAM_NOTIFICATION 表示系统提示，STREAM_VOICE_CALL 表示语音电话。

direction 参数，调节音量大小。有下述几种调节类型：ADJUST_LOWER 表示可调小一个单位，ADJUST_RAISE 表示可调大一个单位，ADJUST_SAME 表示保持之前的音量。

flag 参数，一个或多个标志。有如下几种标志：FLAG_ALLOW_RINGER_MODES 表示更改音量时是否包括振铃模式作为可能的选项，FLAG_PLAY_SOUND 表示是否在改变音量时播放声音，FLAG_REMOVE_SOUND_AND_VIBRATE 表示删除可能在队列中或正在播放的任何声音/振动(与更改音量有关)，FLAG_SHOW_UI 表示显示包含当前音量的进度条，FLAG_VIBRATE 表示进入振动振铃模式时是否振动。

使用 AudioManage 来管理音量，有下列几段关键代码。

(1) 创建 AudioManage 对象和获取音量。

```
//音量控制，初始化定义
AudioManager mAudioManager = (AudioManager) getSystemService(Context.AUDIO_SERVICE);
//获取最大音量
int maxVolume = mAudioManager.getStreamMaxVolume(AudioManager.STREAM_MUSIC);
//获取当前音量
int currentVolume = mAudioManager.getStreamVolume(AudioManager.STREAM_MUSIC);
```

(2) 直接控制音量的大小。

```
if(isSilent){
    mAudioManager.setStreamVolume(AudioManager.STREAM_MUSIC, 0, 0);
}else{
    mAudioManager.setStreamVolume(AudioManager.STREAM_MUSIC, tempVolume, 0);
                                                    //tempVolume 为音量绝对值
}
```

(3) 以步长为单位逐渐增加或逐渐降低地控制音量，并显示音量控制进度条。

```
//降低音量，调出系统音量控制
if(flag == 0){
    mAudioManager.adjustStreamVolume(AudioManager.STREAM_MUSIC,
                AudioManager.ADJUST_LOWER, AudioManager.FLAG_SHOW_UI);
}
//增加音量，调出系统音量控制
else if(flag == 1){
    mAudioManager.adjustStreamVolume(AudioManager.STREAM_MUSIC,
                AudioManager.ADJUST_RAISE, AudioManager.FLAG_SHOW_UI);
}
```

下面通过一个案例来说明运用 MediaPlayer 和 AudioManager 实现播放和控制歌曲音量的应用。

【案例 11.1】 使用一组按钮来控制一首歌曲的播放、暂停、停止、静音、音量调节等操作。歌曲文件存放于本地。

说明：在界面中设计 5 个普通按钮来控制播放、暂停等状态以及音量调节，用 1 个开关按钮 ToggleButton 来控制静音/非静音,一共有 6 个按钮。使用 RelativeLayout 来布局这 6 个按钮。并设计按钮在被按下或弹起的动态效果。

用一个 Service 来实现对音频的控制,这样,当应用页面跳转后仍然可以保持音乐的播放。

开发步骤及解析:过程如下。

1) 创建项目

在 Android Studio 中创建一个名为 Activity_AudioPlayer 的项目。其包名为 ee.example.activity_audioplayer。

2) 准备颜色资源

编写 res/values 目录下的 colors.xml 文件,分别声明 green_light、grey 等颜色。

3) 准备音频资源

在 res 目录下创建 raw 目录,将音频文件 thesamesong.mp3 复制到 res/raw 目录下。

4) 准备图片背景素材

将准备好的按钮图片复制到 res/drawable 目录中。这些图片用于为 6 个按钮提供两组背景图资源。为了突出按钮的按下、弹起动态效果,分别为 6 个按钮的正常/按下状态定义背景描述文件,分别为 btnplay_selector.xml、btnpause_selector.xml、btnstop_selector.xml、tbtn_selector.xml、btninv_selector.xml 和 btndev_selector.xml。

以播放按钮为例,该按钮单击效果背景描述文件为 btnplay_selector.xml,其代码如下。

```
<?xml version = "1.0" encoding = "utf-8"?>
<selector xmlns:android = "http://schemas.android.com/apk/res/android">
    <item android:state_pressed = "false" android:drawable = "@drawable/music_plays"/>
    <item android:state_pressed = "true" android:drawable = "@drawable/music_plays_down"/>
</selector>
```

对于开/关按钮,其背景设置文件代码有些区别,这里,tbtn_selector.xml 就是本案例中开/关按钮的背景设置文件,其代码如下。

```
<?xml version = "1.0" encoding = "utf-8"?>
<selector xmlns:android = "http://schemas.android.com/apk/res/android">
    <item android:state_checked = "true"
        android:drawable = "@drawable/music_mutes" /> <!-- 开状态背景图 -->
    <item android:state_checked = "false"
        android:drawable = "@drawable/music_ons" /> <!-- 关状态背景图 -->
</selector>
```

5) 设计布局

编写 res/layout 目录下的 activity_main.xml 文件。在布局中设计一个 TextView 控件,用于显示播放的歌曲名信息;在该控件的下方设计一排按钮,其中 ID 分别为 btn_play、btn_pause、btn_stop 的 3 个按钮靠左;ID 分别为 tb_mute、btn_upper、btn_lower 的 3 个按钮靠右。并且 ID 为 tb_mute 的按钮是开关按钮 ToggleButton,其余按钮是 Button 按钮。

6) 开发逻辑代码

在 java/ee.example.activity_audioplayer 包下有两个代码文件,一个是 Activity 的子

类代码实现，另一个是 Service 的子类代码实现。

MainActivity.java 是实现 Activity 的子类定义，代码如下。

```
package ee.example.activity_audioplayer;

import androidx.appcompat.app.AppCompatActivity;

import android.os.Bundle;
import android.content.Intent;
import android.view.View;
import android.widget.Button;
import android.widget.CompoundButton;
import android.widget.ToggleButton;

public class MainActivity extends AppCompatActivity {
    private Button btnPlay, btnPause,btnStop,btnUpper, btnLower;
    private ToggleButton tbMute;
    private Intent intent;

    @Override
    protected void onCreate(Bundle savedInstanceState) {
        super.onCreate(savedInstanceState);
        setContentView(R.layout.activity_main);
        btnPlay = (Button)findViewById(R.id.btn_play);
        btnPause = (Button)findViewById(R.id.btn_pause);
        btnStop = (Button)findViewById(R.id.btn_stop);
        btnUpper = (Button)findViewById(R.id.btn_upper);
        btnLower = (Button)findViewById(R.id.btn_lower);
        btnPlay.setOnClickListener(new BtnOnClk());
        btnPause.setOnClickListener(new BtnOnClk());
        btnStop.setOnClickListener(new BtnOnClk());
        btnUpper.setOnClickListener(new BtnOnClk());
        btnLower.setOnClickListener(new BtnOnClk());

        tbMute = (ToggleButton)findViewById(R.id.tb_mute);
        tbMute.setOnCheckedChangeListener(new CompoundButton.OnCheckedChangeListener(){
            @Override
            public void onCheckedChanged(CompoundButton buttonView, boolean isChecked) {
                if(buttonView.getId() == R.id.tb_mute) {
                    Intent intent = new Intent("audioMng");   //创建 Intent
                    if(isChecked)
                      intent.putExtra("CHECKED", 1);           //正常播放设置为 1
                  else
                      intent.putExtra("CHECKED", 0);           //静音设置为 0
                  sendBroadcast(intent);                       //发送广播
                }
            }
        });

        Intent sintent = new Intent(this, AudioService.class); //创建服务的 Intent
        startService(sintent);                                 //启动后台 Service
    }

    private class BtnOnClk implements View.OnClickListener{
```

```
52          @Override
53          public void onClick(View v) {
54              Intent intent = new Intent("audioMng");                //创建 Intent
55              switch (v.getId()) {                                   //分支判断
56                  case R.id.btn_play:                                //按下播放按钮
57                      intent.putExtra("ACTION", 1);                  //存放数据
58                      togglebtnGoPlay(tbMute);
59                      sendBroadcast(intent);                         //发送广播
60                      break;
61                  case R.id.btn_pause:                               //按下暂停按钮
62                      intent.putExtra("ACTION", 2);
63                      sendBroadcast(intent);
64                      break;
65                  case R.id.btn_stop:                                //按下停止按钮
66                      intent.putExtra("ACTION", 3);
67                      sendBroadcast(intent);
68                      break;
69                  case R.id.btn_upper:                               //按下调高音量按钮
70                      intent.putExtra("ACTION", 4);
71                      togglebtnGoPlay(tbMute);
72                      sendBroadcast(intent);
73                      break;
74                  case R.id.btn_lower:                               //按下降低音量按钮
75                      intent.putExtra("ACTION", 5);
76                      togglebtnGoPlay(tbMute);
77                      sendBroadcast(intent);
78                      break;
79              }
80          }
81      }
82
83      private void togglebtnGoPlay(ToggleButton tbtn){
84          if(!tbtn.isChecked()){               //如果按钮是非静音背景,恢复静音背景
85              tbtn.setChecked(true);
86          }
87      }
88
89      @Override
90      protected void onDestroy() {                                   //释放时被调用
91          super.onDestroy();
92          Intent intent = new Intent(this, AudioService.class);      //创建 Intent
93          stopService(intent);                                       //停止后台的 Service
94      }
95  }
```

(1) 第 26～30 行,分别设置 Button 的状态变化监听。回调监听的响应方法在第 53～80 行中定义。在该方法中为 5 个按钮的编号传入 Intent 的 Extra 的 ACTION 中,该 Intent 的 action 是 audioMng,然后发送广播。其中,第 58、71、76 行,设置当相应的按钮被单击时确保开关按钮处于静音状态。

(2) 第 33～45 行,设置 ToggleButton 按钮对象 tbMute 的状态变化监听器。重写

onCheckedChanged()方法，在该方法中实现将 ToggleButton 按钮的开、关状态以整数形式传入 Intent 的另一 Extra 的 CHECKED 中，该 Intent 的 action 是 audioMng，然后发送广播。

(3) 第 48 行，启动后台服务。本案例对音频的所有控制在后台服务中实现。

AudioService.java 是继承 Service 类的实现，在该服务子类里，定义了一个广播接收器，用于对音频资源的播放和控制实现，代码如下。

```
package ee.example.activity_audioplayer;

import android.app.Service;
import android.content.BroadcastReceiver;
import android.content.Context;
import android.content.Intent;
import android.content.IntentFilter;
import android.media.AudioManager;
import android.media.MediaPlayer;
import android.os.IBinder;

public class AudioService extends Service{

    MediaPlayer mymp = null;                              //音频播放器
    private AudioManager myam = null;                     //音频管理器
    ServiceReceiver myserviceRec;
    int status = 1;              //当前的状态,1 没有声音播放,2 正在播放声音,3 暂停

    @Override
    public IBinder onBind(Intent intent) {                //重写的 onBind()方法
        return null;
    }
    @Override
        public void onCreate() {
        myserviceRec = new ServiceReceiver();             //创建 BroadcastReceiver
        IntentFilter filter = new IntentFilter();         //创建过滤器
        filter.addAction("audioMng");                     //添加 Action
        registerReceiver(myserviceRec, filter);           //注册 BroadcastReceiver
        super.onCreate();
    }
    @Override
    public void onDestroy() {                             //重写的 onDestroy()方法
        unregisterReceiver(myserviceRec);                 //取消注册
        super.onDestroy();
    }

    public class ServiceReceiver extends BroadcastReceiver{  //自定义 BroadcastReceiver
        @Override
        public void onReceive(Context context, Intent intent) {   //重写的响应方法
            int action = intent.getIntExtra("ACTION", -1);        //得到需要的数据
            switch(action){
                case 1:                                           //播放声音
                    if(status == 1){                              //当前没有声音播放
                        mymp = MediaPlayer.create(context, R.raw.thesamesong);
                        mymp.setLooping(true);                    //设置循环播放
```

```
46                         mymp.start();                        //播放声音
47                         status = 2;
48                     }
49                     else if(status == 2){                    //暂停声音状态
50                         mymp.start();                        //播放
51                     }
52                     break;
53                 case 2:                                      //暂停声音
54                     mymp.pause();                            //暂停播放声音
55                     status = 2;                              //改变状态
56                     break;
57                 case 3:                                      //停止声音
58                     mymp.stop();                             //停止播放声音
59                     status = 1;
60                     break;
61                 case 4:                                      //调高音量
62                     myam.adjustStreamVolume(
63                         AudioManager.STREAM_MUSIC,      //第一个参数:声音类型
64                         AudioManager.ADJUST_RAISE,      //第二个参数:调整音量的方向为调高
65                         AudioManager.FLAG_SHOW_UI);     //第三个参数:可选的标志位
66                     break;
67                 case 5:                                 //调低音量
68                     myam.adjustStreamVolume(
69                         AudioManager.STREAM_MUSIC,
70                         AudioManager.ADJUST_LOWER,      //第二个参数:调整音量的方向为调低
71                         AudioManager.FLAG_SHOW_UI);
72                     break;
73             };
74
75             myam = (AudioManager)getSystemService(Service.AUDIO_SERVICE);
                                                                //创建实例
76             int checked = intent.getIntExtra("CHECKED", -1);      //得到需要的数据
77             switch(checked) {
78                 case 0:                                          //静音
79                     myam.setStreamMute(AudioManager.STREAM_MUSIC, true);    //设置静音
80                     break;
81                 case 1:                                          //正常
82                     myam.setStreamMute(AudioManager.STREAM_MUSIC, false);   //设置正常
83                     break;
84             };
85         }
86     }
87
88 }
```

(1) 第 26、27 行，创建一个 IntentFilter，来接收 Intent 的 action 为 audioMng 中传过来的信息。

(2) 第 28 行，注册一个广播接收器，其 onReceive()方法在第 37～86 行中重新定义。可以看到在整个 Service 子类中，对音频的播放和控制，都在该 onReceive()方法中实现。

(3) 第 62～65 行，使用 AudioManager 的 adjustStreamVolume()方法实现调高音量操作。其中参数 AudioManager. STREAM_MUSIC 指明是声音类型，参数 AudioManager. ADJUST_RAISE 表示向上调高一个单位的音量，参数 AudioManager. FLAG_SHOW_UI

表示在调整时显示音量进度条。

(4) 第 75 行，创建 AudioManager 的一个实例 myam。第 77～84 行，根据接收到的 Extra CHECKED 中的数值，确定将 myam 实例设置为静音或正常播放。

7）注册 Service

在 AndroidManifest.xml 文件中声明一个 Service 组件即可。

运行结果：在 Android Studio 支持的模拟器上，运行 Activity_AudioPlayer 项目，运行结果如图 11-2 所示。启动应用的初始界面如图 11-2(a)所示，在此状态可以随时单击“暂停”或“停止”播放按钮。单击“静音”按钮，则听不到音乐声但音乐仍在播放中，如图 11-2(b)所示，如果再次单击该按钮，则可听到延续中的音乐声。单击“增大音量”按钮，可听到音量变大，而且显示音量调节进度条，如图 11-2(c)所示。单击“降低音量”按钮，则降低音量并出现音量调节进度条，如图 11-2(d)所示。

(a) 初始界面

(b) 单击“静音”按钮

(c) 单击“音量增大”按钮

(d) 单击“降低音量”按钮

图 11-2 音乐播放与音量控制应用

3. 使用 SoundPool 类

SoundPool 类也是用于播放音频文件的，位于 android.media.SoundPool 包中。相对于 MediaPlayer，SoundPool 具有对象占用 CPU 资源较少、延迟时间短、可以同时播放多个音频文件等优点，这些优点是 MediaPlayer 不具备的。所以 SoundPool 更适合播放短促、密集的音效，或较短的声音文件。在对音频文件的播放控制方面与 MediaPlayer 类基本相同，此外，SoundPool还支持对声音品质、音量、播放比率等参数设置，支持指定循环播放的次数。使用SoundPool类播放音频的步骤为：创建对象、加载音频、播放控制。下面分别介绍。

1）创建SoundPool对象

不同的Android版本，创建SoundPool对象的方式会有所不同。在应用编程中为了更好的兼容性，创建SoundPool对象时应考虑到版本问题。

在Android 5.0（API 21）之前，直接使用SoundPool的构造方法即可完成创建。创建方法如下。

```
SoundPool(int maxStream, int streamType, int srcQuality);
```

参数说明如下。

maxStream，指定同时播放的流的最大数目。

streamType，指定流的类型，一般为STREAM_MUSIC（具体的流类型在AudioManager类中列出）。

srcQuality，指定采样率转化质量，默认值为0，表示不指定采样率转化质量。

例如，创建一个最多支持3个流同时播放的，且类型标记为音乐的SoundPool对象，其代码如下。

```
SoundPool mysoundPool = new SoundPool(3, AudioManager.STREAM_MUSIC, 0);
```

在Android 5.0及以上的版本中，SoundPool对象需要使用SoundPool.Builder来创建。使用Builder模式进行构造Builder可以设置多个参数，设置方法如下。

setMaxStreams(int maxStreams)，用于设置播放的流的最大数目，与前面SoundPool()构造方法的参数maxStream含义一致。

setAudioAttributes(AudioAttributes attributes)，用于设置音频媒体的属性集。参数attributes为属性集，该值要么不设置，要么设置不能为null，否则会导致异常产生。如果不设置，系统会创建一个默认属性，即usage设置为USAGE_MEDIA。

attributes属性集主要指定用途和内容类型两类属性，使用setUsage()设置用途，例如播放的音频是用于游戏还是媒体；使用setContentType来设置内容类型，例如播放的内容是视频还是音乐。这些属性值都是int类型，通常用符号常量表示。常用的属性值见表11-3。

表11-3　SoundPool.Builder的常用属性值及说明

属　性　值	说　　明
USAGE_MEDIA	用于播放音频媒体，比如音乐、电影原声
USAGE_ALARM	用于闹钟铃声
USAGE_VOICE_COMMUNICATION	用于语音通信提醒，比如电话语音或VoIP
USAGE_NOTIFICATION	用于通知提醒
USAGE_NOTIFICATION_RINGTONE	用于来电铃声
USAGE_NOTIFICATION_COMMUNICATION_INSTANT	用于即时通知提醒，比如聊天、短信
USAGE_NOTIFICATION_COMMUNICATION_REQUEST	用于通信出、入的请求，比如VoIP通信或视频会议
USAGE_GAME	用于游戏配音和音效

续表

属　性　值	说　　明
USAGE_UNKNOWN	用于未知用途的音频
CONTENT_TYPE_MOVIE	电影或电视的原声、特色伴奏等内容类型
CONTENT_TYPE_MUSIC	音乐内容类型
CONTENT_TYPE_SONIFICATION	音效内容类型，比如某个动作发声、击键声效或游戏收到奖励的声音类型
CONTENT_TYPE_SPEECH	语音内容类型
CONTENT_TYPE_UNKNOWN	未知类型或多重定义的类型

例如，创建一个最多支持3个流同时播放的，且类型标记为音乐的SoundPool对象，其代码如下。

```
AudioAttributes audioAttributes = new AudioAttributes.Builder()
                                  .setUsage(AudioAttributes.USAGE_MEDIA)
                                  .setContentType(AudioAttributes.CONTENT_TYPE_MUSIC)
                                  .build();
SoundPool mysoundPool = new SoundPool.Builder()
                                  .setMaxStreams(3)
                                  .setAudioAttributes(audioAttributes)
                                  .build();
```

2）加载音频资源

SoundPool可以通过load()方法来加载一个音频资源，load()方法有四种加载方式，它们分别如下。

int load(AssetFileDescriptor afd, int priority)，通过一个AssetFileDescriptor对象描述的文件加载音频。

int load(Context context, int resId, int priority)，通过一个资源ID号加载音频。该资源文件存放在res/raw目录下。

int load(String path, int priority)，通过指定的路径加载音频。

int load(FileDescriptor fd, long offset, long length, int priority)，加载FileDescriptor对象指定的音频文件从offset开始、长度为length的声音。

一个SoundPool对象能同时加载多个音频，所以可以通过多次调用load()函数来加载。每次加载成功将返回一个非0的声音soundID，之后在播放时就是根据这个ID找到对应音频的。在编程中一般采用HashMap来存储并管理SoundPool对象中的多个声音。

例如，上面创建的mysoundPool对象，加载本地资源dingdong.ogg文件，并且以优先级为1传入HashMap对象中，其代码片段如下。

```
HashMap<Integer, Integer> soundPoolMap = new HashMap<Integer,Integer>();
soundPoolMap.put(1, mysoundPool.load(this, R.raw.dingdong, 1));
```

其中，第一个"1"是播放流的ID号，第二个"1"是优先级。

3）播放控制

由于一个SoundPool对象可以加载多个音频，所以在播放控制时需要指定soundID号。SoundPool提供的播放方法如下。

```
int play(int soundID, float leftVolume, float rightVolume, int priority, int loop, float rate)
```

返回值为 streamID,是一个播放流的 ID 号。

参数说明如下。

soundID,指定播放音频的 ID 号,该 ID 由 load()方法返回。

leftVolume,左声道的音量。

rightVolume,右声道的音量。

priority,流的优先级,值越大优先级越高。当同时播放数量超出了最大支持数时,这个优先级会决定 SoundPool 对该流的处理。

loop,循环播放的次数,−1 表示无限循环,0 表示不循环,其他值为播放 loop+1 次。

rate,播放的速率,范围为 0.5～2.0。0.5 表示播放速率减慢一半,1.0 表示正常速率,2.0 表示播放速率加快到两倍。

除了 play()方法外,SoundPool 还提供了其他控制方法。常用的控制音频播放方法见表 11-4。

表 11-4　SoundPool 常用的播放控制方法及说明

属　性　值	说　　明
pause(int streamID)	暂停指定播放流的音效,参数 streamID 是 play()的返回值,以毫秒为单位
resume(int streamID)	继续播放指定播放流的音效,参数 streamID 是 play()的返回值
stop(int streamID)	终止指定播放流的音效,参数 streamID 是 play()的返回值
setLoop(int streamID, int loop)	设置指定播放流的循环
setVolume(int streamID, float leftVolume, float rightVolume)	设置指定播放流的音量
setPriority(int streamID, int priority)	设置指定播放流的优先级
setRate(int streamID, float rate)	设置指定播放流的速率
unload(int soundID)	卸载一个指定的音频资源,卸载成功返回 true
release()	释放 SoundPool 中的所有音频资源

这里,play()方法传递的是一个 load()返回的 soundID,它指向一个被记载的音频资源。同一个 soundID 可以通过多次调用 play()而获得多个不同的 streamID ,当然不要超出同时播放的最大数目。pause()、resume()和 stop()方法是针对播放流操作的,其流的 ID 来自 play()的返回值。

需要在 API 中指出,即使调用相关的方法时,使用了无效的 soundID /streamID 也不会导致错误中断。在程序退出时,需要终止播放并释放资源。

下面通过一个案例来说明 SoundPool 的应用。

【案例 11.2】 使用 SoundPool 播放 MID、OGG、WAV、MP3 多种音频文件,并且可以实现同时播放的效果。

说明: 本案例将准备四种不同格式的音频文件,均存放于本地。通过单击按钮来分别加载音频,并使用 HashMap 来保存多个音频的 streamID,以便指定播放的音频文件。

开发步骤及解析: 过程如下。

1）创建项目

在 Android Studio 中创建一个名为 Activity_SoundPool 的项目。其包名为 ee.example.activity_soundpool。

2）准备颜色资源

编写 res/values 目录下的 colors.xml 文件，分别声明 green_light、grey 等颜色。

3）准备音频资源

在 res 目录下创建 raw 目录，将多个音频文件复制到 res/raw 目录下。本案例准备了四个音频文件，分别是 dingdong.ogg、dingling.wav、guzhang.mp3、backsound.mid。

4）设计布局

编写 res/layout 目录下的 activity_main.xml 文件。在布局中设计一个 TextView 控件，用于显示信息；在该控件的下方设计 4 个并排按钮，ID 分别为 btn_sp1、btn_sp2、btn_sp3 和 btn_sp4，分别用于播放不同的音频文件。

5）开发逻辑代码

在 java/ee.example.activity_soundpool 包下编写 Activity 子类的实现代码 MainActivity.java，代码如下。

```
package ee.example.activity_soundpool;

import androidx.appcompat.app.AppCompatActivity;

import android.os.Bundle;
import android.annotation.TargetApi;
import android.media.AudioAttributes;
import android.media.AudioManager;
import android.media.SoundPool;
import android.os.Build;
import android.view.View;
import android.widget.Button;
import java.util.HashMap;

public class MainActivity extends AppCompatActivity {

    SoundPool mysp;
    HashMap<Integer, Integer> soundPoolMap;
    private Button button1,button2,button3,button4;

    @Override
    protected void onCreate(Bundle savedInstanceState) {
        super.onCreate(savedInstanceState);
        setContentView(R.layout.activity_main);

        button1 = (Button) this.findViewById(R.id.btn_sp1);
        button2 = (Button) this.findViewById(R.id.btn_sp2);
        button3 = (Button) this.findViewById(R.id.btn_sp3);
        button4 = (Button) this.findViewById(R.id.btn_sp4);
        button1.setOnClickListener(new BtnOnClk());
        button2.setOnClickListener(new BtnOnClk());
        button3.setOnClickListener(new BtnOnClk());
        button4.setOnClickListener(new BtnOnClk());
```

```
34
35          createSoundPool();                //创建 SoundPool 对象
36
37          soundPoolMap = new HashMap<Integer,Integer>();
38          soundPoolMap.put(1,mysp.load(this,R.raw.dingdong,1));
39          soundPoolMap.put(2,mysp.load(this,R.raw.dingling,1));
40          soundPoolMap.put(3,mysp.load(this,R.raw.guzhang,1));
41          soundPoolMap.put(4,mysp.load(this,R.raw.backsound,1));
42      }
43
44      @TargetApi(Build.VERSION_CODES.LOLLIPOP)
45      private void createSoundPool() {
46          if (mysp == null) {
47              //Android 5.0 及之后版本
48              if (android.os.Build.VERSION.SDK_INT >=
                                android.os.Build.VERSION_CODES.LOLLIPOP) {
49                  AudioAttributes audioAttributes = null;
50                  audioAttributes = new AudioAttributes.Builder()
51                          .setUsage(AudioAttributes.USAGE_MEDIA)
52                          .setContentType(AudioAttributes.CONTENT_TYPE_MUSIC)
53                          .build();
54                  mysp = new SoundPool.Builder()
55                          .setMaxStreams(10)
56                          .setAudioAttributes(audioAttributes)
57                          .build();
58              } else { //Android 5.0 以前版本
59                  mysp = new SoundPool(10, AudioManager.STREAM_MUSIC, 0);  //创建对象
60              }
61          }
62      }
63
64      private class BtnOnClk implements View.OnClickListener{
65          @Override
66          public void onClick(View v) {
67              switch (v.getId()) {              //分支判断
68                  case R.id.btn_sp1:
69                      mysp.play(soundPoolMap.get(1),1,1,0,0,1.2f);
70                      break;
71                  case R.id.btn_sp2:
72                      mysp.play(soundPoolMap.get(2),1,1,0,0,1);
73                      break;
74                  case R.id.btn_sp3:
75                      mysp.play(soundPoolMap.get(3),1,1,0,0,0.8f);
76                      break;
77                  case R.id.btn_sp4:
78                      mysp.play(soundPoolMap.get(4),1,1,0,0,1);
79                      break;
80              }
81          }
82      }
83
84  }
```

(1) 第 35 行，调用自定义方法 createSoundPool()来创建 SoundPool 对象。由于创建方

法因Android的版本不同而不同,所以,在第45～62行,定义了创建SoundPool对象的方法,在第48行,检查版本号是否大于Android 5.0的版本号,即Android昵称为LOLLIPOP、API 21之后的版本。如果版本高于Android 5.0则使用new SoundPool.Builder()的方法创建对象,注意,创建对象之前需要先创建AudioAttributes实例,设置SoundPool对象需要的属性集,为创建对象提供参数。如果版本低于Android 5.0则使用new SoundPool()方法创建对象。

(2) 第38～41行,向HashMap<Integer,Integer>数据结构压入音频播放流ID号,音频流由SoundPool对象调用load()方法实现。

(3) 第44行,在方法前面加一个@TargetApi(),该方法在低版本SDK上运行不会报错。

(4) 第66～81行,重写onClick()回调方法,定义多个Button的onClick()事件。在其中,定义每一个按钮被单击时回调play()方法播放音频,其音频流ID号取自于HashMap对象。第69行定义播放音频会比正常速率快1.2倍,第75行定义播放的音频速率减慢0.8倍。

图 11-3　播放多个音频音效的应用

运行结果:在Android Studio支持的模拟器上,运行Activity_SoundPool项目,运行结果如图11-3所示。单击每一个按钮都可以发出一种声音,同时可以发出多种声音。

11.1.2　视频播放

Android中播放视频可以通过两种方式来实现。一种是通过VideoView组件,该种方式实现起来比较简单容易,但是其可控性不强,可以完成简单的播放任务;另一种通过MediaPlayer在SurfaceView进行播放,该种方式实现起来比较麻烦,但是可控性极强。所以在实际的项目中可以根据不同的需求进行使用。

Android支持的视频格式有mp4(MPEG-4低比特率)、3gp、avi、flv、h.263、h.264(avc)等,在相同存储容量下h.264比h.263画质更好。由于视频格式的文件都比较大,一般都存放在SD卡中。

1. 使用VideoView类

VideoView类位于android.widget.VideoView包中,它是Android自带的、专业化的显示视频的视图控件。VideoView类可从各种源(如资源或内容提供者)加载视频,并且可提供各种显示选项设置。常用的方法见表11-5。

表 11-5　VideoView常用的播放控制方法及说明

方　法	说　明
setVideoPath(String path)	加载path路径指定的视频文件
setVideoURI(Uri uri)	加载Uri地址所对应的视频资源

续表

方　　法	说　　明
setMediaController(MediaController controller)	设置媒体控制条的对象
setOnPreparedListener(MediaPlayer.OnPreparedListener l)	设置预备播放监听器。在监听器内需要实现 OnPreparedListener 的 onPrepared()方法，该方法在准备播放时调用
setOnCompletionListener(MediaPlayer.OnCompletionListener l)	设置结束播放监听器。在监听器内需要实现 OnCompletionListener 的 onCompletion()方法，该方法在结束播放时调用
setOnErrorListener(MediaPlayer.OnErrorListener l)	设置播放异常监听器。在监听器内需要实现 OnErrorListener 的 onError()方法，该方法在播放出现异常时调用
start()	开始播放视频
pause()	暂停播放视频
resume()	恢复播放视频
suspend()	结束播放并释放资源
seekTo(int msec)	视频到指定进度开始播放，msec 是指定的时间，单位为毫秒
getDuration()	获得视频的总时长
getCurrentPosition()	获得当前的播放时间位置
isPlaying()	判断是否正在播放
getBufferPercentage()	获得已缓冲的比例，返回值为 0～1

由于 VideoView 只是一个展示视频的控件，就相当于 ImageView 控件只显示图片一样。所以在实际应用中，要结合 MediaController 类来完成对视频播放的控制，结合 SeekBar 对象来显示播放的进度或拖动进度条确定播放的位置。

下面通过一个案例来学习如何使用 VideoView 播放视频。

【案例 11.3】 利用 VideoView 来播放来自 SD 卡上的 mp4 视频文件，并且显示进度条。可以实现前进、后退、暂停播放，可以拖动进度条实现从任意位置开始播放的效果。

说明：本案例将联合使用 VideoView ＋ MediaController ＋ SeekBar 完成编程，将使用线程来控制播放进度条的显示。

开发步骤及解析：过程如下。

1）创建项目

在 Android Studio 中创建一个名为 Activity_VideoViewPlayer 的项目。其包名为 ee.example.activity_videoviewplayer。

2）准备图片资源

将准备好的按钮图片复制到 res/drawable 目录中。

3）准备视频资源

将准备好的视频文件上传到 SD 卡的 Movies 目录中。单击 Android Studio 窗口右下角的 Device File Explorer 调出设备文件管理器，在 sdcard/Movies 目录下，单击右键弹出菜单，选择 Upload 选项，如图 11-4 所示。然后将视频文件 bordercollie.3gp 上传到 SD 卡的 Movies 目录中。

4）设计布局

编写 res/layout 目录下的 activity_main.xml 文件，在其中添加< VideoView >和

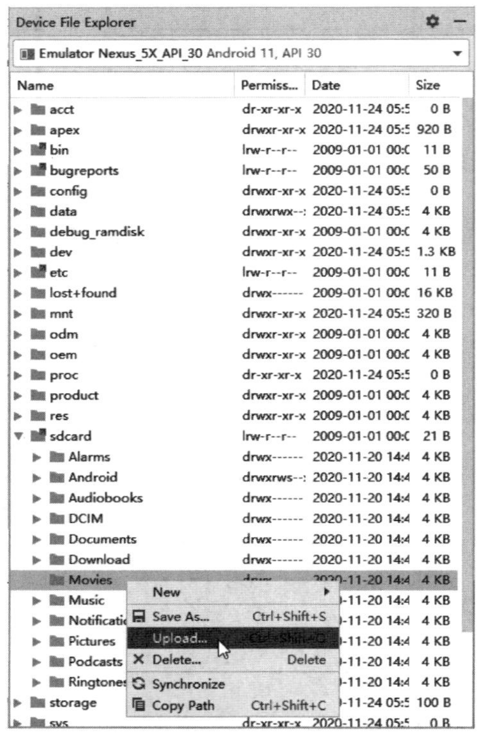

图 11-4　上传视频文件到 SD 卡的 Movies 目录中

<SeekBar>控件，<SeekBar>控件可以显示视频播放进度，并指定进度标志点的图片资源ID，代码片段如下。

```
<VideoView
    android:id = "@ + id/vv_play"
    android:layout_width = "match_parent"
    android:layout_height = "wrap_content"
    android:layout_gravity = "center"
    android:layout_marginStart = "2dp"
    android:layout_marginLeft = "2dp"
    android:layout_marginTop = "40dp"
    android:layout_marginEnd = "2dp"
    android:layout_marginRight = "2dp" />

<SeekBar
    android:id = "@ + id/sb_play"
    android:layout_width = "match_parent"
    android:layout_height = "wrap_content"
    android:layout_marginTop = "20dp"
    android:max = "100"
    android:thumb = "@drawable/seekbar_points" />
```

5）声明权限

播放存储在 SD 卡上的视频文件时，需要有读外存的权限，所以在 AndroidManifest.xml 中需要声明读外部存储器文件权限，声明代码如下。

```
<uses-permission android:name="android.permission.READ_EXTERNAL_STORAGE" />
```

6）开发逻辑代码

在 java/ee.example.activity_videoviewplayer 包下编写程序文件，代码如下。

```
package ee.example.activity_videoviewplayer;

import androidx.appcompat.app.AppCompatActivity;

import android.os.Bundle;
import android.Manifest;
import android.app.Activity;
import android.content.pm.PackageManager;
import android.os.Build;
import android.os.Environment;
import android.os.Handler;
import android.widget.MediaController;
import android.widget.SeekBar;
import android.widget.SeekBar.OnSeekBarChangeListener;
import android.widget.VideoView;
import androidx.annotation.RequiresApi;
import androidx.core.app.ActivityCompat;

public class MainActivity extends AppCompatActivity implements OnSeekBarChangeListener {

    private VideoView vv_play;
    private SeekBar sb_play;
    MediaController myctrl;
    private static final int REQUEST_EXTERNAL_STORAGE = 1;

    @RequiresApi(api = Build.VERSION_CODES.M)
    @Override
    protected void onCreate(Bundle savedInstanceState) {
        super.onCreate(savedInstanceState);
        setContentView(R.layout.activity_main);

        vv_play = (VideoView) findViewById(R.id.vv_play);
        sb_play = (SeekBar) findViewById(R.id.sb_play);
        sb_play.setOnSeekBarChangeListener(this);
        sb_play.setEnabled(false);
        myctrl = new MediaController(this);
        vv_play.setMediaController(myctrl);
        myctrl.setMediaPlayer(vv_play);

        verifyStoragmissions(this);          //检查是否有对 SD 卡的读写授权
        play_mp4();
    }

    private void verifyStoragmissions(Activity activity) {
        if (android.os.Build.VERSION.SDK_INT >= android.os.Build.VERSION_CODES.M) {
            int permission_read = ActivityCompat.checkSelfPermission(activity,
                                            "android.permission.READ_EXTERNAL_STORAGE");
            if ( permission_read != PackageManager.PERMISSION_GRANTED) {  //检查是否有权限
                if (ActivityCompat.shouldShowRequestPermissionRationale(
```

```
                                activity, Manifest.permission.READ_EXTERNAL_STORAGE)) {
49                      } else {
50                          //动态申请权限
51                          ActivityCompat.requestPermissions(activity,
52                                  new String[]{Manifest.permission.READ_EXTERNAL_STORAGE},
53                                  REQUEST_EXTERNAL_STORAGE);
54                      }
55                  }
56              }
57          }
58
59          private void play_mp4() {
60              vv_play.setVideoPath(Environment.getExternalStorageDirectory().getPath()
                                                    + "/Movies/bordercollie.3gp");
61              vv_play.requestFocus();
62              vv_play.start();
63              sb_play.setEnabled(true);
64              mHandler.post(mRefresh);
65          }
66
67          private Handler mHandler = new Handler();
68          private Runnable mRefresh = new Runnable() {
69              @Override
70              public void run() {
71                  sb_play.setProgress(100 * vv_play.getCurrentPosition()/vv_play.getDuration());
72                  mHandler.postDelayed(this, 500);
73              }
74          };
75
76          @Override
77          public void onProgressChanged(SeekBar seekBar, int progress,boolean fromUser) {
78          }
79
80          @Override
81          public void onStartTrackingTouch(SeekBar seekBar) {
82          }
83
84          @Override
85          public void onStopTrackingTouch(SeekBar seekBar) {
86              int pos = seekBar.getProgress() * vv_play.getDuration() / 100;
87              vv_play.seekTo(pos);
88          }
89
90      }
```

(1) 第 26 行，注解@RequiresApi()指定了版本是 Android 6.0(VERSION_CODES.M)，说明在代码中要对运行版本做检查，对高于该版本和低于该版本的运行设备，会有区分地实现代码。

(2) 第 36～38 行，为 VideoView 对象 vv_play 设置一个媒体控制条，用于视频播放时对前进、后退、暂停及播放进度的控制。同时，MediaController 对象 myctrl 也要指定 VideoView 对象。

(3) 第 40 行，调用自定义方法 verifyStoragmissions()，用于动态为 App 授权。第 44～

57 行定义该自定义方法，其中第 46、47 行，检查是否已授权外部存储器的读权限；如果判断没有授权则进行授权(见第 51～53 行)，此时系统会给出提示授权对话框。关于对访问 SD 卡文件的动态授权，在第 9 章的 9.3.2 小节中有详细介绍。

(4) 第 59～65 行，自定义播放视频方法 play_mp4()。在该方法中，首先要从指定的路径处获取视频文件(见第 60 行)，在调用 start()方法之前，VideoView 对象要获得焦点，即调用 requestFocus()。第 64 行，向 mHandler 发送 Runnable 对象 mRefresh。

(5) 第 67 行，定义 Handler 对象 mHandler。第 68 行，定义 Runnable 对象 mRefresh。第 70～73 行，重写了 Runnable 的 run()方法，在其中每隔 0.5 秒执行一次 Runnable 对象，重写进度条当前位置。

(6) 第 85～88 行，是监听进度条拖动停止时触发调用的方法，在该方法中控制视频从进度标志位置开始播放。

运行结果：在 Android Studio 支持的模拟器上，首次运行 Activity_VideoViewPlayer 项目时，会出现是否允许授权的系统制式对话框，如图 11-5 所示。这时请选择 Allow(即允许授权)，否则将不能播放存储在 SD 卡上的视频文件。如果选择了 Allow，以后运行该项目都不会再次出现该对话框了。

图 11-5　App 访问外部存储卡的授权对话框

再次运行项目会正常执行，运行结果如图 11-6 所示。启动应用会自动播放视频，如果拖动进度条，可以改变播放的时间位置，如图 11-6(a)所示。单击视频窗口，会短暂浮现视频控制按钮条，此时可以对视频进行前进、后退或暂停播放操作，如图 11-6 (b)所示。当改变手机方向时视频会跟随横屏，但是播放比例失真，如图 11-6(c)所示。

(a) 初始播放界面

(b) 使用菜单按钮控制播放

图 11-6　VideoView 播放视频与控制播放应用

(c) 改变设备方向播放

图 11-6 （续）

从案例代码来看，使用 VideoView 播放视频比较简单易学。结合 MediaController 提供的播放控制按钮和进度条可以很容易地控制 VideoView 的播放管理。但是对于控制屏幕上的播放位置，以及播放窗口的大小控制还难以实现。下面介绍的 SurfaceView 与 MediaPlayer 配合实现视频播放可以解决窗口调整问题。

2. 使用 MediaPlayer 和 SurfaceView 类

从前面介绍 MediaPlayer 类中可知，MediaPlayer 对象既可以加载音频文件，也可以加载视频文件。但是 MediaPlayer 主要用于播放音频，不提供视频图像的输出界面。这时必须借助于 SurfaceView 控件，将它与 MediaPlayer 结合起来，才能达到音频、视频的同时输出。

SurfaceView 类位于 android. view. SurfaceView 包中，主要用于视频的屏幕输出。SurfaceView 可以像其他控件一样在 XML 布局文件中声明，其常用的方法见表 11-6。

表 11-6 SurfaceView 常用的播放控制方法及说明

方 法	说 明
getHolder()	得到一个 SurfaceHolder 对象来管理 SurfaceView。返回值是一个 SurfaceHolder 对象
setVisibility(int visibility)	设置是否可见，visibility 取值可以是 VISIBLE、INVISIBLE、GONE
setAlpha(float alpha)	设置播放透明度。取值为 0～1，0 为完全透明，1 为完全不透明

SurfaceHolder 是一个接口，用于管理 SurfaceView，SurfaceHolder 类也位于 android. view. SurfaceHolder 包中，它的内部接口 SurfaceHolder. Callback 用于接收预览界面变化的信息，可通过重写以下三个抽象方法来实现。

当 SurfaceView 创建时触发方法如下。

```
public abstract void surfaceCreated(SurfaceHolder holder);
```

当 SurfaceView 改变（如预览界面的格式和大小发生改变）时触发方法如下。

```
public abstract void surfaceChanged(SurfaceHolder holder, int format, int width, int height);
```

当 SurfaceView 销毁时触发方法如下。

```
public abstract void surfaceDestroyed(SurfaceHolder holder);
```

使用 MediaPlayer 和 SurfaceView 的实现步骤如下。

(1) 在界面布局文件中定义 SurfaceView 控件。

(2) 创建 MediaPlayer 对象,并设置加载的视频文件(使用 setDataSource()方法)。

(3) 创建 SurfaceView 子类实现 SurfaceHolder. Callback 接口。在其中重定义三个抽象方法 surfaceCreated()、surfaceChanged()、surfaceDestroyed()。

(4) 通过 MediaPlayer. setDisplay(SurfaceHolder mysh)来指定视频画面输出到 SurfaceView 之上。

(5) 通过 MediaPlayer 对象的 start()、pause()、stop()等方法来控制播放视频。

下面通过一个视频播放案例来说明其用法。

【案例 11.4】 使用 MediaPlayer 和 SurfaceView 播放一个存放在 SD 上的视频文件,并可以随窗口的变化而调整。

说明:使用一个 RelativeLayout,在其中设计一个 SurfaceView。使用 SurfaceView 的尺寸和视频的原始尺寸及屏幕尺寸来计算横屏和竖屏时的视频播放窗口尺寸。在视频播放窗口的下面布局按钮,其中播放/暂停按钮是开关按钮,播放时显示"暂停";暂停时显示"播放"。

开发步骤及解析:过程如下。

1) 创建项目

在 Android Studio 中创建一个名为 Activity_SurfaceView 的项目。其包名为 ee. example. activity_surfaceview。

2) 准备颜色资源

编写 res/values 目录下的 colors. xml 文件,分别声明 black 等颜色。

3) 准备视频资源

将准备好的视频文件 bordercollie. 3gp 上传到 SD 卡的 Movies 目录中。

4) 设计布局

编写 res/layout 目录下的 activity_main. xml 文件。在布局中设计一个 RelativeLayout,在其中添加一个 SurfaceView;在 RelativeLayout 的下面添加两个按钮,ID 分别为 parent_play 和 btnstop。其中声明< RelativeLayout >的代码片段如下。

```
<RelativeLayout
    android:id = "@ + id/parent_play"
    android:layout_width = "match_parent"
    android:layout_height = "300dp"
    android:background = "@color/black">
    <!-- 添加一个 SurfaceView 用于播放视频 -->
    <SurfaceView
        android:id = "@ + id/surfaceView"
        android:layout_width = "wrap_content"
        android:layout_height = "wrap_content"
        android:layout_centerInParent = "true" /> <!-- 设置居中于 RelativeLayout 中 -->
</RelativeLayout>
```

5) 声明权限

在 AndroidManifest. xml 中添加对外部存储器文件的读权限。

6）开发逻辑代码

在 java/ee. example. activity_ surfaceview 包下编写 MainActivity. java 文件，代码如下。

```
package ee.example.activity_surfaceview;

import androidx.appcompat.app.AppCompatActivity;

import android.os.Bundle;
import androidx.annotation.RequiresApi;
import androidx.core.app.ActivityCompat;
import android.Manifest;
import android.os.Build;
import android.media.AudioManager;
import android.media.MediaPlayer;
import android.os.Environment;
import android.view.SurfaceHolder;
import android.view.SurfaceView;
import android.view.View;
import android.view.View.OnClickListener;
import android.widget.Button;
import android.content.pm.PackageManager;
import android.content.pm.ActivityInfo;
import android.net.Uri;
import android.widget.RelativeLayout;

public class MainActivity extends AppCompatActivity implements OnClickListener{

    private static final int REQUEST_EXTERNAL_STORAGE = 1;
    private static String[] PERMISSIONS_STORAGE = {
        Manifest.permission.READ_EXTERNAL_STORAGE};

    Button btn_playorpause, btn_stop;
    boolean isPlayed = false;
    RelativeLayout mParent;

    SurfaceView surfaceView;
    SurfaceHolder surfaceHolder;
    MediaPlayer mediaPlayer;

    @RequiresApi(api = Build.VERSION_CODES.M)
    @Override
    protected void onCreate(Bundle savedInstanceState) {
        super.onCreate(savedInstanceState);
        setContentView(R.layout.activity_main);
        btn_playorpause = (Button) findViewById(R.id.btnplayorpause);
        btn_stop = (Button) findViewById(R.id.btnstop);
        surfaceView = (SurfaceView) findViewById(R.id.surfaceView);
        mParent = findViewById(R.id.parent_play);
        btn_playorpause.setOnClickListener(this);
        btn_stop.setOnClickListener(this);

        surfaceView = (SurfaceView) findViewById(R.id.surfaceView);
```

```
            mediaPlayer = new MediaPlayer();

            surfaceHolder = surfaceView.getHolder();
            surfaceHolder.setKeepScreenOn(true);
            surfaceHolder.addCallback(new SurfaceHolder.Callback() {
                @Override
                public void surfaceCreated(SurfaceHolder holder) {
                if (android.os.Build.VERSION.SDK_INT >= android.os.Build.VERSION_CODES.M) {
                    //对 SD 卡读写权限的检测与授权,仅首次运行需要检测
                    if (ActivityCompat.checkSelfPermission(
                            MainActivity.this, Manifest.permission.READ_EXTERNAL_STORAGE) ==
                                PackageManager.PERMISSION_GRANTED) {
                            readyPlay();        //加载播放
                        } else {
                            ActivityCompat.requestPermissions(MainActivity.this,
                                PERMISSIONS_STORAGE, REQUEST_EXTERNAL_STORAGE);
                        }
                    } else {
                        readyPlay();            //加载播放
                    }
                }
                @Override
                public void surfaceChanged(SurfaceHolder holder, int format, int width, int height) {
                }
                @Override
                public void surfaceDestroyed(SurfaceHolder holder) {
                }
            });

        mediaPlayer.setOnVideoSizeChangedListener(new MediaPlayer.OnVideoSizeChangedListener() {
            @Override
            public void onVideoSizeChanged(MediaPlayer mp, int width, int height) {
                changeVideoSize();
            }
        });
    }

    @Override
    public void onRequestPermissionsResult(int requestCode, String permissions[], int[]
                                            grantResults) {
        if (requestCode == REQUEST_EXTERNAL_STORAGE) {
            if (grantResults.length > 0 && grantResults[0] ==
                                            PackageManager.PERMISSION_GRANTED) {
                readyPlay();                    //加载播放
            }
        }
    }

    //改变视频的尺寸自适应
    public void changeVideoSize() {
        int videoWidth = mediaPlayer.getVideoWidth();
        int videoHeight = mediaPlayer.getVideoHeight();
        int surfaceWidth = surfaceView.getWidth();
        int surfaceHeight = surfaceView.getHeight();
```

```
        //根据视频尺寸去计算->视频可以在sufaceView中放大的最大倍数
        float maxsize;
        if (getResources().getConfiguration().orientation ==
                                    ActivityInfo.SCREEN_ORIENTATION_PORTRAIT) {
            //竖屏模式下按视频宽度计算放大倍数值
            maxsize = (float) videoWidth / (float) surfaceWidth;
        } else {
            //横屏模式下按视频高度计算放大倍数值
            maxsize = (float) videoHeight / (float) surfaceHeight;
        }
        //视频宽高分别除以最大倍数值 计算出放大后的视频尺寸
        videoWidth = (int) Math.ceil((float) videoWidth / maxsize);
        videoHeight = (int) Math.ceil((float) videoHeight / maxsize);
        //无法直接设置视频尺寸,将计算出的视频尺寸设置到surfaceView,让视频自动填充
        RelativeLayout.LayoutParams params =
                            new RelativeLayout.LayoutParams(videoWidth, videoHeight);
        params.addRule(RelativeLayout.CENTER_HORIZONTAL,mParent.getId());
        surfaceView.setLayoutParams(params);
    }

    //视频加载并播放
    public void readyPlay() {
         String url = Environment.getExternalStorageDirectory().getPath() + "/Movies/
                      bordercollie.3gp";
        //String path = Environment.getExternalStorageDirectory().getPath() + "/Movies/
                      bordercollie.3gp";
        mediaPlayer.setAudioStreamType(AudioManager.STREAM_MUSIC);
        try {
            mediaPlayer.setDataSource(this, Uri.parse(url));
            //mediaPlayer.setDataSource(path);
        } catch (Exception e) {
            e.printStackTrace();
        }
        mediaPlayer.setLooping(true);
        mediaPlayer.setDisplay(surfaceHolder);  //把视频画面输出到SurfaceView
        mediaPlayer.prepareAsync();             //通过异步的方式加载媒体资源
        mediaPlayer.setOnPreparedListener(new MediaPlayer.OnPreparedListener() {
            @Override
            public void onPrepared(MediaPlayer mp) {
                play();                         //装载完毕回调播放
                isPlayed = true;
            }
        });
    }

    private void play() {
        if (mediaPlayer != null) {
            if (mediaPlayer.isPlaying()) {
                mediaPlayer.pause();
                btn_playorpause.setText("播放");
            } else {
                mediaPlayer.start();
                btn_playorpause.setText("暂停");
            }
```

```
149         }
150     }
151
152     public void onClick(View v) {
153         switch (v.getId()) {
154             case R.id.btnplayorpause:
155                 if(!isPlayed) {
156                     isPlayed = true;
157                     readyPlay();                //首次播放要加载视频资源
158                 }else{
159                     play();
160                 }
161                 break;
162             case R.id.btnstop:
163                 mediaPlayer.stop();
164                 btn_playorpause.setText("播放");
165                 mediaPlayer.reset();
166                 isPlayed = false;
167                 break;
168             }
169         }
170
171         @Override
172         protected void onDestroy() {
173         if (mediaPlayer != null) {
174             if (mediaPlayer.isPlaying()) {
175                 mediaPlayer.stop();
176             }
177             mediaPlayer.release();
178             mediaPlayer = null;
179         }
180         super.onDestroy();
181         }
182     }
```

(1) 第 26、27 行，定义 PERMISSIONS_STORAGE 字符串数组常量，用于表示外存储设备文件的读权限。

(2) 第 53 行，设置 Surface 所播放的视频持续出现在屏幕上。

(3) 第 54～74 行，定义了 SurfaceHolder.Callback 子类的实现。在实现中分别重写了 surfaceCreated()、surfaceChanged()、surfaceDestroyed()抽象方法。本案例只实现了 surfaceCreated()方法(第 56～67 行)，在重写的方法中实现：如果运行设备的版本需要(VERSION_CODES.M 及以上版本)，对 App 访问 SD 卡的权限进行检测，如果已经授权，或如果在低版本设备上运行，则调用 readyPlay()方法准备播放视频；如果没有授权，则动态授权，如果授权成功则自动回调 onRequestPermissionsResult()方法(第 85～91 行)。

(4) 第 76～81 行，设置 MediaPlayer 对象对视频尺寸变化的监听器，当视频尺寸发生变化时，调用 changeVideoSize()自定义方法(第 94～115 行)，该方法主要计算横屏和竖屏模式时的视频尺寸自适应调整。

(5) 第 118～138 行，定义视频加载方法 readyPlay()。由于是从外部存储设备上读取视频文件，此操作可能会受到其他不可预知的因素干扰，所以调用 setDataSource()方法，一般会使用 try...catch 来处理。这里是使用 URL 的方法从网络上获取资源的 setDataSource() 方式，也可以使用路径方式获取，代码如下。

```
String path = Environment. getExternalStorageDirectory ( ). getPath ( ) + "/Movies/
              bordercollie. 3gp";
mediaPlayer. setDataSource(path);
```

视频文件普遍比较大，加载过程相对来说比较慢，一般采用异步方式 prepareAsync() 加载不会造成线程阻塞。只有在视频资源准备好之后，才能调用播放方法实现视频播放。

(6) 第 140～150 行，定义播放方法 play()，该方法是要实现视频的播放或暂停功能和播放/暂停按钮的状态切换。

运行结果：在 Android Studio 支持的模拟器上，首次运行 Activity_SurfaceView 项目时，会出现是否允许授权的系统制式对话框。这时请选择 Allow（即允许授权），然后再次运行该项目。运行结果如图 11-7 所示。启动应用会自动播放视频，如图 11-7(a)所示。当单击“暂停”按钮时，视频暂停播放，此时按钮变为“播放”，如图 11-7(b)所示。如果将模拟器转向横屏，视频窗口会随动，并且窗口做自适应调整，不会变形，仍然保持在屏幕中央，如图 11-7(c)所示。

(a) 初始播放界面

(b) 单击“暂停”按钮后

(c) 横屏后的播放效果

图 11-7　SurfaceView 播放视频与自适应播放窗口

11.2　声音数据采集

现在的移动便携设备都配置有麦克风、录音等功能，可以非常方便地进行声音的采集。Android 提供了 MediaRecorder 类，可以实现录音、录像等操作。下面主要介绍录音应用。

Android 的声音采集就是通常说的录音，可使用 MediaRecorder 类来实现。MediaRecorder 类位于 android.media.MediaRecorder 包中，它包含对 Audio 和 Video 的媒体录制、编码、压缩等功能，被称作媒体录制器。

11.2.1　MediaRecorder 的常用方法

MediaRecorder 既可以录音，也可以录视频。下面主要针对声音采集方面的常用方法进行介绍，见表 11-7。表中的方法均无返回值，即返回 void。

表 11-7　MediaRecorder 常用于录音的方法及说明

方　法	说　明
prepare()	准备录制
start()	开始录制
stop()	停止录制
reset()	重置 MediaRecorder
release()	释放 MediaRecorder 占用的资源
setAudioChannels(int numChannels)	设置音频的声道数。1 表示单声道，2 表示双声道
setAudioEncoder(int audio_encoder)	设置音频录制的编码格式
setAudioEncodingBitRate(int bitRate)	设置音频每秒录制的字节数。数值越大音频越清晰
setAudioSamplingRate(int samplingRate)	设置音频的采样率。单位千赫兹(kHz)
setAudioSource(int audioSource)	设置音频录制的音频源，一般使用 AudioSource.MIC，指麦克风录制声音
setOutputFormat(int output_format)	设置录制媒体文件的输出格式
setOutputFile(File file)	媒体输出文件的路径

注意：在开发中，setAudioEncoder()方法必须在 setOutputFormat()之后执行，即应该先设置输出格式，再设置编码格式，否则程序将会抛出 IllegalStateException 异常。

setOutputFormat()方法常见的输出格式见表 11-8。

表 11-8　setOutputFormat()方法常用的音频输出格式及说明

输出格式	输出文件的扩展名	说　明
AMR_NB	.amr	窄带格式
AMR_WB	.mp3	宽带格式
AAC_ADTS	.aac	高级音频传输流格式
MPEG_4	.m4a	MPEG4 的音频格式
WEBM	.webm	WEBM 容器中的 VP8 / VORBIS 数据

常见的编码格式有 ACC、ACC-ELD、AMR_NB、AMR_WB、DEFAULT 等。

11.2.2 使用 MediaRecorder 的步骤

使用 MediaRecorder 进行录音应用开发比较简单，通常遵循下列步骤进行编程就可以实现。

(1) 创建 MediaRecorder 对象。例如，创建 recorder 对象如下。

```
MediaRecorder recorder = new MediaRecorder();
```

(2) 设置声音来源。例如，设置录音源为麦克风，代码如下。

```
recorder.setAudioSource(MediaRecorder.AudioSource.MIC);
```

(3) 设置输出格式。例如，设置为 MPEG_4 的输出格式，这时的文件扩展名为.m4a。

```
recorder.setOutputFormat(MediaRecorder.OutputFormat.MPEG_4);
```

(4) 设置音频编码格式。例如，采用高级音频编码格式，代码如下。

```
recorder.setAudioEncoder(MediaRecorder.AudioEncoder.AAC);
```

(5) 设置编码位率。

(6) 设置采样率。

(7) 设置音频通道。

(8) 设置音频文件保存路径，例如，将录制的音频文件保存在 SD 卡的 myrec 目录下，代码如下。

```
recorder.setOutputFile("/sdcard/myrec/myrecord01.m4a");
```

(9) 准备录制。例如：recorder.prepare();。

(10) 开始录制。例如：recorder.start();。

注意：以上步骤中，(3)和(4)的顺序不能颠倒。否则程序就会抛出异常。步骤(5)～(7)不是必须的，在编程中可以选择设置或不设置。

11.2.3 申请权限

图片、音频、视频文件占用存储空间较大，一般都保存在外部存储器 SD 卡中。录音操作需要拥有录音权限和对 SD 存储卡的写入权限。如果还要求播放录制的声音，则还需要拥有对 SD 卡的读权限。因此，必须在 Androidmanifest.xml 文件的“<manifest>”元素内添加下列权限元素。

```
<uses-permission android:name="android.permission.RECORD_AUDIO"/>
<uses-permission android:name="android.permission.WRITE_EXTERNAL_STORAGE"/>
<uses-permission android:name="android.permission.READ_EXTERNAL_STORAGE"/>
```

注意：对于 Android 6.0 以上的版本，在编程中还需要对 App 进行动态授权，才能保证程序的正常运行。

下面通过一个案例来说明录音程序的设计。

【案例 11.5】 使用 MediaRecorder 实现一个录音器,该录音器可以进行录音、停止、播放、删除功能,且将录制的文件存放在 SD 卡中。

说明:使用 Java 的文件读写操作,在 SD 卡指定目录中存储录制的声音文件,如果该目录不存在则创建。使用 ListView 列出已录制的音频文件列表,对录制的声音可以进行播放和删除。

使用 SD 卡保存文件,首先要判断移动设备(手机或模拟器)上是否有 SD 卡,使用下列方法。

```
Environment.getExternalStorageState().equals(android.os.Environment.MEDIA_MOUNTED)
```

如果返回的值为 true,则设备上有 SD 卡,否则没有 SD 卡。

开发步骤及解析:过程如下。

1) 创建项目

在 Android Studio 中创建一个名为 Activity_VoiceRecord 的项目。其包名为 ee.example.activity_voicerecord。

2) 准备颜色资源

编写 res/values 目录下的 colors.xml 文件,分别声明 black、white、lightgreen、lightpurple 等颜色。

3) 准备图片资源

将准备好的按钮图片复制到 res/drawable 目录中。

4) 设计布局

编写 res/layout 目录下的 activity_main.xml 文件。在布局中并排添加 4 个 ImageButton 图片按钮,ID 号分别为 ImgBtn_record、ImgBtn_stop、ImgBtn_play 和 ImgBtn_del;在按钮的下方添加一个文本框控件,ID 号为=TextView01,用于显示当前的操作信息;在文本框下面添加一个 ListView 控件,ID 号为 ListView01,用于显示录音文件名称列表。

在 res/layout 目录下创建另一个布局文件 my_list_item.xml,用于设计 ListView 列表项的显示样式布局,使用<CheckedTextView>元素。当选择列表框的某一项时,选中项使用该布局。该文件代码如下。

```
<?xml version = "1.0" encoding = "utf - 8"?>
< CheckedTextView
    xmlns:android = "http://schemas.android.com/apk/res/android"
    android:id = "@ + id/myCheckedTextView1"
    android:layout_width = "match_parent"
    android:layout_height = "match_parent"
    android:textSize = "16dp"
    android:textColor = "@color/white"/>
```

5) 声明权限

在 AndroidManifest.xml 中声明录音权限和对外部存储器文件的读写权限。即在<manifest>元素内添加权限如下。

```
< uses - permission android:name = "android.permission.RECORD_AUDIO"/>
```

```
<uses-permission android:name="android.permission.WRITE_EXTERNAL_STORAGE"/>
<uses-permission android:name="android.permission.READ_EXTERNAL_STORAGE"/>
```

6）开发逻辑代码

在 java/ee. example. activity_voicerecord 包下编写 MainActivity. java 文件，代码如下。

```
package ee.example.activity_voicerecord;

import androidx.annotation.RequiresApi;
import androidx.appcompat.app.AppCompatActivity;
import androidx.core.app.ActivityCompat;
import androidx.core.content.ContextCompat;

import java.io.File;
import java.io.IOException;
import java.util.ArrayList;
import java.util.Calendar;
import java.util.Locale;

import android.app.Activity;
import android.Manifest;
import android.content.pm.PackageManager;
import android.media.MediaPlayer;
import android.os.Build;
import android.os.Bundle;
import android.media.MediaRecorder;
import android.os.Environment;
import android.text.format.DateFormat;
import android.util.Log;
import android.view.View;
import android.widget.AdapterView;
import android.widget.ArrayAdapter;
import android.widget.CheckedTextView;
import android.widget.ImageButton;
import android.widget.ListView;
import android.widget.TextView;
import android.widget.Toast;

public class MainActivity extends AppCompatActivity implements View.OnClickListener,
                                                      AdapterView.OnItemClickListener{

    private static final String LOG_TAG = "AudioRecordTest";
    private ImageButton myBtnrecord,myBtnstop,myBtnplay,myBtndel;
    private ListView myListView1;
    private File myRecAudioFile;
    private File myRecAudioDir;
    private File myPlayFile;
    private MediaRecorder myMediaRecorder;
    private ArrayList<String> recordFiles;
    private ArrayAdapter<String> adapter;
    private TextView myTextView1;
    private boolean sdCardExit;
    private boolean isStopRecord;

```

```
    @RequiresApi(api = Build.VERSION_CODES.M)
    @Override
    protected void onCreate(Bundle savedInstanceState) {
        super.onCreate(savedInstanceState);
        setContentView(R.layout.activity_main);
        checkandgrantedPermissions(MainActivity.this);          //检测并授权

        //判断 SD 卡是否插入
        sdCardExit = Environment.getExternalStorageState().equals(
                                            Environment.MEDIA_MOUNTED);
        //取得 SD 卡路径作为录音的文件存储位置
        if (sdCardExit){
            myRecAudioDir = new File(Environment.getExternalStorageDirectory() + "/test/");
            if (!myRecAudioDir.exists())
                myRecAudioDir.mkdir();                          //创建目录
        }else {
            Toast.makeText(this, "没有检测到 SD 卡,请插入.", Toast.LENGTH_LONG).show();
            System.exit(0);
        }

        myBtnrecord = (ImageButton) findViewById(R.id.ImgBtn_record);
        myBtnstop = (ImageButton) findViewById(R.id.ImgBtn_stop);
        myBtnplay = (ImageButton) findViewById(R.id.ImgBtn_play);
        myBtndel = (ImageButton) findViewById(R.id.ImgBtn_del);
        myListView1 = (ListView) findViewById(R.id.ListView01);
        myTextView1 = (TextView) findViewById(R.id.TextView01);
        myBtnstop.setEnabled(false);
        myBtnplay.setEnabled(false);
        myBtndel.setEnabled(false);

        getRecordFiles();                           //取得 SD 卡目录里的所有.m4a 文件

        adapter = new ArrayAdapter<String>(this,R.layout.my_list_item, recordFiles);

        myListView1.setAdapter(adapter);        //将 ArrayAdapter 添加到 ListView 对象中
        myTextView1.setText("已有的录音文件:");

        myBtnrecord.setOnClickListener(this);  //录音
        myBtnstop.setOnClickListener(this);    //停止
        myBtnplay.setOnClickListener(this);    //播放
        myBtndel.setOnClickListener(this);     //删除
        myListView1.setOnItemClickListener(this);       //音频文件列表监听
    }

    private static final int REQUEST_EXTERNAL_STORAGE = 1;
    private void checkandgrantedPermissions(Activity activity) {   //检测并授权方法
        String[] requestpermissions = {
                Manifest.permission.WRITE_EXTERNAL_STORAGE,
                Manifest.permission.READ_EXTERNAL_STORAGE,
                Manifest.permission.RECORD_AUDIO};
        boolean permissiongranted = true;           //被授权为 true

        if (Build.VERSION.SDK_INT >= Build.VERSION_CODES.M) {
            for (int i = 0; i < requestpermissions.length; i++) {//检查是否有了权限
```

```
                if (ContextCompat.checkSelfPermission(activity, requestpermissions[i]) !=
                                        PackageManager.PERMISSION_GRANTED) {
                    permissiongranted = false;
                    break;
                }
            }
            if ( permissiongranted ) { //permissiongranted 为 true,表示都授予了权限
            }else {                               //请求权限方法
                //动态申请权限
                ActivityCompat.requestPermissions(activity, requestpermissions,
                                        REQUEST_EXTERNAL_STORAGE);
            }
        }
    }

    @Override
    protected void onStop(){
        if (myMediaRecorder != null && !isStopRecord){
            //停止录音
            myMediaRecorder.stop();
            myMediaRecorder.release();
            myMediaRecorder = null;
        }
        super.onStop();
    }

    @Override
    public void onClick(View v) {
        switch (v.getId()) {
            case R.id.ImgBtn_record:                //录音
                if(myMediaRecorder == null)
                    myMediaRecorder = new MediaRecorder();
                try {
                    //创建录音频文件
                    String FileName = "RecVoice_" + DateFormat.format("yyyyMMdd_HHmmss",
                                        Calendar.getInstance(Locale.CHINA)) + ".m4a";
                    myRecAudioFile = new File( myRecAudioDir.getPath() + "/" + FileName);
                    myMediaRecorder.setOutputFile(myRecAudioFile.getAbsolutePath());
                                        //设置录音文件的路径

                    //设置录音来源、输出格式、编码格式和双声道
                    myMediaRecorder.setAudioSource(MediaRecorder.AudioSource.MIC);
                    myMediaRecorder.setOutputFormat(MediaRecorder.OutputFormat.MPEG_4);
                    myMediaRecorder.setAudioEncoder(MediaRecorder.AudioEncoder.AAC);
                    myMediaRecorder.setAudioChannels(2);

                    myMediaRecorder.prepare();
                    myMediaRecorder.start();

                    myTextView1.setText("录音中");
                    myBtnstop.setEnabled(true);
                    myBtnplay.setEnabled(false);
                    myBtndel.setEnabled(false);
                    isStopRecord = false;
```

```
                } catch (IOException e) {
                    e.printStackTrace();
                }
                break;
            case R.id.ImgBtn_stop:                //停止录音
                try {
                    myMediaRecorder.stop();
                    myMediaRecorder.release();
                    myMediaRecorder = null;
                    adapter.add(myRecAudioFile.getName());  //将录音频文件名给 adapter
                    myTextView1.setText("停止:" + myRecAudioFile.getName());
                    myBtnstop.setEnabled(false);
                    isStopRecord = true;
                } catch (RuntimeException e) {
                    myMediaRecorder.reset();
                    myMediaRecorder.release();
                    myMediaRecorder = null;
                    if (myRecAudioFile.exists())
                        myRecAudioFile.delete();
                }
                break;
            case R.id.ImgBtn_play:                //播放录音
                if (myPlayFile != null && myPlayFile.exists()) {
                    MediaPlayer mPlayer = new MediaPlayer();
                    try{
                        mPlayer.setDataSource(myPlayFile.getAbsolutePath());
                        mPlayer.prepare();
                        mPlayer.start();
                    } catch (IOException e) {
                        e.printStackTrace();
                        Log.e(LOG_TAG,"播放失败");
                    }
                }
                break;
            case R.id.ImgBtn_del:                 //删除指定录音
                if (myPlayFile != null) {
                    if (myPlayFile.exists())
                        myPlayFile.delete();     //删除 SD 卡中的文件
                    adapter.remove(myPlayFile.getName());  //删除 adapter 列表中的文件名项
                    myTextView1.setText("完成删除");
                    if (adapter.isEmpty()) {     //判断 adapter 为空时按钮的状态
                        myBtnplay.setEnabled(false);
                        myBtndel.setEnabled(false);
                    }
                }
                break;
        }
    }

    @Override
    public void onItemClick(AdapterView<?> parent, View view, int position, long id) {
        //当有单击列表中的文件名时,将"删除"及"播放"按钮置为 Enable
        myBtnplay.setEnabled(true);
```

```
202         myBtndel.setEnabled(true);
203         myPlayFile = new File(myRecAudioDir.getAbsolutePath()
                    + File.separator
                    + ((CheckedTextView) view).getText());
204         myTextView1.setText("你选的是:"
                    + ((CheckedTextView) view).getText());
205     }
206     //文件读操作,向 ArrayList 对象中添加 SD 卡中的所有.m4a 文件
207     private void getRecordFiles(){
208         recordFiles = new ArrayList<String>();
209         if (sdCardExit){
210             File files[] = myRecAudioDir.listFiles();
211             if (files != null){
212                 for (int i = 0; i < files.length; i++){
213                     if (files[i].getName().indexOf(".") >= 0){
214                         //只取.m4a 文件
215                         String fileS = files[i].getName().substring(files[i].getName().
                                    indexOf("."));
216                         if (fileS.toLowerCase().equals(".m4a"))
217                             recordFiles.add(files[i].getName());
218                     }
219                 }
220             }
221         }
222     }
223
224 }
```

(1) 第 54 行，调用自定义方法 checkandgrantedPermissions()，该方法对于 Android 6.0 以上的版本有效。该方法在第 92～110 行定义实现，用于检测应用项目需要的权限，如果没有授权则请求权限。本案例需要三个权限，第 94 行，定义了一个字符串数组 requestpermissions，用于存放这三个权限。第 98～103 行，检查这三个权限是否已授权，如果查出有一个权限未予授权，则标记变量 permissiongranted 为 false。当 permissiongranted 为 false 时，则动态申请权限（见第 107 行）。

(2) 第 57～66 行，判断 SD 卡是否插入设备中。如果有 SD 卡，则获取录音文件存储的路径。本案例设置的路径是 sdcard 下的 test 目录，如果 test 目录不存在则创建之。其中 android.os.Environment.MEDIA_MOUNTED 是系统环境常量，表示 SD 卡已插入。方法 Environment.getExternalStorageDirectory()的功能是取 SD 卡所在目录路径。这里，用变量 myRecAudioDir 保存录音文件的存储路径，以便后面用于保存录音文件时使用。

如果 SD 卡没有插入设备中，则提示用户插入，并退出运行。

(3) 第 78 行，调用自定义方法 getRecordFiles()，该方法在第 207～220 行定义实现。按照本案例中指定的文件存储目录（由变量 myRecAudioDir 给出）中所有的扩展名为.m4a 的文件名添加到 ArrayList<String>（列表数组）recordFiles 中，该对象就是文件名列表。

(4) 第 80 行，实例化一个 ArrayAdapter 对象 adapter，该对象为文件名列表框 myListView1 适配数据，其样式由 my_list_item.xml 布局文件定义，数据内容由 recordFiles 列表数组提供，该数组由自定义方法 getRecordFiles()提供。

(5) 第 124～196 行，实现 onClick()方法。其中分别定义了四个按钮的单击响应操作。其中第 126～152 行，实现录音操作，这段代码定义了录音文件名的命名规则，是以 RecVoice_开头，按日期+时间的形式命名，扩展名为.m4a。然后按照 MediaRecorder 的编程步骤，分别实现录音来源、输出格式、编码格式、双声道、准备、开始录音编程。第 153～169 行，实现停止录音操作，停止录音后将新生成的录音文件名添加到文件名列表适配器对象中(见第 158 行)。第 170～182 行，实现播放选择的音频文件，使用 MediaPlayer 对象 mPlayer 完成音频的播放，这个播放文件由 myPlayFile.getAbsolutePath()指定。第 183～194 行，实现删除操作，在这里不仅要删除指定的文件，而且还要把文件名列表适配器对象中的该项同时删除(见第 187 行)。

(6) 第 199～205 行，实现 onItemClick()方法。该方法定义了对文件名列表框每一项单击的响应操作，当单击到某一个文件名项，则即刻获取该文件的完整路径 myPlayFile(见第 203 行)，供用户单击播放按钮或删除按钮时对指定的文件进行相应操作。

运行结果：在 Android Studio 支持的模拟器上，运行前先来看看指定的 SD 卡中 test 目录。单击 Android Studio 窗口右下角的 Device File Explorer 调出设备文件管理器，如图 11-8 所示。

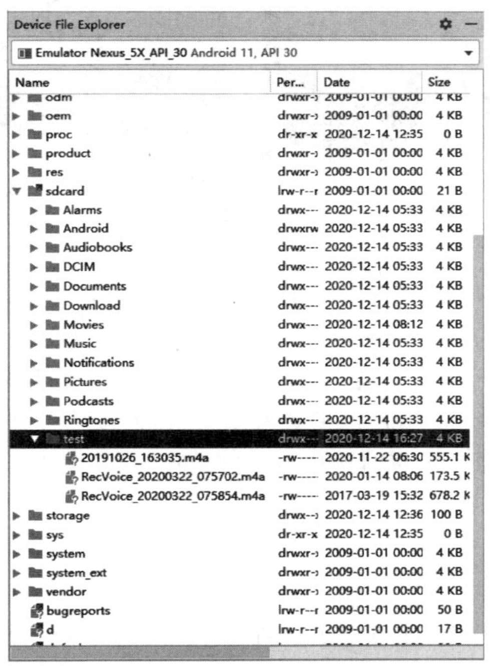

图 11-8 SD 卡的 test 目录中的音频文件

从图 11-8 中可以看到 test 目录下已经有了三个音频文件。这三个文件是为了测试应用项目而事先上传的三个音频文件。注意，如果首次运行项目，该目录中应该没有任何音频文件。首次运行 Activity_VoiceRecord 项目时，会出现是否允许授权的系统制式对话框。这时请选择 Allow(即允许授权)，然后再次运行该项目，运行结果如图 11-9 所示。启动应用即可以看到文件名列表框中有三个音频文件名，如图 11-9(a)所示。单击“录音”按钮，开始录制声音，如图 11-9(b)所示。单击“停止”，即可在文件名列表中看到新增的文件名项，如图 11-9(c)所示。在文件名列表中单击一个文件名项，该文件为选中状态，如果单击“播放”按钮，即可播放该文件对应的录制声音，如图 11-9(d)所示；如果单击“删除”按钮，则删除该选中文件，同时文件名列表中也会清除该文件名项。

(a) 初始界面

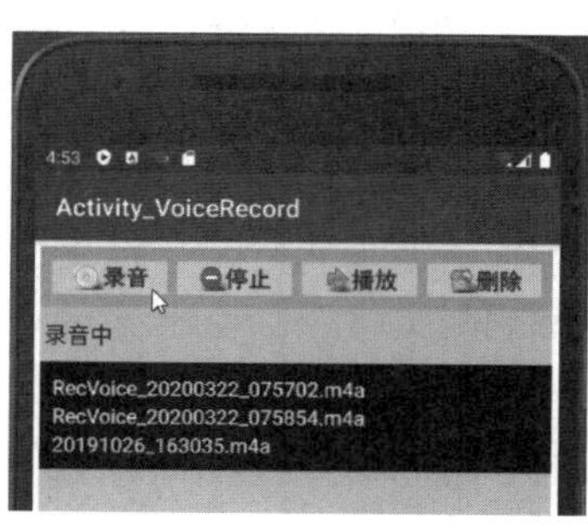

(b) 单击“录音”按钮后

(c) 单击“停止”按钮后

(d) 单击“播放”按钮后

图 11-9　录音器的运行效果

注意：在模拟器上运行时，是不能实现真正的录音测试的。因为模拟器不支持麦克风功能。如果要测试这一功能，必须在真机上运行。

在 Android Studio 中，可以连接手机来调试应用项目，步骤如下。

(1) 设置手机属性：打开手机“开发者权限”。如果没有打开，请在手机的“设置”中查找到“版本号”，在版本号项上连续多次单击，打开“开发者权限”。

(2) 设置手机允许“USB 调试”。

(3) 确保在 Android Studio 中勾选了 Google USB Driver 项。操作流程为依次选择 Android Studio 的菜单：file→Settings...→ Android SDK→SDK Tools，如图 11-10 所示。

(4) 检查 Android Studio 的 SDK Platforms 中是否包含测试手机的 Android 版本号。如果不包含则需要添加。

(5) 连接手机到电脑，即可以在 App 运行设备列表中看到该手机型号选项了。本案例选择的真机是华为 Mate 9，在运行设备列表中显示为 HUAWEI MHA-AL00，如图 11-11 所示。

选择该手机为测试设备，然后单击运行，即可实现在手机上运行应用项目。

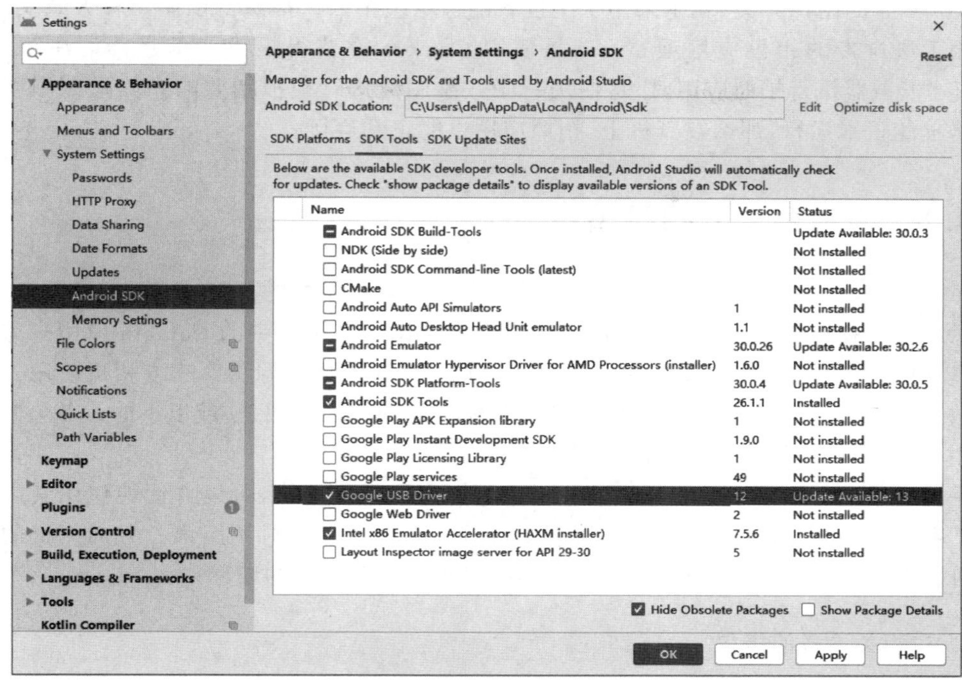

图 11-10　Android Studio 的 Settings 对话框

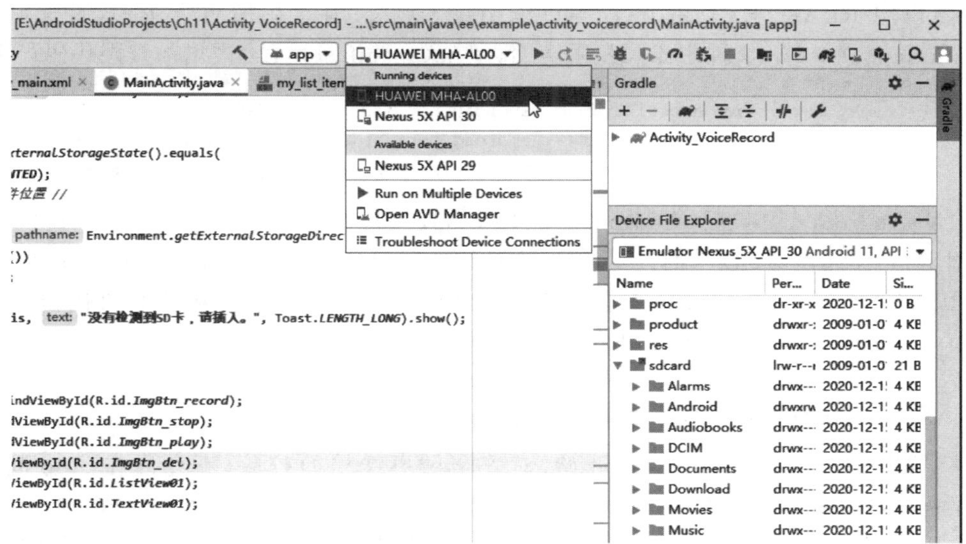

图 11-11　Android Studio 中 App 运行设备列表选项

11.3　图像数据采集

现在的智能手机或其他移动便携设备同时都是照相机、摄像机，使用手机拍照、摄像是非常重要的应用需求。下面主要介绍拍照应用。

Android 的图像采集，就是采用手机设备的摄像头拍照。拍照可以用两种方式实现，一是利用手机自带的相机应用项目；二是使用 Camera 类获得摄像头信息。请注意，使用 Android 模拟器执行拍照应用时，是使用一张三维图片进行模拟拍照的。如果应用项目在模拟器上能正常运行，那么在真机上，拍照应用也能正常运行。

11.3.1 调用第三方相机拍照

每个安装 Android 系统的智能手机，都会有一个自带相机的内置 Camera 应用，实现拍照摄像功能。程序员可以在自己的应用项目中直接调用系统自带的 Camera 应用来完成图像采集功能。但是必须指出，从 Android 11(API 30)开始禁止调用第三方相机应用。目前大多数 Android 设备都是 Android 10 或以下的，如果在 Android 10 或以下的设备上开始拍照应用，仍然可以使用系统自带的拍照应用。

手机自带的相机应用是第三方的相机应用。调取手机相机，可以方便地使用其内置的各种拍照属性设置，在应用开发中运用非常普遍。下面本书仍对调用第三方的相机的编程应用进行介绍。调取手机相机进行开发，需要实现以下几方面的编程。

1. 对相关设备的授权

拍照操作需要使用摄像头，需要声明 CAMERA 权限，拍照的图像文件一般都会保存在 SD 存储卡上，需要声明 WRITE_EXTERNAL_STORAGE 权限，如果要从相册中选取图片，需要声明 READ_EXTERNAL_STORAGE 权限。因此在拍照应用中，首先必须在 AndroidManifest.xml 文件中添加权限声明，即在<manifest>元素内增加权限代码如下。

```
<uses-permission android:name="android.permission.CAMERA"/>
<uses-permission android:name="android.permission.WRITE_EXTERNAL_STORAGE"/>
<uses-permission android:name="android.permission.READ_EXTERNAL_STORAGE"/>
```

自从 Android 6.0 以后，Android 对一些设备使用提供了安全性保护，这里对摄像头设备的调用、外存储设备 SD 卡的读写访问，在使用前需要进行动态授权。这种授权只需执行一次，以后再次运行该应用将不会重复要求授权。检查权限是否授予的代码应该在 Activity 的 onCreate()方法中调用。

2. 指定 App 间的文件共享存储空间

从 Android 7.0 (Android N)开始，对安全做更严格的校验，不允许在 App 间使用 file://的方式传递一个文件，否则会抛出 FileUriExposedException 异常。如果应用项目是利用设备自带的相机来拍照，就涉及自定义 App 与设备相机 App 之间、以及与设备上自带的相册 App 之间的文件传递问题。为解决必要的 App 文件传递，Android 提供了 FileProvider 机制，即通过 content://模式替换 file://模式，并且将项目的 targetSdkVersion 升级到 24 以上即可。

在 Android 7.0 及以下版本，FileProvider 由 android support v4.content 包提供；在 Android 8.0 及以上版本，FileProvider 由 androidx.core.content 包提供，是 ContentProvider 的子类，用于应用项目之间私有文件的传递。

在 Android 7.0 和 Android 9 之间的版本(含 Android 7.0 和 Android 9),使用 FileProvider 向外界传递 URI 信息,传递形式为 content ://Uri。

1) 设置清单文件

使用 FileProvider 向外界传递 URI 信息,需要在 AndroidManifest.xml 文件中注册,注册代码在< application >元素内添加,其代码如下。

```
1    <provider
2        android:name = "androidx.core.content.FileProvider"
3        android:authorities = " ${applicationId}.fileprovider"
4        android:exported = "false"
5        android:grantUriPermissions = "true">
6        < meta - data
7            android:name = "android.support.FILE_PROVIDER_PATHS"
8            android:resource = "@xml/file_paths" />
9    </provider >
```

第 3 行,android:authorities 表示授权者,其中,${applicationId}表示当前项目的包名,例如当前项目的包名为 ee.example.activity_takepic,那么 ${applicationId}.fileprovider 等同于 ee.example.activity_takepic.fileprovider。

第 4 行,代表是否可以输出被外部程序使用,一般为 false。

第 5 行,grantUriPermissions 表示是否允许为文件授予临时权限,必须为 true。

第 8 行,说明定义图片文件路径的描述文件的所在位置。android:resource="@xml/file_paths"含义是该描述文件位于应用项目的 res 目录下的 xml 子目录中,描述文件名为 file_paths.xml。

2) 编写文件路径的描述文件

文件路径的描述文件常用的元素如下。

```
<?xml version = "1.0" encoding = "utf - 8"?>
< paths >
    < files - path
        name = "file_uri"
        path = "file_path"/>
    < cache - path
        name = "cache_uri"
        path = "cache_path"/>
    < external - path
        name = "SD_uri"
        path = "SD_path"/>
    < external - files - path
        name = "SDfile_uri"
        path = "SDfile_path"/>
    < external - cache - path
        name = "SDcache_uri "
        path = " SDcache_path"/>
    ...
</paths >
```

在< paths >元素内,声明子元素,每一个子元素指向一个路径,其含义分别如下。

（1）< files-path >相当于调用 getFilesDir()方法，用于获取/data/data/< application package >/files 路径。

（2）< cache-path >相当于调用 getCacheDir()方法，用于获取/data/data/< application package >/cache 路径。

（3）< external-path >相当于调用 Environment. getExternalStorageDirectory()方法，用于获取设备上的 SD 卡的根路径，该路径为 mnt/sdcard。

（4）< external-files-path >相当于调用 getExternalFilesDir()方法，用于获取到 SD 卡中的 files-path 路径，其完整路径为 mnt/sdcard/Android/data/< application package >/files，一般存放较长时间保存的文件。

（5）< external-cache-path >相当于调用 getExternalCacheDir()方法，用于获取到 SD 卡中的 cache 路径，其完整路径为 mnt/sdcard/Android/data/< application package >/cache，一般存放临时缓存文件。注意，如果应用项目被用户卸载后，mnt/sdcard/Android/data/< application package >/这个目录下的所有文件都会被删除，不会留下垃圾信息。

这些子元素只有两个属性，一个是 name，一个是 path。

（1）name 的值替代 path 指定的路径，生成一个 Uri。

（2）path 的值表示该子元素指定的路径下的子目录，生成文件的存储路径。如果 path 的值只是一个点，即 path=“.”，则表示不改变之前自定义的路径。

例如< files-path name="file_uri" path="file_path"/>，表示 data/data/< application package >/files/file_path 路径下的文件是可共享的。在生成 Uri 时，name 的值 file_uri 会替代上面的路径，即 data/data/< application package >/files/file_path/部分将由 file_uri 代替。例如应用项目的< application package >是 ee. example. activity_takepic，将指定路径下的 test. jpg 文件是共享的，那么 test. jpg 的完整路径是 data/data/ee. example. activity_takepic/files/file_path/test. jpg，Uri 可表示为：content:// ee. example. activity_takepic/file_uri /test. jpg。

3）新版本的不同访问方式

请注意，在 Android 10（即 Android Q，API 29）及以上版本，不支持直接使用文件保存路径方式，而是使用 MediaStore 生成对应文件的 Uri。其中 FileProvider 类的 getUriForFile()方法提供了从文件路径格式转换为 Uri 格式的功能。方法的使用格式如下。

```
FileProvider.getUriForFile(<上下文>,<清单文件中 authorities 的值>,<共享的文件>);
```

3. 开启系统应用 App

开启系统自带应用项目，只需要创建 Intent，然后调用 startActivityForResult()方法。从系统应用项目返回时，会回调 onActivityResult()方法。

启动相机应用，使用常量 MediaStore. ACTION_IMAGE_CAPTURE 作为 Intent 的 action 来开启这个 Activity。

启动相册应用，使用常量 Intent. ACTION_GET_CONTENT 作为 Intent 的 action 来开启这个 Activity。

下面通过一个案例来说明如何使用手机自带的相机开发图像采集应用。

【案例 11.6】 通过一个简单界面调用手机内置的相机实现拍照，并且可以从手机相册中浏览并选择需要的图片。

说明：在用户界面上设计两个按钮，一个用来调用自带相机完成拍照功能，在拍照中可以使用与拍照相关的调节操作，拍照完毕后，向用户界面返回照片。另一个用来调用相册，浏览已有的图片，并可以选择图片展示在界面中。

当用户界面调用系统应用时，该用户的 Activity 将会转入后台，而当相机应用完毕后，用户的 Activity 将重新被激活。因为用户的 Activity 在其后台期间，有可能会因系统内存吃紧被销毁，所以在程序中需要使用生命周期方法 onSaveInstanceState ( ) 和 onRestoreInstanceState()来保存状态。

开发步骤及解析：过程如下。

1）创建项目

在 Android Studio 中创建一个名为 Activity_TakePic 的项目。其包名为 ee. example. activity_takepic。

2）准备文件路径描述文件

在 res/下创建子目录 xml，在 res/xml 目录下新建文件路径描述文件 file_path. xml，指定共享路径为 SD 卡根目录下 Pictures 子目录，其代码如下。

```
<?xml version = "1.0" encoding = "utf - 8"?>
< paths >
    < external - path
        name = "images"
        path = "Pictures" />
</paths >
```

3）声明权限和文件共享空间

在 AndroidManifest. xml 中的< manifest >元素内添加 CAMERA 权限、WRITE_EXTERNAL_STORAGE 权限和 READ_EXTERNAL_STORAGE 权限。在< application >元素内添加< provider >元素，指明相应共享权限和共享空间路径，添加< provider >元素的代码片段如下。

```
< provider
    android:name = "androidx. core. content. FileProvider"
    android:authorities = "ee. example. activity_takepic. fileprovider"
    android:exported = "false"
    android:grantUriPermissions = "true">
    < meta - data
        android:name = "android. support. FILE_PROVIDER_PATHS"
        android:resource = "@xml/file_paths" />
 </provider >
```

4）设计布局

编写 res/layout 目录下的 activity_main. xml 文件。在布局中添加一个 ImageView，ID 为 ivPic，用于浏览拍照图像；添加两个按钮，ID 为 bt_take_picture 的按钮用于启动相机拍照，ID 为 bt_choose_from_album 的按钮用于浏览相册照片。

5）开发逻辑代码

在 java/ee. example. activity_takepic 包下编写 MainActivity. java 程序文件，代码如下。

```
package ee.example.activity_takepic;

import java.io.File;
import java.io.IOException;
import java.text.SimpleDateFormat;
import java.util.Date;
import java.util.Locale;

import android.Manifest;
import android.content.ContentValues;
import android.content.Intent;
import android.content.pm.PackageManager;
import android.net.Uri;
import android.os.Bundle;
import android.os.Environment;
import android.provider.MediaStore;
import android.view.View;
import android.widget.Button;
import android.widget.ImageView;
import android.os.Build;
import android.widget.Toast;

import androidx.annotation.Nullable;
import androidx.appcompat.app.AppCompatActivity;
import androidx.core.app.ActivityCompat;
import androidx.core.content.ContextCompat;
import androidx.core.content.FileProvider;
import androidx.core.os.EnvironmentCompat;

public class MainActivity extends AppCompatActivity {
    private ImageView ivPicture;
    private Button btCamera;
    private Button btPhoto;

    private static final int CAMERA_REQUEST_CODE = 1;        //拍照的请求码
    private static final int ALBUM_REQUEST_CODE = 2;         //打开相册的请求码
    private static final int PERMISSION_CAMERA_REQUEST_CODE = 3;    //申请相机权限的请求码
    private static final int PERMISSION_ALBUM_REQUEST_CODE = 4;    //申请相册读写权限请求码
    private boolean u_taken = false;
    private static final String PHOTO_TAKEN = "photo_taken";

    private Uri myCameraImageUri;                              //用于保存拍照图片的 Uri
    private boolean isUpAndroidQ = Build.VERSION.SDK_INT >= Build.VERSION_CODES.Q;
                                                          //是否是 Android 10 以上手机

    @Override
    protected void onCreate(Bundle savedInstanceState) {
        super.onCreate(savedInstanceState);
        setContentView(R.layout.activity_main);
        ivPicture = findViewById(R.id.ivPic);
        btCamera = findViewById(R.id.bt_take_picture);
```

```
        btPhoto = findViewById(R.id.bt_choose_from_album);
        btCamera.setOnClickListener(new View.OnClickListener() {
            @Override
            public void onClick(View view) {
                checkPermissionAndCamera();
            }
        });

        //打开系统图库
        btPhoto.setOnClickListener(new View.OnClickListener() {
            @Override
            public void onClick(View view) {
                checkPermissionAndPhotoAlbum();
            }
        });
    }

    //检查权限,授权后调用相机拍照
    private void checkPermissionAndCamera() {
        int hasCameraPermission =
             ContextCompat.checkSelfPermission(getApplication(),Manifest.permission.CAMERA);
        if (hasCameraPermission == PackageManager.PERMISSION_GRANTED) {
            //有权限,调用相机拍照
            openCamera();
        } else {
            //没有权限,申请权限
            ActivityCompat.requestPermissions(this,
                    new String[]{Manifest.permission.CAMERA,
                            Manifest.permission.WRITE_EXTERNAL_STORAGE},
                    PERMISSION_CAMERA_REQUEST_CODE);
        }
    }

    //检查权限,授权后调用相册选择图片
    private void checkPermissionAndPhotoAlbum(){
        int hasPhotoAlbumPermission = ContextCompat.checkSelfPermission(
                        getApplication(),Manifest.permission.WRITE_EXTERNAL_STORAGE);
    if (hasPhotoAlbumPermission == PackageManager.PERMISSION_GRANTED) {
        //有权限,调用相册浏览
        openPhotoAlbum();
    } else {
        //没有权限,申请权限
        ActivityCompat.requestPermissions(this,
                    new String[]{Manifest.permission.WRITE_EXTERNAL_STORAGE,
                                    Manifest.permission.READ_EXTERNAL_STORAGE},
                    PERMISSION_ALBUM_REQUEST_CODE);
    }
}

    //处理请求权限的回调方法
    @Override
    public void onRequestPermissionsResult(int requestCode, String[] permissions, int[] grantResults) {
        if (requestCode == PERMISSION_CAMERA_REQUEST_CODE) {
            if (grantResults.length > 0
```

```
                    && grantResults[0] == PackageManager.PERMISSION_GRANTED) {
                //允许权限,则调用相机拍照
                openCamera();
            } else {
                //拒绝权限,弹出提示框
                Toast.makeText(this,"拍照权限被拒绝",Toast.LENGTH_LONG).show();
            }
        }else if (requestCode == PERMISSION_ALBUM_REQUEST_CODE) {
            if (grantResults.length > 0
                        && grantResults[0] == PackageManager.PERMISSION_GRANTED) {
                //允许权限,则调用相册浏览
                openPhotoAlbum();
            } else {
                //拒绝权限,弹出提示框
                Toast.makeText(this,"相册读写权限被拒绝",Toast.LENGTH_LONG).show();
            }
        }
    }

    @Override
    protected void onActivityResult(int requestCode, int resultCode, @Nullable Intent data) {
        super.onActivityResult(requestCode, resultCode, data);
        if (requestCode == CAMERA_REQUEST_CODE) {
            if (resultCode == RESULT_OK) {
                ivPicture.setImageURI(myCameraImageUri); // 显示拍照的照片
            } else {
                Toast.makeText(this,"取消拍照",Toast.LENGTH_LONG).show();
            }
        } else if (requestCode == ALBUM_REQUEST_CODE) {
            if (resultCode == RESULT_OK) {
                ivPicture.setImageURI(data.getData());   // 显示相册中选择的图片
            }
        }
    }

    //生命周期方法:当系统要销毁 Activity 之前调用
    @Override
    protected void onSaveInstanceState( Bundle outState ) {
        super.onSaveInstanceState(outState);
        outState.putBoolean(this.PHOTO_TAKEN, u_taken);
    }
    //生命周期方法:如果 Activity 在后台没有因为运行内存吃紧被清理,
    //则切换回时会触发 onRestoreInstanceState 方法
    @Override
    protected void onRestoreInstanceState( Bundle savedInstanceState){
        super.onRestoreInstanceState(savedInstanceState);
        if( savedInstanceState.getBoolean( this.PHOTO_TAKEN ) ) {
            openCamera();
        }
    }

    //打开系统相册
    private void openPhotoAlbum() {
        Intent intent = new Intent(Intent.ACTION_GET_CONTENT);
```

```
            intent.setType("image/ * "); //选择图片
            startActivityForResult(intent, ALBUM_REQUEST_CODE);
    }

    //打开系统相机拍照并保存
    private void openCamera() {
            u_taken = true;
            Intent captureIntent = new Intent(MediaStore.ACTION_IMAGE_CAPTURE);
            //判断是否有相机
            if (captureIntent.resolveActivity(getPackageManager()) != null) {
                File photoFile = null;
                Uri photoUri = null;
                if (isUpAndroidQ) {
                    //适配 Android 10 以上
                    photoUri = createImageUri();
                } else {
                    try {
                        photoFile = createImageFile();
                    } catch (IOException e) {
                        e.printStackTrace();
                    }
                    if (photoFile != null) {
                        if (Build.VERSION.SDK_INT >= Build.VERSION_CODES.N) {
                            //适配从 Android 7.0 到 Android 9 的文件权限,通过 FileProvider
                            //创建一个 Uri
                            photoUri = FileProvider.getUriForFile(MainActivity.this,
                                "ee.example.activity_takepic.fileprovider", photoFile);
                        } else {     //适配 Android 7.0 以下, 如:file:
                            photoUri = Uri.fromFile(photoFile);
                        }
                    }
                }
                myCameraImageUri = photoUri;
                if (photoUri != null) {
                    captureIntent.putExtra(MediaStore.EXTRA_OUTPUT, photoUri);
                    captureIntent.addFlags(Intent.FLAG_GRANT_WRITE_URI_PERMISSION);
                    startActivityForResult(captureIntent, CAMERA_REQUEST_CODE);
                }
            }
    }

    //创建图片地址 Uri,用于保存拍照后的照片,Android 10 以后使用这种方法
    private Uri createImageUri() {
        //设置保存参数到 ContentValues 中
        ContentValues contentValues = new ContentValues();
        //兼容 Android Q 和以上版本
        if (Build.VERSION.SDK_INT >= Build.VERSION_CODES.Q) {
            //Android Q 中不再使用 DATA 字段,而用 RELATIVE_PATH 代替
            contentValues.put(MediaStore.Images.Media.RELATIVE_PATH, "Pictures/AndQ");
        }
        //设置图片文件名
        contentValues.put(MediaStore.Images.Media.DISPLAY_NAME,getPicturefileName());
        //执行 insert 操作,向系统文件夹中添加文件
        // EXTERNAL_CONTENT_URI 代表外部存储器,该值不变
```

```
204            Uri uri;
205            //判断是否有 SD 卡
206            String status = Environment.getExternalStorageState();
207            if (status.equals(Environment.MEDIA_MOUNTED)) {
208                //有 SD 卡,则使用 SD 卡存储
209                uri = getContentResolver()
                        .insert(MediaStore.Images.Media.EXTERNAL_CONTENT_URI, contentValues);
210        }else {// 当没有 SD 卡时使用手机存储
211            uri = getContentResolver()
                    .insert(MediaStore.Images.Media.INTERNAL_CONTENT_URI, contentValues);
212        }
213        return uri;
214    }
215
216    //创建保存图片的文件路径
217    private File createImageFile() throws IOException {
218        String imageName = getPicturefileName();
219        File storageDir = new File(Environment.getExternalStorageDirectory().getAbsolutePath()
                    + File.separator + "Pictures" + File.separator + "AndPie");
220        if (!storageDir.exists()) storageDir.mkdirs();
221        File tempFile = new File(storageDir, imageName);
222        if (!Environment.MEDIA_MOUNTED.equals(EnvironmentCompat.getStorageState(tempFile))) {
223            return null;
224        }
225        return tempFile;
226    }
227
228    //创建保存图片的文件名
229    private String getPicturefileName(){
230        String picName = "IMG_" + new SimpleDateFormat("yyyyMMdd_HHmmss",
                                Locale.getDefault()).format(new Date()) + ".jpg";
231        return picName;
232    }
233 }
```

(1) 第 69～78 行，自定义方法实现检查拍照和写入 SD 卡权限，如果授权则打开相机执行拍照方法，否则进行授权。

(2) 第 81～90 行，自定义方法实现检查读写 SD 卡权限，如果授权则打开相册浏览图片的方法，否则进行授权。

(3) 第 94～114 行，是请求权限的回调方法，在执行授权之后会自动回调该方法。

(4) 第 156～189 行，自定义一个打开系统相机的方法 openCamera()，其中第 158 行创建一个 Intent 实例 captureIntent，该 Intent 是调用相机的 App。如果设备的相机存在，根据不同的版本获取拍照后照片存放的公共空间的 Uri 实例 photoUri。第 184 行将这个 photoUri 加入 captureIntent 的 EXTRA 中。

(5) 第 192～214 行，自定义方法 createImageUri()，创建拍照图片保存地址的 Uri，该方法用于 Android 10 及以上版本，指定保存地址为 sdcard/Pictures/AndQ 目录下。

(6) 第 217～226 行，自定义方法 createImageFile()，创建拍照图片保存地址的文件路

径。该方法用于 Android 9 及以下的版本，指定保存地址为 sdcard/Pictures/AndPie 目录下。

(7) 第 229～232 行，自定义方法 getPicturefileName()，用于创建拍照图片的文件名。在方法中采用当前日期时间为图片命名，扩展名为.jpg。

(8) 第 134～137 行，是 onSaveInstanceState()方法；第 141～146 行，是 onRestoreInstanceState()方法。这两个方法是为了保证该应用的数据完整性。防止该应用 App 在调用系统相机或相册而移入后台时，因系统内存不足被销毁而造成数据丢失。

运行结果：在 Android Studio 支持的模拟器上，运行 Activity_TakePic 项目。首次运行 App 时会有权限请求对话框出现，请选择允许授权。

在授权通过之后，运行结果如图 11-12 所示。进入初始运行界面，如图 11-12(a)所示。单击“拍照”按钮，打开设备自带的相机 App。首次进入相机时会出现一个选择对话框，如图 11-12(b)所示。选择“GOT IT”即可进入相机拍照界面，如图 11-12(c)所示。单击相机图标进行拍照，得到拍照预览界面，如图 11-12(d)所示。选择打钩，确定拍照的图片，返回首界面，此时可以在首界面中看到刚刚拍照所得照片，如图 11-12(e)所示。

单击“浏览”按钮，打开设备的相册，即可看到设备中所有的图片，如图 11-12(f)所示。单击其中一个图片，即可选中该图片，返回首界面，如图 11-12(g)所示。

可以打开 Android Studio 的 Device File Explore，查看图片保存结果，如图 11-13 所示。在图中可以看到保存在 sdcard/Pictures/AndQ 路径中的图片文件 IMG_20200419_052532.jpg。

(a) 初始界面

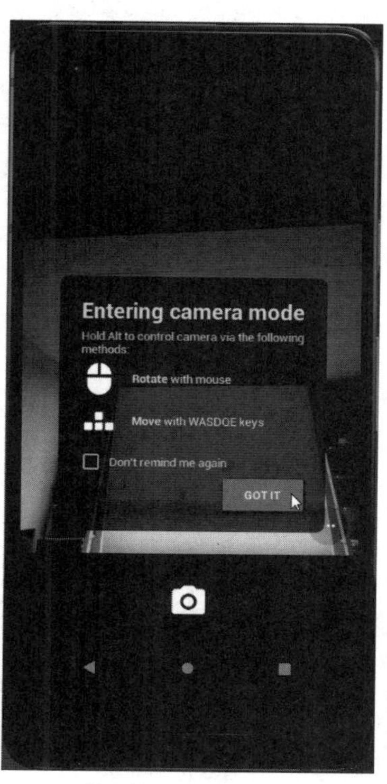

(b) 首次单击“拍照”进入相机

图 11-12　打开相机和相册的应用

(c) 进入相机界面

(d) 拍照后预览确认界面

(e) 拍照确认返回首界面

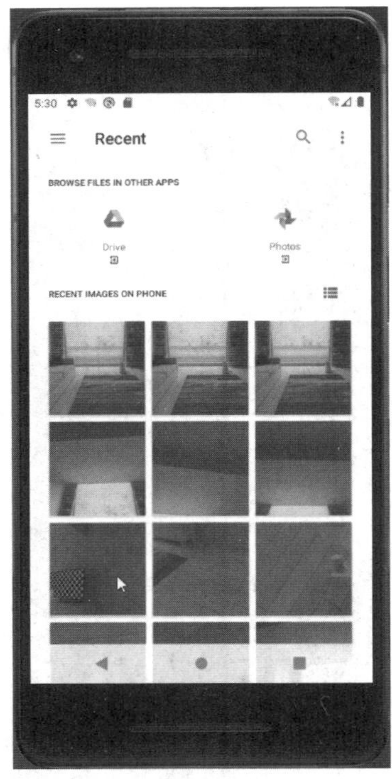

(f) 进入相册界面

(g) 选择相册图片返回首界面

图 11-12 （续）

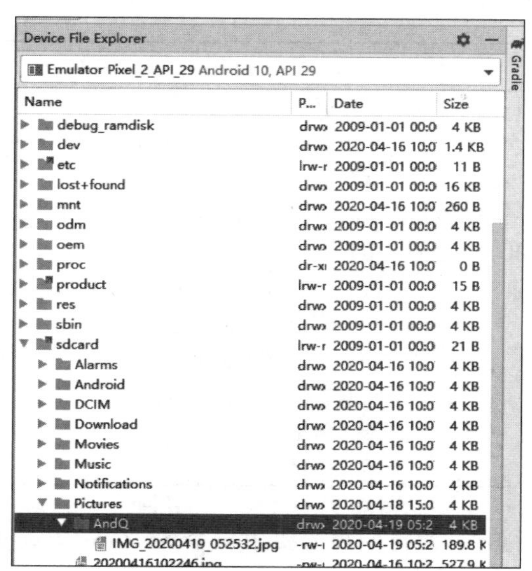

图 11-13 拍照照片的存储位置

该应用项目在模拟器上运行时，因受内存占用问题的影响，不同的 API 版本模拟器可能会出现一些异常表现，所以推荐在真机上运行，效果更好、更稳定。

调用设备自带相机拍照，应用项目相对简单，而且在拍照过程中可以调用相机固有的调节功能。但是，如果需要根据应用项目的具体需求改变拍照界面就不能做到，因为由系统定制的相机应用项目界面是不能被修改的。另外，从 Android 11 之后，Google 强制所有的拍照应用使用系统提供的类实现。接下来将学习 Camera 类来完成拍照编程。

11.3.2 使用 Android 提供的类实现拍照

1. 使用 Camera 类

Android 的 Camera 类位于 android. hardware. Camera 包中，它主要用于摄像头捕获图片、启动/停止预览图片、拍照、获取视频帧等，是直接对设备上的摄像头硬件进行管理的工具类。Camera 类常用的方法见表 11-9。

表 11-9 Camera 类常用的方法及说明

方法	说明
getNumberOfCameras()	获取设备的摄像头数目
open(int cameraId)	打开摄像头。默认为后置摄像头。参数 0 为后置摄像头，1 为前置摄像头
getParameters()	获取摄像头的拍照参数，返回 Camera. Parameters 对象
setParameters(Camera. Parameters params)	设置摄像头的拍照参数。例如： Camera. Parameters. setPreviewSize，设置预览界面的尺寸； Camera. Parameters. setPictureSize，设置保存图片的尺寸； Camera. Parameters. setPictureFormat，设置图片的格式

续表

方　法	说　明
setPreviewDisplay(SurfaceHolder holder)	设置预览界面的 SurfaceHolder 对象
startPreview()	开始预览，该方法必须在 setPreviewDisplay()方法之后调用
setDisplayOrientation(int degrees)	设置预览的角度
takePicture (Camera. ShutterCallback shutter, Camera. PictureCallback raw, Camera. PictureCallback jpeg)	开始拍照，并设置拍照回调事件。 参数： shutter，快门回调接口，回调方法 onShutter()在快门关闭后立即触发； raw，原始图像的回调接口。图片原始数据通过 byte[]传入回调方法； jpeg，JPG 图像的回调接口。压缩后的图像数据通过 byte[]传入回调方法，可在该接口的 onPictureTaken()方法中获得
stopPreview()	停止预览
release()	释放摄像头

使用 Camera 类进行拍照，预览照片需要用一个 SurfaceView 视图来呈现，在编程中需要实现 SurfaceHolder. Callback 接口。

使用 Camera 类拍照的主要操作是通过调用 takePicture()方法实现。它将以异步的方式从 Camera 中获取图像，其参数为回调类接口。参数 raw、jpeg 都是 Camera. PictureCallback 的回调，通常人们会关注 jpeg，因为拍照后获得的图像保存数据就是从该参数中获取的，其他参数都可以是 null。在回调接口 Camera. PictureCallback 中需要实现 onPictureTaken(byte[] data，Camera camera) 方法，其中 data 为图像数据。

下面通过案例说明如何使用 Camera 类开发自己的图像采集应用。

【案例 11.7】 使用 Camera 类设计一个简单的拍照应用，实现拍照、保存和预览功能。拍照文件存放在 sdcard/Pictures 目录中，文件名以日期时间命名。

说明：设计 3 个类的实现代码，一个是主 Activity 类的代码，在该活动类中触发拍照操作；第二个是拍照预览活动类的实现；第三个是 SurfaceView 类的派生子类的实现。拍照预览的图像视图由 SurfaceView 子类实例提供。

开发步骤及解析：过程如下。

1）创建项目

在 Android Studio 中创建一个名为 Activity_APICamera1 的项目，其包名为 ee. example. activity_apicamera1。

2）准备图片资源

在首次运行本应用时，在指定的目录中是没有图像的，所以预先准备好一个图片文件在首次运行时显示在界面上。将准备好的图片复制到 res/drawable 目录中。

3）设计布局

编写 res/layout 目录下的 activity_main. xml 文件。在布局中添加一个 ImageView，ID 为 main_image，用于浏览拍照图像；添加两个按钮，ID 为 bt_open_camera 的按钮用于拍照，ID 为 bt_quit 的按钮用于退出应用。

在 res/layout 目录下新建 activity_camera. xml 文件，该文件使用一个自定义的控件来显示拍照预览视图，该视图来自本案例自定义的子类 CameraSurfaceView，继承于

SurfaceView 类,在元素中需要给出包括包名的子类名全称。自定义控件的声明代码片段如下。

```
<ee.example.activity_apicamera1.CameraSurfaceView
    android:id="@+id/camera_preview"
    android:layout_width="match_parent"
    android:layout_height="match_parent"
    android:layout_gravity="center_vertical|center_horizontal" />
```

4) 开发逻辑代码

在 java/ee.example.activity_apicamera1 包下有 3 个 Java 代码,下面分别介绍。

MainActivity.java 文件,是主 Activity 类的实现代码,代码如下。

```
package ee.example.activity_apicamera1;

import java.io.File;

import android.Manifest;
import android.content.pm.PackageManager;
import android.database.Cursor;
import android.graphics.BitmapFactory;
import android.net.Uri;
import android.os.Build;
import android.os.Bundle;
import android.app.Activity;
import android.content.Intent;
import android.graphics.Bitmap;
import android.provider.MediaStore;
import android.view.View;
import android.view.View.OnClickListener;
import android.widget.Button;
import android.widget.ImageView;

import androidx.appcompat.app.AppCompatActivity;
import androidx.core.app.ActivityCompat;
import androidx.core.content.ContextCompat;

public class MainActivity extends AppCompatActivity implements OnClickListener{
    private Button cameraButton;                          //拍照按钮
    private Button quitButton;                            //退出按钮
    private ImageView imageView;                          //图片显示
    private File[] files;
    private static final int CAMERA_REQUEST_CODE = 1;     //拍照的请求码
    private boolean isAndroidQ = Build.VERSION.SDK_INT >= Build.VERSION_CODES.Q;
                                                          //是否是 Android 10 以上手机

    @Override
    protected void onCreate(Bundle savedInstanceState) {
        super.onCreate(savedInstanceState);
        setContentView(R.layout.activity_main);
        checkandgrantedPermissions(this);                 //检测并授权

        cameraButton = (Button) findViewById(R.id.bt_open_camera);
```

```
        quitButton = (Button) findViewById(R.id.bt_quit);
        cameraButton.setOnClickListener(MainActivity.this);
        quitButton.setOnClickListener(MainActivity.this);
        imageView = (ImageView) findViewById(R.id.main_image);
    }

    private static final int REQUEST_EXTERNAL_STORAGE = 1;
    private void checkandgrantedPermissions(Activity activity) {    //检测并授权方法
        String[] requestPermissions = {
                Manifest.permission.CAMERA,
                Manifest.permission.WRITE_EXTERNAL_STORAGE,
                Manifest.permission.READ_EXTERNAL_STORAGE,
        };
        boolean permissiongranted = true;                  //被授权为 true

        if (android.os.Build.VERSION.SDK_INT >= android.os.Build.VERSION_CODES.M) {
            for (int i = 0; i < requestPermissions.length; i++) {  //检查是否有了权限
                if (ContextCompat.checkSelfPermission(activity, requestPermissions[i])
                                        != PackageManager.PERMISSION_GRANTED) {
                    permissiongranted = false;
                    break;
                }
            }
            if ( permissiongranted ) {//permissiongranted 为 ture,表示都授予了权限
            }else {                         //请求权限方法
                                            //动态申请权限
                ActivityCompat.requestPermissions(activity,
                                        requestPermissions,
                                        REQUEST_EXTERNAL_STORAGE);
            }
        }
    }

    @Override
    protected void onResume(){
        super.onResume();
        initViews();
    }

    private void initViews() {
        //从本地取图片(在 SD 卡中获取文件)
        String picFilePath = null;
        Bitmap mybitmap = null;

        Uri mImageUri = MediaStore.Images.Media.EXTERNAL_CONTENT_URI;

        String[] projection = {MediaStore.Images.Media._ID,
             MediaStore.Images.Media.DATA,
             MediaStore.Images.Media.DATE_ADDED
        };
        //指定目录下的图片
        String where = MediaStore.Images.Media.DATA + " like ? and ("
             + MediaStore.Images.Media.MIME_TYPE + " = ? or "
             + MediaStore.Images.Media.MIME_TYPE + " = ?)";
```

```
86          //指定过滤条件值
87          String picPath = "/storage/emulated/0/Pictures";
88          String[] whereArgs = {picPath + "%","image/jpeg","image/jpg"};
89          //查询
90          Cursor mCursor = this.getContentResolver()
                                    .query(mImageUri, projection, where, whereArgs,
                                          MediaStore.Images.Media.DATE_MODIFIED + " desc");
91          if (mCursor != null && mCursor.getCount()> 0) {
92              mCursor.moveToFirst();
93              //获取图片的路径
94              int pathIndex = mCursor.getColumnIndex(MediaStore.Images.Media.DATA);
95              picFilePath = mCursor.getString(pathIndex);
96              mCursor.close();
97              //加载图片到 ImageView 控件上
98              mybitmap = BitmapFactory.decodeFile(picFilePath);
99              imageView.setImageBitmap(mybitmap);
100         }
101     }
102
103     @Override
104     public void onClick(View v) {
105         switch (v.getId()) {
106             case R.id.bt_open_camera:
107                 startActivity(new Intent(MainActivity.this, CameraActivity.class));
108                 break;
109             case R.id.bt_quit:
110                 MainActivity.this.finish();
111                 break;
112             default:
113                 break;
114         }
115     }
116
117 }
```

(1) 第 47～68 行，定义对权限检测和授权的方法，其中第 56～61 行对所要求赋予的权限进行授权检测，这里通过数组 requestPermissions 来存储所要求的权限值(见第 49～51 行)。当所有权限已授权，则 permissiongranted 为 true。

(2) 第 71～74 行，在 Activity 的生命周期中，每当该 Activity 从暂停状态转换到激活状态时调用 onResume()方法。通常用于页面刷新。

(3) 第 76～101 行，自定义方法 initViews()，用于为 ImageView 控件赋予图像并显示。在此使用 BitmapFactory 先获取 BitMap 对象，然后使用 setImageBitmap()方法把图像显示到 ImageView 控件中(见第 98、99 行)。此时没有使用 setImageURI()方法来显示图像，主要原因是使用该方法只是显示被压缩了的缩略图。

(4) 第 83、85、88 行，分别定义一个查询集的查询选项、查询过滤条件和查询过滤值。第 90 行生成一个查询集，并且是按文件修改日期的降序排序的。该查询集的结果包括在 SD 卡的 Pictures 目录下的所有 JPG 和 JPEG 格式的图片文件。

CameraActivity.java 文件，是预览拍照界面 Activity 的实现代码，代码如下。

```
package ee.example.activity_apicamera1;
```

```
import android.content.ContentValues;
import android.content.Intent;
import android.content.res.Configuration;
import android.graphics.Bitmap;
import android.graphics.BitmapFactory;
import android.graphics.Matrix;
import android.hardware.Camera;
import android.net.Uri;
import android.os.Bundle;
import android.os.Environment;
import android.provider.MediaStore;
import android.view.View;
import android.view.View.OnClickListener;
import android.widget.Button;
import android.widget.Toast;

import androidx.appcompat.app.AppCompatActivity;

import java.io.File;
import java.io.FileOutputStream;
import java.io.OutputStream;
import java.text.SimpleDateFormat;
import java.util.Date;

public class CameraActivity extends AppCompatActivity {
    private CameraSurfaceView myCameraSurfaceView;

    //拍照快门的回调
    private Camera.ShutterCallback ShutterCallback = new Camera.ShutterCallback() {
        @Override
        public void onShutter() {
        }
    };

    //拍照完成之后返回原始数据的回调
    private Camera.PictureCallback rawPictureCallback = new Camera.PictureCallback() {
        @Override
        public void onPictureTaken(byte[] data, Camera camera) {
        }
    };

    //拍照完成之后返回压缩数据的回调
        final private Camera.PictureCallback jpegPictureCallback = new Camera.
            PictureCallback() {
        @Override
        public void onPictureTaken(byte[] data, Camera camera) {
            saveFile(data);
        }
    };

    @Override
    protected void onCreate(Bundle savedInstanceState) {
        super.onCreate(savedInstanceState);
```

```
            setContentView(R.layout.activity_camera);

            Button btn_camera_save = (Button) findViewById(R.id.camera_save);
            myCameraSurfaceView = (CameraSurfaceView) findViewById(R.id.camera_preview);

            btn_camera_save.setOnClickListener(new OnClickListener() {
                @Override
                public void onClick(View v) {
                    takepicture();
                }
            });
        }

        public void takepicture() {
            myCameraSurfaceView.takePicture(ShutterCallback, rawPictureCallback,
            jpegPictureCallback);
        }

        //保存图片到硬盘
        public void saveFile(byte[] data) {
            Bitmap mybitmap;
            Uri myuri;

            SimpleDateFormat format = new SimpleDateFormat("yyyyMMddHHmmss");
            String fileName = format.format((new Date())) + ".jpg";
            //获取当前图片
            mybitmap = BitmapFactory.decodeByteArray(data, 0, data.length);
            //如果是纵屏拍照,则将图片旋转 90 度
            if (CameraActivity.this.getResources().getConfiguration().orientation
                    == Configuration.ORIENTATION_PORTRAIT)
                mybitmap = rotateBitmap(mybitmap, 90);

            if (android.os.Build.VERSION.SDK_INT < android.os.Build.VERSION_CODES.Q) {
                File storageDir = new File(Environment.getExternalStorageDirectory().getAbsolutePath()
                                    + File.separator + "Pictures");
                if (!storageDir.exists()) storageDir.mkdirs();
                File picoutFile = new File(storageDir , fileName);

                myuri = Uri.fromFile(picoutFile);
                try {
                    FileOutputStream fos = new FileOutputStream(picoutFile.getPath());
                    mybitmap.compress(Bitmap.CompressFormat.JPEG, 100, fos);
                    fos.flush();
                    fos.close();
                } catch (Exception e) {
                    e.printStackTrace();
                }
          } else {
              ContentValues contentValues = new ContentValues();
              contentValues.put(MediaStore.Images.Media.RELATIVE_PATH, "Pictures");
              contentValues.put(MediaStore.Images.Media.DISPLAY_NAME, fileName);
              contentValues.put(MediaStore.Images.Media.DESCRIPTION, fileName);
              contentValues.put(MediaStore.Images.Media.MIME_TYPE, "image/jpeg");
              myuri = CameraActivity.this.getContentResolver().insert(
```

```
                            MediaStore.Images.Media.EXTERNAL_CONTENT_URI, contentValues);
107                try {
108                    OutputStream cos = CameraActivity.this.getContentResolver()
                                                                .openOutputStream(myuri);
109                    mybitmap.compress(Bitmap.CompressFormat.JPEG, 100, cos);
110                    cos.flush();
111                    cos.close();
112                } catch (Exception e) {
113                     e.printStackTrace();
114                }
115            }
116            //将拍的照片添加到相册
117            Intent mediaScanIntent = new Intent(Intent.ACTION_MEDIA_SCANNER_SCAN_FILE);
118            mediaScanIntent.setData(myuri);
119            sendBroadcast(mediaScanIntent);
120
121            Toast.makeText(CameraActivity.this, "拍照成功", Toast.LENGTH_SHORT).show();
122            CameraActivity.this.finish() ;
123        }
124
125        //旋转图片,使图片保持正确的方向
126        private Bitmap rotateBitmap(Bitmap bitmap, int degrees) {
127            if (degrees == 0 || null == bitmap) {
128                return bitmap;
129            }
130            Matrix matrix = new Matrix();
131            matrix.setRotate(degrees, bitmap.getWidth() / 2, bitmap.getHeight() / 2);
132            Bitmap bmp = Bitmap.createBitmap(bitmap, 0, 0, bitmap.getWidth(), bitmap.getHeight(),
                                                    matrix, true);
133            bitmap.recycle();
134            return bmp;
135        }
136
137    }
```

（1）第 28 行，声明 CameraSurfaceView 子类的对象 myCameraSurfaceView。这个 CameraSurfaceView 子类继承 SurfaceView 类，在 CameraSurfaceView.java 中定义。

（2）第 31～35 行、第 38～42 行、第 45～50 行，分别定义 takePicture()三个参数的回调方法。当拍照后回传图像压缩数据给 jpegPictureCallback 时，回调方法 onPictureTaken()，在该方法中调用 saveFile()方法。

（3）第 73～123 行，定义 saveFile()方法。该方法主要完成 5 大操作：首先给要保存的图像文件命名(见第 77、78 行)；然后调整图片的角度方向；第三将图片写入指定的位置；第四通过扫描图片发出广播信息以更新设备的相册，将拍照图片添加到相册中(见第 117～119 行)；最后给出拍照成功提示并结束本页面，返回到首页面。

（4）第 80～84 行，从 data 参数中获取图片数据，并加载到 Bitmap 对象中，然后判断设备的方向，通过 getResources().getConfiguration().orientation 获取设备的方向，系统常量 ORIENTATION_PORTRAIT 表示纵向，ORIENTATION_LANDSCAPE 表示横向。因为 Camera 默认横屏拍照为正，竖屏拍照图像会旋转 90 度。所以在此要检测设备的方向，如果是竖屏，则需要把图像回转 90 度才能得到正向的 mybitmap。

（5）第 86～115 行，分 Android 版本完成保存图片文件到指定的目录 sdcard/Pictures 中。由于 Android 10 以上版本不再支持使用文件路径，编程时需要区分处理。其中第 86～99 行针对 Android 10（不包含 Android 10）以下版本的保存图片文件处理；第 101～114 行针对 Android 10 以上版本，在保存图片文件时，需要用 contentValues 来携带图片文件相关信息。

（6）第 126～135 行，实现将图片旋转指定角度的方法。其中，第 131 行的方法完成以屏幕为中心点进行旋转的操作。

CameraSurfaceView.java 文件，是定义继承 SurfaceView 类的子类实现文件，代码如下。

```
package ee.example.activity_apicamera1;

import android.content.Context;
import android.content.res.Configuration;
import android.graphics.ImageFormat;
import android.hardware.Camera;
import android.util.AttributeSet;
import android.view.SurfaceHolder;
import android.view.SurfaceView;

import java.io.IOException;
import java.util.List;

public class CameraSurfaceView extends SurfaceView implements SurfaceHolder.Callback {
    private SurfaceHolder mHolder;
    private Camera myCamera;

    public CameraSurfaceView(Context context, AttributeSet attrs) {
        super(context, attrs);
        mHolder = getHolder();
        mHolder.addCallback(this);
    }

    public void takePicture(Camera.ShutterCallback mShutterCallback,
                            Camera.PictureCallback rawPictureCallback,
                            Camera.PictureCallback jpegPictureCallback) {
        myCamera.takePicture(mShutterCallback, rawPictureCallback, jpegPictureCallback);
    }

    public void startPreview() {
        try {
            myCamera.setPreviewDisplay(mHolder);
            myCamera.startPreview();
        } catch (IOException e) {
            e.printStackTrace();
        }
    }

    public void setCamera(Camera camera) {
        myCamera = camera;
        Camera.Parameters parameters = camera.getParameters();      //获取 camera 的参数对象
```

```
            Camera.Size largestSize = getBestSupportedSize(parameters.getSupportedPreviewSizes());
            parameters.setPreviewSize(largestSize.width, largestSize.height);
                                                                     //设置预览图片尺寸
            largestSize = getBestSupportedSize(parameters
                    .getSupportedPictureSizes());                    //设置捕捉图片尺寸
            parameters.setPictureSize(largestSize.width, largestSize.height);
            parameters.setPictureFormat(ImageFormat.JPEG);           //设置格式
            parameters.set("jpeg-quality", 100);                     //设置照片质量
            myCamera.setParameters(parameters);
        }

        @Override
        public void surfaceCreated(SurfaceHolder holder) {
            if (myCamera == null) {
                myCamera = Camera.open(0);
            }
            followScreenOrientation(getContext(),myCamera);
            try {
                myCamera.setPreviewDisplay(holder);
            } catch (IOException e) {
                e.printStackTrace();
                myCamera.release();
                myCamera = null;
            }
        }

        @Override
        public void surfaceChanged(SurfaceHolder holder, int format, int width, int height) {
            setCamera(myCamera);
            myCamera.startPreview();
        }

        @Override
        public void surfaceDestroyed(SurfaceHolder holder) {
            if (myCamera != null) {
                myCamera.setPreviewCallback(null);
                myCamera.stopPreview();
                myCamera.release();
                myCamera = null;
            }
        }

        //设置手机横屏/纵屏摄像头方向
        public static void followScreenOrientation(Context context, Camera camera){
            final int orientation = context.getResources().getConfiguration().orientation;
            if(orientation == Configuration.ORIENTATION_LANDSCAPE) {
                camera.setDisplayOrientation(180);
            }else if(orientation == Configuration.ORIENTATION_PORTRAIT) {
                camera.setDisplayOrientation(90);
            }
        }

        //取能适用的最大的 Size
        private Camera.Size getBestSupportedSize(List<Camera.Size> sizes) {
```

```
93             Camera.Size largestSize = sizes.get(0);
94             int largestArea = sizes.get(0).height * sizes.get(0).width;
95             for (Camera.Size s : sizes) {
96                 int area = s.width * s.height;
97                 if (area > largestArea) {
98                     largestArea = area;
99                     largestSize = s;
100                }
101            }
102            return largestSize;
103        }
104
105    }
```

(1) 第 18～22 行，是该子类的构造方法。

(2) 第 37～48 行，自定义方法 setCamera()，实现对 Camera 设备的属性设置。其中 getBestSupportedSize()是自定义方法(见 92～103 行)，用于获得适应手机屏幕的最大尺寸。

(3) 第 51～63 行，重写 surfaceCreated()方法，在该方法中打开 Camera 的后置摄像头设备；调用 followScreenOrientation()方法，根据设备的方向调整摄像头方向，使得拍照预览时图像保持正向；并设置显示拍照预览图片的 SurfaceHolder 对象。

(4) 第 66～69 行，重写 surfaceChanged()方法，在该方法中对 Camera 设备进行属性设置，并开始预览。

(5) 第 72～79 行，重写 surfaceDestroyed()方法，在该方法中需要对 Camera 设备所占用的资源进行释放。

5) 声明权限和相关属性

本应用使用拍照操作，需要在 AndroidManifest.xml 中添加 CAMERA 权限、WRITE_EXTERNAL_STORAGE 权限和 READ_EXTERNAL_STORAGE 权限。并且，本例定义了两个 Activity 类，需要在< application >元素内添加一个< activity >声明，代码如下。

```
<activity android:name = ".CameraActivity">
```

更重要的是，从 Android Q(Android 10，API 29)版本开始，默认开启沙箱模式。因此，对于 Android Q 及以上版本，会导致读写外部存储设备上的文件操作失败。如果要把存储在 SD 卡上的图像文件显示出来，需要在清单文件的< application >元素的属性中添加此项属性设置，以请求为本应用开启使用旧的存储模式。其代码片段如下。

```
<application
    ...
    android:requestLegacyExternalStorage = "true"
    ...>
```

运行结果：在 Android Studio 支持的模拟器上，运行 Activity_ APICamera1 项目。首次运行 App 时会有权限请求对话框出现，请选择允许授权。

在授权通过之后，运行结果如图 11-14 所示。进入初始运行界面，如果此手机在指定的保存照片目录中没有任何图片，会出现一个默认的图片在屏幕中，如图 11-14(a)所示。单击"拍照"按钮，如果是首次进入相机预览，会出现一个选择对话框，如图 11-14(b)所示。选择

GOT IT 即可进入相机拍照界面，如图 11-14(c)所示。单击“拍照”按钮即进行拍照，并保存照片到指定的目录 sdcard/Pictures 中，返回到首页面，如图 11-14(d)所示。

(a) 进入拍照初始界面

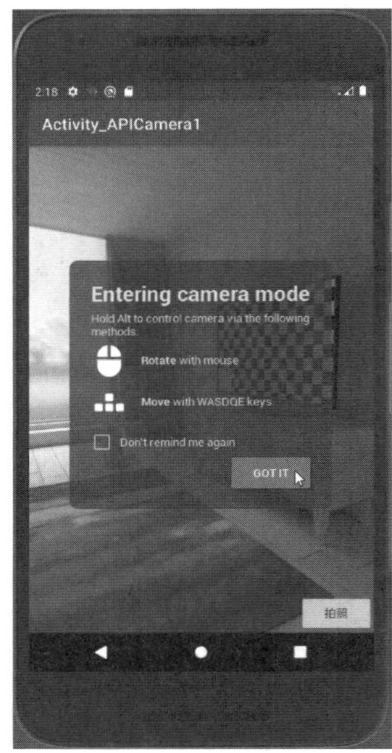

(b) 首次进入拍照预览界面

(c) 再次进入拍照预览界面

(d) 拍照完成返回首界面

图 11-14　使用 Camera 类自定义拍照应用

这时，可以打开 Android Studio 的 Device File Explore，可以看到保存在 sdcard/Pictures 下的图片文件，如图 11-15 所示。

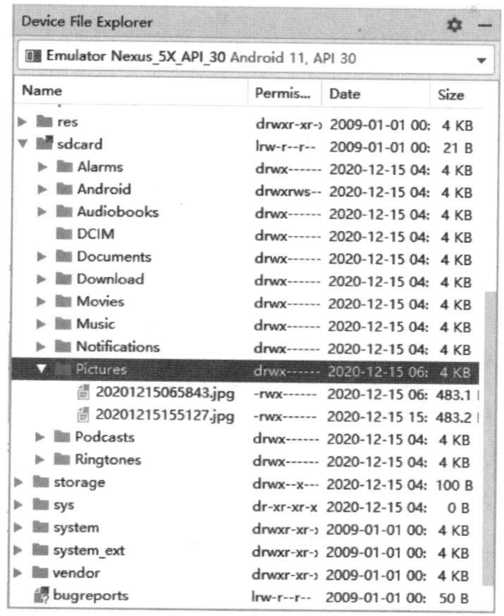

图 11-15 拍照照片的存储位置

Camera 类对拍照功能的控制不多，无法开发出应对现在日益强大的相机功能需求。所以自 Android 5.0 开始，Google 废弃了 Camera 类，引入了一套全新的相机框架 Camera2。对于应用在 Android 5.0 以上版本的设备，建议使用新的 Camera2 架构来编程实现拍照应用。

2. 使用 Camera2 类

Camera2 类位于 android. hardware. camera2 包中，它提供了许多高级特性，可以构建出高质量的相机应用项目，主要的新特性如下。

(1) 在开启相机之前就可以检查相机信息。

(2) 支持高速高质量连拍，每秒可有 30 帧的全高清连拍，甚至在连拍 30 张图片中可以使用不同曝光时间。

(3) 支持帧之间的手动设置，可以一次拍摄多张不同格式和尺寸的图片。

(4) 支持更多格式的图片，例如 RAW 图片格式，RAW 格式是无损照片的原生格式。

(5) 支持快门 0 延迟以及电影速拍。

(6) 支持相机其他方面的手动控制，例如控制噪音消除的级别、控制曝光时间等。

Camera2 在相机框架结构上做了大幅改造。Google 采用了管道(Pipeline)的概念，将 Camera Device(相机设备)和 Android Device(安卓设备)连接起来，Android Device 通过管道发送 CaptureRequest 拍照请求给 Camera Device，Camera Device 通过管道返回 CameraMetadata 数据给 Android Device，这一切的操作是在 CameraCaptureSession 的会话中进行的。为此 Camera2 可以提供数码相机拥有的强大功能，需要调用的 API 及回调也增加许多，在使用上比原来的 Camera 会复杂许多。

1）几个重要的管理类

为支持 Camera2 的强大功能，Android 提供了以下几个重要的管理类。下面对这些管理类及相应功能进行介绍。

（1）CameraManager。

摄像头管理类，其对象从系统服务 CAMERA_SERVICE 获取。主要功能包括：获取摄像头的 ID；获取摄像头的特征信息，比如摄像头前后位置信息、其支持的分辨率等信息；打开指定 ID 的摄像头；打开或关闭闪光灯等。常用的方法如下。

① getCameraIdList()，功能是获取摄像头列表，返回一组 String 类型的数组。通常返回两条字符串内容，一条是后置摄像头，另一条是前置摄像头。

② getCameraCharacteristics(String cameraId)，功能是获取摄像头的参数信息，例如摄像头支持级别、照片的尺寸等。

③ openCamera(String cameraId, CameraDevice. StateCallback callback, Handler handler)，功能是打开指定的摄像头。其中第一个参数是指定的摄像头 ID，第二个参数是设备状态监听器，该监听器需要实现接口 CameraDevice. StateCallback 的 onOpened()方法，在该方法内部调用 CameraDevice 对象的 createCaptureRequest()方法。

④ setTorchMode(String cameraId, boolean enabled)，功能是在不打开摄像头的情况下，开启或关闭闪光灯。true 表示开启闪光灯，false 表示关闭闪光灯。

（2）CameraDevice。

摄像头设备类。主要功能包括创建获取数据请求类 CaptureRequest. Builder，创建获取预览数据会话通道，创建获取拍照数据会话通道，关闭摄像头，等等。常用的方法如下。

① createCaptureRequest(int templateType)，功能是获取一个 CaptureRequest. Builder 对象。其中的 int 取值为系统常量，常见的系统常量如下。

TEMPLATE_PREVIEW：用于创建一个相机预览请求。相机会优先保证高帧率而不是高画质。适用于所有相机设备。

TEMPLATE_STILL_CAPTURE：用于创建一个拍照请求。相机会优先保证高画质而不是高帧率。适用于所有相机设备。

TEMPLATE_RECORD：用于创建一个录像请求。相机会使用标准帧率，并设置录像级别的画质。适用于所有相机设备。

② createCaptureSession(List < Surface > outputs, CameraCaptureSession. StateCallback callback, Handler handler)，功能是获取一个 CameraCaptureSession 对象。其中第一个参数是 Surface 列表集，用于为 addTarget()提供添加内容，作为请求的输出显示目标；第二个参数是会话状态监听器，该监听器需要实现会话状态回调接口 CameraCaptureSession. StateCallback 的 onConfigured()方法，将预览影像输出到屏幕。

③ close()，功能是关闭摄像头。

（3）CameraDevice. StateCallback。

摄像头状态接口回调类。主要是负责回调摄像头的开启、断开、异常、销毁。使用 CameraManager 打开指定 ID 的摄像头时需要添加这个回调。

（4）CameraCaptureSession. StateCallback。

获取数据会话的状态接口回调类。在创建相机的预览图像、拍照、录像时，需要通过这个回调来获取数据会话的通道状态是否配置成功。它还负责回调一个重要的CameraCaptureSession提供操作。

(5) CameraCaptureSession. CaptureCallback。

获取数据会话的数据接口回调类。负责回调获取数据的生命周期，比如开始、进行中、完成、失败等回调。它是一个创建预览图像、拍照、录像的数据入口，但不是出口，即这些数据不是从这里返回的。在回调方法中，获取的拍照或者录像的数据不仅包含其生命周期信息，还包含其他相关信息，比如图片的尺寸、分辨率等。

(6) CaptureRequest. Builder。

获取数据请求配置类。这个类很重要，由CameraDevice类创建。它主要负责设置返回数据的surface，surface是用于显示图像的控件；配置预览、拍照、录制的拍照参数，例如自动对焦、自动曝光、拍照自动闪光、设置Hz值、颜色校正等相机参数；完成数据配置后交给CameraCaptureSession会话类，从CameraCaptureSession中获得需要的数据，例如图像预览、拍照照片或者录制视频等。常用的方法如下。

① addTarget(Surface outputTarget)，添加一个请求的输出目标，该输出目标是一个Surface，并且这个Surface必须包含在CameraDevice. createCaptureSession()方法设置的输出Surface集合中。

② set(Key key, T value)，设置CaptureRequest. Builder对象的属性值，其中key是属性名。常用的属性名如下。

CONTROL_AE_MODE：相机自动曝光程序所需的模式。

CONTROL_AF_MODE：自动对焦(AF)当前是否启用，以及设置为何种模式。

CONTROL_AF_TRIGGER：相机设备是否会为该请求触发自动对焦。

COLOR_CORRECTION_ABERRATION_MODE：色差校正算法的操作模式。

CONTROL_MODE：整个3A控制程序模式。

JPEG_ORIENTATION：JPEG图像的方向。

JPEG_QUALITY：JPEG图像的压缩质量。

JPEG_THUMBNAIL_QUALITY：JPEG缩略图的压缩质量。

STATISTICS_FACE_DETECT_MODE：人脸检测单元的操作模式。

(7) CameraCaptureSession。

获取数据会话类。这个类很重要，由CameraCaptureSession. StateCallback接口回调方法回调提供。它主要负责创建/停止预览、拍照、录像的操作。常用的方法如下。

① getDevice()，获得该会话的摄像头设备对象。

② capture(CaptureRequest request, CameraCaptureSession. CaptureCallback listener, Handler handler)，拍照并输出到指定目标。输出目标为CaptureRequest对象时，表示显示在屏幕上；输出目标为ImageReader时，表示可以保存为图片文件。

③ setRepeatingRequest(CaptureRequest request, CameraCaptureSession. CaptureCallback listener, Handler handler)，设置连续不断的拍照请求并输出到指定目标。输出目标为CaptureRequest对象时，表示显示在屏幕上；输出目标为ImageReader时，表示可以保存为图片文件。

④ stopRepeating()，停止连续请求。

(8) ImageReader。

图片读取类。它不属于 Camera2Api 的类，但是是拍照功能重要的类。它用于对照片的数据流进行缓存，从缓存中提取图像数据流可保存为图片文件或者直接显示到 ImageView 控件中。常用的方法如下。

① newInstance(int width, int height, int format, int maxImages)，创建一个 ImageReader 对象，携带其尺寸、格式及最多可获取的图像流个数。

② setOnImageAvailableListener(ImageReader.OnImageAvailableListener listener, Handler handler)，为图像数据的可用注册监听器。该监听器需要实现图像数据可用的回调接口 ImageReader.OnImageAvailableListener 的 onImageAvailable()方法。

2) 管理类与回调之间的工作流程

使用 Camera2 对拍照应用进行编程实现中，CameraManager 是所有相机设备(CameraDevice)的管理者，而每个 CameraDevice 自己会负责建立 CameraCaptureSession 以及建立 CaptureRequest。在 CaptureRequest 中不断地发出请求 request，其中定义了照相效果的一些参数，并且必须使用 addTarget()函数为这个 request 添加一个 target surface；在 CameraCaptureSession 中调用 capture，不停地捕获画面，当捕获到画面内容后，CameraDevice 将返回的数据送到这个 target surface 中。这个 target surface 可以是 Surface View，或是 Surface Texture，这时返回的数据传递到预览界面中；也可以是 ImageReader 或 MediaRecorder，这时返回的数据传给这两个类，进行进一步处理，形成图片文件或者视频文件。拍照应用的工作流程如图 11-16 所示。

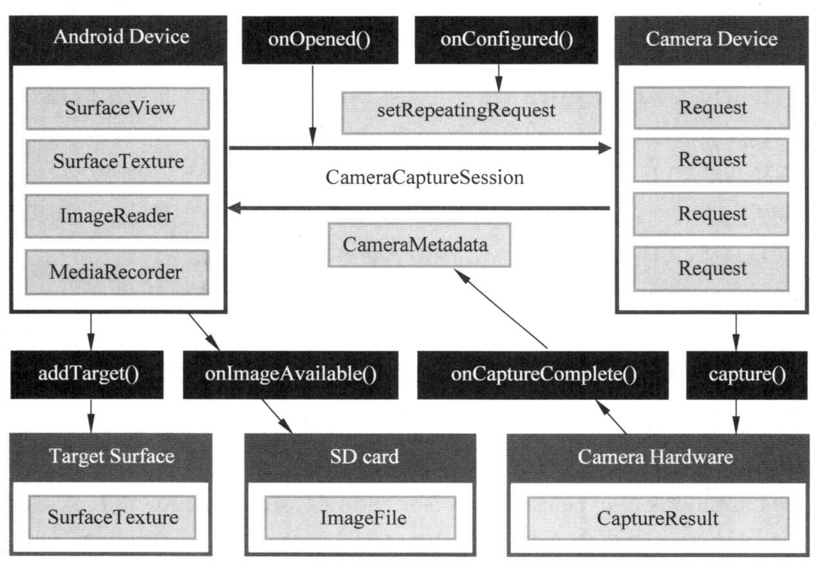

图 11-16 Camera2 拍照应用工作流程

3) 拍照应用的实现步骤

运用 Camera2 框架来实现拍照应用，编程实现步骤如下。

(1) 判断摄像头是否存在。

(2) 定义 TextureView 作为相机预览界面。

(3) 设置相机参数,并打开相机。

(4) 打开相机预览。

(5) 拍照,保存图片。

这里的第(2)步,是使用 TextureView 来作为拍照的预览窗口。TextureView 是纹理视图,它与 SurfaceView 均继承于 android. view. View,与其他 View 不同的是,两者都能在独立的线程中进行绘制和渲染,不会影响到主线程。但是 SurfaceView 不支持平移、缩放变换和嵌套,在使用中受到局限;而 TextureView 更像是一般的 View,像 TextView 那样能被缩放、平移,也能加上动画,但是 TextureView 消耗的内存比 SurfaceView 要多,并且伴随着 1～3 帧的延迟,所以必须在开启了硬件加速的 Windows 中使用。综合拍照应用的功能需求,TextureView 更方便控制,适合作为相机的预览显示控件。

TextureView 通过 TextureView. setSurfaceTextureListener 注册监听器,在子线程中来监控表面纹理的状态变化,从而更新 UI。在该监听器中,需要重写以下 4 个方法。

onSurfaceTextureAvailable(),在表面纹理可用时触发,可在此方法中进行打开相机等操作。

onSurfaceTextureSizeChanged(),在表面纹理尺寸发生变化时触发。

onSurfaceTextureDestroyed(),在表面纹理销毁时触发。

onSurfaceTextureUpdated(),在表面纹理更新时触发。

注意:使用 Camera2 进行拍照,同样需要在清单文件 AndroidManifest. xml 中声明相关权限。通常一个拍照应用都会需要保存图片到 SD 卡上,所以不例外地要加上对外存储设备的写权限。一般需要添加的权限如下。

```
<uses-permission android:name = "android.permission.CAMERA" />
<uses-feature android:name = "android.hardware.camera2.full" />
<uses-permission android:name = "android.permission.WRITE_EXTERNAL_STORAGE"/>
```

使用 Camera2 编程,必须加上 android:name="android. hardware. camera2. full"特性权限。官方文档指出:如果没有该权限,则 Camera2 暴露的特性与 Camera 的 API 没有区别。

下面通过案例来说明如何使用 Camera2 类来开发自己的图像采集应用。

【案例 11.8】 使用 Camera2 类设计一个简单的拍照应用,执行效果同案例 11.7,拍照文件存放在 SD 卡的 DCIM/CameraV2 目录中,图片文件以 IMG_开头后接日期时间来命名。

说明:设计两个 Activity,其中在首界面中设计两个按钮,一个是拍照,另一个是退出应用。在拍照预览界面用一个 TextureView 来作为拍照的预览窗口,在实现其监听器时必须重写 SurfaceTexture 的 4 个方法。

开发步骤及解析:过程如下。

1) 创建项目

在 Android Studio 中创建一个名为 Activity_APICamera2 的项目。其包名为 ee. example. activity_apicamera2。

2) 准备图片资源

将准备好的图片文件 ic_camera. jpg 复制到 res/drawable 目录中。

3）声明权限和相关属性

在 AndroidManifest. xml 中，除了声明权限外还要声明一些特性，其余与案例 11. 7 相似。声明权限和特性的代码如下。

```
<uses-permission android:name="android.permission.CAMERA" />
<uses-permission android:name="android.permission.WRITE_EXTERNAL_STORAGE"/>
<uses-permission android:name="android.permission.READ_EXTERNAL_STORAGE"/>
<uses-feature android:name="android.hardware.camera" />
<uses-feature android:name="android.hardware.camera.autofocus" />
<uses-feature android:name="android.hardware.camera2.full" />
```

4）设计布局

编写 res/layout 目录下的 activity_main. xml 文件，该文件设计首页面布局，其代码与案例 11. 7 的首页面布局文件一致，在此不做赘述。

在 res/layout 目录下新建 activity_camera2. xml 文件，该文件用来显示拍照预览视图，预览视图使用 TextureView 控件，声明该控件的代码片段如下。

```
<TextureView
    android:id="@+id/textureView"
    android:layout_width="match_parent"
    android:layout_height="match_parent"
    android:layout_centerHorizontal="true"
    android:layout_centerVertical="true"/>
```

5）开发逻辑代码

在 java/ee. example. activity_apicamera2 包下编写 MainActivity. java 程序文件，这个程序代码与案例 11. 7 大部分都相同，initViews()方法中只有两处与案例 11. 7 的不同，并且多了一个 loadBitmap()方法，这两个方法的代码如下。

```
private void initViews() {
    //从本地取图片(在 SD 卡中获取文件)
    String picFilePath = null;
    Bitmap mybitmap = null;

    Uri mImageUri = MediaStore.Images.Media.EXTERNAL_CONTENT_URI;

    String[] projection = {MediaStore.Images.Media._ID,
                           MediaStore.Images.Media.DATA,
                           MediaStore.Images.Media.DATE_ADDED
    };
    //指定目录下的图片
    String where = MediaStore.Images.Media.DATA + " like ? and ("
                 + MediaStore.Images.Media.MIME_TYPE + " = ? or "
                 + MediaStore.Images.Media.MIME_TYPE + " = ?)";
    //指定过滤条件值
    String picPath = "/storage/emulated/0/DCIM/CameraV2";
    String[] whereArgs = {picPath + "%","image/jpeg","image/jpg"};
    //查询
    Cursor mCursor = this.getContentResolver()
                          .query(mImageUri, projection, where, whereArgs,
                              MediaStore.Images.Media.DATE_MODIFIED + " desc");
```

```
        if (mCursor != null && mCursor.getCount()> 0) {
            mCursor.moveToFirst();
            //获取图片的路径
            int pathIndex = mCursor.getColumnIndex(MediaStore.Images.Media.DATA);
            picFilePath = mCursor.getString(pathIndex);

            mCursor.close();
            mybitmap = loadBitmap(picFilePath);
            imageView.setImageBitmap(mybitmap);
        }
    }
    ...
    //从给定的路径加载图片,并指定是否自动旋转方向
    public static Bitmap loadBitmap(String imgpath) {
        Bitmap bm = BitmapFactory.decodeFile(imgpath);
        int degree = 0;
        ExifInterface exif = null;
        try {
            exif = new ExifInterface(imgpath);
        } catch (IOException e) {
                e.printStackTrace();
                exif = null;
        }
        if (exif != null) {
            //读取图片中相机方向信息
            int ori = exif.getAttributeInt(ExifInterface.TAG_ORIENTATION,
                                           ExifInterface.ORIENTATION_UNDEFINED);
            //计算旋转角度
            switch (ori) {
                case ExifInterface.ORIENTATION_ROTATE_90:
                    degree = 90;
                    break;
                case ExifInterface.ORIENTATION_ROTATE_180:
                    degree = 180;
                    break;
                case ExifInterface.ORIENTATION_ROTATE_270:
                    degree = 270;
                    break;
                  default:
                    degree = 0;
                    break;
            }
        }
        if (degree != 0) {
                                                                    // 旋转图片
            Matrix m = new Matrix();
            m.postRotate(degree);
            bm = Bitmap.createBitmap(bm, 0, 0, bm.getWidth(),bm.getHeight(), m, true);
        }
        return bm;
    }
```

(1) 对Camera权限和SD卡读写权限的动态请求授权方法,相关代码同案例11.7的MainActivity.java文件,在此没有给出代码。

(2) 第 12、23 行，是本案例与案例 11.7 代码不同的地方，其中第 12 行是指定的图片文件保存目录，本案例与上一案例的保存目录不同；第 23 行根据文件保存的路径调用自定义方法 loadBitmap()，该方法将判断图片文件的方向属性，如果方向不正即进行旋转调正。

(3) 第 32～38 行，定义一个 ExifInterface 对象 exif，附加到指定的图像文件 imgpath 上。ExifInterface 是 Android 提供的一个支持库，它可以定义一个可交换图像文件格式，是专门为数码相机的照片设定的，可以记录数码照片的属性信息和拍摄数据。由它定义的对象 exif，可以附加于 JPEG、TIFF、RIFF 等文件之中，包括的属性信息有分辨率、旋转方向、感光度、白平衡、拍摄的光圈、焦距、分辨率、相机品牌、型号、GPS 等。

(4) 第 41 行，读取指定图像文件的旋转方向属性值。

(5) 第 43～57 行，根据图像的旋转方向信息，计算需要旋转的角度值存入变量 degree 中。

(6) 第 58～63 行，按 degree 给的角度，对图像进行旋转，并传入 Bitmap 对象 bm 中。

Camera2Activity.java 是使用 Camera2 实现预览拍照操作活动类的实现代码文件，本案例对 Android 5.0 及以上的版本有效，代码如下。

```
package ee.example.activity_apicamera2;

import android.Manifest;
import android.content.ContentValues;
import android.content.Context;
import android.content.Intent;
import android.content.pm.PackageManager;
import android.graphics.ImageFormat;
import android.graphics.SurfaceTexture;
import android.hardware.camera2.CameraAccessException;
import android.hardware.camera2.CameraCaptureSession;
import android.hardware.camera2.CameraCharacteristics;
import android.hardware.camera2.CameraDevice;
import android.hardware.camera2.CameraManager;
import android.hardware.camera2.CameraMetadata;
import android.hardware.camera2.CaptureRequest;
import android.hardware.camera2.CaptureResult;
import android.hardware.camera2.TotalCaptureResult;
import android.hardware.camera2.params.StreamConfigurationMap;
import android.media.Image;
import android.media.ImageReader;
import android.net.Uri;
import android.os.Build;
import android.os.Bundle;
import android.os.Environment;
import android.os.Handler;
import android.os.HandlerThread;
import android.provider.MediaStore;
import android.util.Size;
import android.util.SparseIntArray;
import android.view.Surface;
import android.view.TextureView;
import android.view.View;
import android.view.Window;
```

```
import android.view.WindowManager;
import android.widget.Button;
import android.widget.Toast;

import androidx.annotation.RequiresApi;
import androidx.appcompat.app.AppCompatActivity;
import androidx.core.app.ActivityCompat;

import java.io.File;
import java.io.FileOutputStream;
import java.io.IOException;
import java.io.OutputStream;
import java.io.ByteBuffer;
import java.text.SimpleDateFormat;
import java.util.ArrayList;
import java.util.Arrays;
import java.util.Collections;
import java.util.Comparator;
import java.util.Date;
import java.util.List;

@RequiresApi(api = Build.VERSION_CODES.LOLLIPOP)
public class Camera2Activity extends AppCompatActivity implements View.OnClickListener {

    private String mCameraId;
    private Size mPreviewSize;
    private Size mCaptureSize;
    private TextureView mTextureView;
    private Button btn_takepicture;
    private HandlerThread mCameraThread;
    private Handler mCameraHandler;
    private CameraDevice mCameraDevice;
    private ImageReader mImageReader;
    private CaptureRequest.Builder mCaptureRequestBuilder;
    private CaptureRequest mCaptureRequest;
    private CameraCaptureSession mCameraCaptureSession;
    private CameraManager manager;
    private static final SparseIntArray ORIENTATION = new SparseIntArray();
    static {
        ORIENTATION.append(Surface.ROTATION_0, 90);
        ORIENTATION.append(Surface.ROTATION_90, 0);
        ORIENTATION.append(Surface.ROTATION_180, 270);
        ORIENTATION.append(Surface.ROTATION_270, 180);
    }

    @Override
    protected void onCreate(Bundle savedInstanceState) {
        super.onCreate(savedInstanceState);
        //全屏无状态栏
        getWindow().setFlags(WindowManager.LayoutParams.FLAG_FULLSCREEN,
                    WindowManager.LayoutParams.FLAG_FULLSCREEN);
        requestWindowFeature(Window.FEATURE_NO_TITLE);
        setContentView(R.layout.activity_camera2);
        mTextureView = (TextureView) findViewById(R.id.textureView);
```

```
        btn_takepicture = (Button) findViewById(R.id.photoButton);
        btn_takepicture.setOnClickListener(this);
    }

    @Override
    protected void onResume() {
        super.onResume();
        startCameraThread();
        if (!mTextureView.isAvailable()) {
            mTextureView.setSurfaceTextureListener(mTextureListener);
        } else {
            startPreview();
        }
    }

    private void startCameraThread() {
        mCameraThread = new HandlerThread("CameraThread");
        mCameraThread.start();
        mCameraHandler = new Handler(mCameraThread.getLooper());
    }

    private TextureView.SurfaceTextureListener mTextureListener = new
                                          TextureView.SurfaceTextureListener() {
        @Override
        public void onSurfaceTextureAvailable(SurfaceTexture surface, int width, int height) {
          //当 SurefaceTexture 可用的时候,设置相机参数并打开相机
          setupCamera(width, height);
        }
        @Override
        public void onSurfaceTextureSizeChanged(SurfaceTexture surface, int width, int height) {
        }
        @Override
        public boolean onSurfaceTextureDestroyed(SurfaceTexture surface) {
            return false;
        }
        @Override
        public void onSurfaceTextureUpdated(SurfaceTexture surface) {
        }
    };

    @RequiresApi(api = Build.VERSION_CODES.LOLLIPOP)
    private void setupCamera(int width, int height) {
        //获取摄像头的管理者 CameraManager
        manager = (CameraManager) getSystemService(Context.CAMERA_SERVICE);

        try {
            String[] cameraList = manager.getCameraIdList();
            if (cameraList.length == 0) {
                return; //没有可用相机
            }
            //遍历所有摄像头
            for (String cameraId : cameraList) {
                CameraCharacteristics characteristics = manager.getCameraCharacteristics
                (cameraId);
```

```
                Integer facing = characteristics.get(CameraCharacteristics.LENS_FACING);
                //此处默认打开后置摄像头
                if (facing != null && facing == CameraCharacteristics.LENS_FACING_FRONT)
                    continue;
                //获取 StreamConfigurationMap,它是管理摄像头支持的所有输出格式和
                //尺寸
                StreamConfigurationMap map = characteristics
                     .get(CameraCharacteristics.SCALER_STREAM_CONFIGURATION_MAP);
                //根据 TextureView 的尺寸设置预览尺寸
                mPreviewSize = getOptimalSize(map.getOutputSizes(SurfaceTexture.class),
                                             width, height);
                //获取相机支持的最大拍照尺寸
                mCaptureSize =
                     Collections.max(Arrays.asList(map.getOutputSizes(ImageFormat.JPEG)),
                      new Comparator<Size>() {
                    @Override
                    public int compare(Size lhs, Size rhs) {
                        return Long.signum(lhs.getWidth() * lhs.getHeight()
                                          - rhs.getHeight() * rhs.getWidth());
                    }
                });
                //此 ImageReader 用于拍照所需
                mCameraId = cameraId;
                setupImageReader();
                break;
            }
        } catch (CameraAccessException e) {
            e.printStackTrace();
        }
        openCamera();
    }

    //选择 sizeMap 中大于并且最接近 width 和 height 的 Size
    private Size getOptimalSize(Size[] sizeMap, int width, int height) {
        List<Size> sizeList = new ArrayList<>();
        for (Size option : sizeMap) {
            if (width > height) {
                if (option.getWidth() > width && option.getHeight() > height) {
                    sizeList.add(option);
                }
            } else {
                if (option.getWidth() > height && option.getHeight() > width) {
                    sizeList.add(option);
                }
            }
        }
        if (sizeList.size() > 0) {
            return Collections.min(sizeList, new Comparator<Size>() {
                @Override
                public int compare(Size lhs, Size rhs) {
                    return Long.signum(lhs.getWidth() * lhs.getHeight()
                                      - rhs.getWidth() * rhs.getHeight());
                }
            });
```

```
        }
        return sizeMap[0];
    }

    private void openCamera() {
        try {
            if (ActivityCompat.checkSelfPermission(this,
                Manifest.permission.CAMERA) != PackageManager.PERMISSION_GRANTED) {
                return;
            }
            manager.openCamera(mCameraId, mStateCallback, mCameraHandler);
        } catch (CameraAccessException e) {
            e.printStackTrace();
        }
    }

    private CameraDevice.StateCallback mStateCallback = new CameraDevice.StateCallback() {
        @Override
        public void onOpened(CameraDevice camera) {
            mCameraDevice = camera;
            startPreview();
        }
        @Override
        public void onDisconnected(CameraDevice camera) {
            camera.close();
            mCameraDevice = null;
        }
        @Override
        public void onError(CameraDevice camera, int error) {
            camera.close();
            mCameraDevice = null;
        }
    };

    private void startPreview() {
        SurfaceTexture mSurfaceTexture = mTextureView.getSurfaceTexture();
        mSurfaceTexture.setDefaultBufferSize(mPreviewSize.getWidth(),
                                        mPreviewSize.getHeight());
        Surface previewSurface = new Surface(mSurfaceTexture);
        try {
            mCaptureRequestBuilder =
                mCameraDevice.createCaptureRequest(CameraDevice.TEMPLATE_PREVIEW);
            mCaptureRequestBuilder.addTarget(previewSurface);
            //自动对焦
            mCaptureRequestBuilder.set(CaptureRequest.CONTROL_AF_MODE,
                        CaptureRequest.CONTROL_AF_MODE_CONTINUOUS_PICTURE);
            mCameraDevice.createCaptureSession(Arrays.asList(previewSurface,
                    mImageReader.getSurface()), new CameraCaptureSession.StateCallback() {
                @Override
                public void onConfigured(CameraCaptureSession session) {
                    try {
                        mCaptureRequest = mCaptureRequestBuilder.build();
                        mCameraCaptureSession = session;
                        mCameraCaptureSession.setRepeatingRequest(mCaptureRequest, null,
```

```
                                                        mCameraHandler);
                        } catch (CameraAccessException e) {
                            e.printStackTrace();
                        }
                    }
                    @Override
                    public void onConfigureFailed(CameraCaptureSession session) {
                                                                    //开启预览会话失败
                    }
                }, mCameraHandler);
            } catch (CameraAccessException e) {
                e.printStackTrace();
            }
        }

        @Override
        public void onClick(View v) {
            takePicture();
        }

        private void takePicture() {
            try {
                mCaptureRequestBuilder.set(CaptureRequest.CONTROL_AF_MODE,
                        CaptureRequest.CONTROL_AF_MODE_CONTINUOUS_PICTURE); // 自动对焦
                mCaptureRequestBuilder.set(CaptureRequest.CONTROL_AF_TRIGGER,
                        CameraMetadata.CONTROL_AF_TRIGGER_START);
                mCameraCaptureSession.capture(mCaptureRequestBuilder.build(),mCaptureCallback,
                                            mCameraHandler);
            } catch (CameraAccessException e) {
                e.printStackTrace();
            }
        }

        private CameraCaptureSession.CaptureCallback mCaptureCallback =
                                            new CameraCaptureSession.CaptureCallback() {
            @Override
            public void onCaptureProgressed(CameraCaptureSession session, CaptureRequest request,
                                        CaptureResult partialResult) {
            }

            @Override
            public void onCaptureCompleted(CameraCaptureSession session, CaptureRequest request,
                                        TotalCaptureResult result) {
                capture();
            }
        };

        private void capture() {
            try {
                final CaptureRequest.Builder mCaptureBuilder =
                 mCameraDevice.createCaptureRequest(CameraDevice.TEMPLATE_STILL_CAPTURE);
                int rotation = getWindowManager().getDefaultDisplay().getRotation();
                                                                //获取手机方向
                mCaptureBuilder.set(CaptureRequest.JPEG_ORIENTATION,
```

```
                    ORIENTATION.get(rotation));  //根据屏幕方向对保存的照片进
                                                 //行旋转
            mCaptureBuilder.addTarget(mImageReader.getSurface());
            CameraCaptureSession.CaptureCallback CaptureCallback =
                        new CameraCaptureSession.CaptureCallback() {
                @Override
                public void onCaptureCompleted(CameraCaptureSession session,
                            CaptureRequest request, TotalCaptureResult result) {
                    Toast.makeText(getApplicationContext(), "照片已保存!",
                            Toast.LENGTH_SHORT).show();
                    unLockFocus();
                    Camera2Activity.this.finish();
                }
            };
            mCameraCaptureSession.stopRepeating();
            mCameraCaptureSession.capture(mCaptureBuilder.build(), CaptureCallback, null);
        } catch (CameraAccessException e) {
            e.printStackTrace();
        }
    }

    private void unLockFocus() {
        try {
            mCaptureRequestBuilder.set(CaptureRequest.CONTROL_AF_TRIGGER,
                            CameraMetadata.CONTROL_AF_TRIGGER_CANCEL);
            mCameraCaptureSession.setRepeatingRequest(mCaptureRequest, null, mCameraHandler);
        } catch (CameraAccessException e) {
            e.printStackTrace();
        }
    }

    @Override
    protected void onPause() {
        super.onPause();
        if (mCameraCaptureSession != null) {
            mCameraCaptureSession.close();
            mCameraCaptureSession = null;
        }
        if (mCameraDevice != null) {
            mCameraDevice.close();
            mCameraDevice = null;
        }
        if (mImageReader != null) {
            mImageReader.close();
            mImageReader = null;
        }
    }

    private void setupImageReader() {
        //2 代表 ImageReader 中最多可以获取两帧图像流
        mImageReader = ImageReader.newInstance(mCaptureSize.getWidth(),
                            mCaptureSize.getHeight(), ImageFormat.JPEG, 2);
        mImageReader.setOnImageAvailableListener(new ImageReader.OnImageAvailableListener() {
            @Override
```

```
                public void onImageAvailable(ImageReader reader) {
                    mCameraHandler.post(new Camera2Activity.imageSave(reader.acquireNextImage()));
                }
            }, mCameraHandler);
        }

        public class imageSave implements Runnable {
            private Image mImage;
            public imageSave(Image image) {
                mImage = image;
            }

            @Override
            public void run() {
                ByteBuffer buffer = mImage.getPlanes()[0].getBuffer();
                byte[] data = new byte[buffer.remaining()];
                buffer.get(data);
                Uri myuri;

                String path = Environment.getExternalStorageDirectory() + "/DCIM/CameraV2/";
                String timeStamp = new SimpleDateFormat("yyyyMMdd_HHmmss").format(new Date());
                String fileName = "IMG_" + timeStamp + ".jpg";

                if (android.os.Build.VERSION.SDK_INT < android.os.Build.VERSION_CODES.Q) {
                //适配 Andriod 9 以下
                    File imageFile = new File(path + fileName);
                    myuri = Uri.fromFile(imageFile);
                    File FileDir = new File(path);
                    if (!FileDir.exists()) {
                        FileDir.mkdir();
                    }

                    FileOutputStream fos = null;
                    try {
                        fos = new FileOutputStream(imageFile);
                        fos.write(data, 0, data.length);
                        fos.flush();
                    } catch (IOException e) {
                        e.printStackTrace();
                    } finally {
                        if (fos != null) {
                            try {
                                fos.close();
                            } catch (IOException e) {
                                e.printStackTrace();
                            }
                        }
                    }
                } else {//适配 Android 10 以上
                    ContentValues contentValues = new ContentValues();
                    contentValues.put(MediaStore.Images.Media.RELATIVE_PATH,
                                "DCIM/CameraV2");
                    contentValues.put(MediaStore.Images.Media.DISPLAY_NAME, fileName);
                    contentValues.put(MediaStore.Images.Media.DESCRIPTION, fileName);
```

```
376                      contentValues.put(MediaStore.Images.Media.MIME_TYPE, "image/jpeg");
377                      myuri = Camera2Activity.this.getContentResolver().insert(
                                  MediaStore.Images.Media.EXTERNAL_CONTENT_URI, contentValues);
378                      OutputStream cos = null;
379                      try {
380                          cos = Camera2Activity.this.getContentResolver().openOutputStream(myuri);
381                          cos.write(data,0,data.length);
382                          cos.flush();
383                      } catch (Exception e) {
384                          e.printStackTrace();
385                      } finally {
386                          if (cos != null) {
387                              try {
388                                  cos.close();
389                              } catch (IOException e) {
390                                  e.printStackTrace();
391                              }
392                          }
393                      }
394                  }
395
396                  //将拍的照片添加到相册
397                  Intent mediaScanIntent = new Intent(Intent.ACTION_MEDIA_SCANNER_SCAN_FILE);
398                  mediaScanIntent.setData(myuri);
399                  sendBroadcast(mediaScanIntent);
400              }
401          }
402  }
```

(1) 第 79 行设置窗口全屏，第 80 行设置窗口无状态栏。

(2) 第 88～96 行实现 Resume()方法，该方法是当该 Activity 被激活时调用。该方法主要实现三个操作：其一，调用自定义方法 startCameraThread()，初始化一个子线程，Camera2 的操作一般放在子线程里进行；其二，创建 SurfaceTextureView 对象的监听器；其三，调用 startPreview()方法对 Camera2 进行初始化设置。

(3) 第 98～102 行实现 startCameraThread()方法，对子线程实例化，并启动线程，提供该子线程的 mCameraHandler 实例。

(4) 第 104～120 行实现 SurfaceTextureView 对象的监听器定义。对其 4 个方法进行重写。本例重点重写 onSurfaceTextureAvailable()方法，在该方法中调用设置相机方法 setupCamera()。

(5) 第 123～159 行实现 setupCamera()方法，设置相机参数。因为 Camera2 是 Android 5.0 之后才推出的，在该方法里使用的某些方法只适用于 Android 5.0 以上的版本，所以在定义方法之前要加上@RequiresApi()，以避免低版本的系统运行高版本的 API 时编译出异常。

该方法主要实现五个操作：其一，创建管理者对象 manager；其二，通过 manager 获得所有摄像头，通过遍历摄像头找到后置摄像头（见第 133～138 行）；其三，使用 StreamConfigurationMap 来管理后置摄像头的相关设置参数（本案例的参数是预览显示尺寸、相机的最大尺寸以及图像的格式是 JPEG），预览尺寸由自定义方法 getOptimalSize()实现(见第 162～184 行)；其四，调用 setupImageReader()方法，初始化 ImageReader；其五，调用自定义方法 openCamera()，打开相机。

(6) 第 186～195 行实现 openCamera()方法，打开相机。注意，在调用 manager.openCamera()方法前必须对 CAMERA 权限是否授权进行检测，只有通过授权才能继续执行程序。在 manager.openCamera()方法中，第二个参数 mStateCallback 将执行 CameraDevice.StateCallback 接口回调类(见第 197～213 行)，在这个接口类的 onOpened()方法里，需要实现创建预览图像请求配置和创建获取数据会话，这个在 startPreview()方法中定义实现；第三个参数 mCameraHandler 表示该 Camera 的操作通信将在子线程中进行。

(7) 第 215～242 行实现 startPreview()方法，创建预览会话。该方法主要实现三个操作：其一，实例化预览显示控件(见第 216～218 行)；其二，创建预览图像请求配置(见第 220～223 行)；其三，创建获取数据会话(见第 224～238 行)。

(8) 第 249～257 行实现 takePicture()方法，进行拍照。在该方法中，首先设置拍照参数，第 251 行启用自动对焦模式，第 252 行设置相机设备为该会话请求触发自动对焦操作。然后在子线程中进行拍照捕获操作。触发拍照操作后，会回调摄像头获取数据会话接口类 CameraCaptureSession.CaptureCallback(见第 259～269 行)。

(9) 第 271～290 行实现 capture()方法，该方法定义拍照操作，主要实现三个操作，其一，创建请求拍照的 CaptureRequest(见第 273 行)；其二，对新创建的请求设置属性，如设置摄像头拍照的旋转方向，保持拍照的摄像头与屏幕方向一致(见第 274、275 行)，添加预览 Surface 目标，此时设置目标为 ImageReader，即将拍照所得图像以文件形式保存；其三，定义会话的回调监听器(见第 277～284 行)，在该监听器内重写 onCaptureCompleted()方法，用于定义拍照捕获完成后的收尾操作。

(10) 第 318～327 行实现 setupImageReader()方法，设置图像捕获后的相关操作。在该方法内定义了一个图像可用监听(见第 321～326 行)，当监听到图像有可用数据时执行保存图像子线程(见第 324 行操作)。

(11) 第 329～401 行实现子线程的 imageSave 类，保存图像到指定的目录中。在该子线程中重写 run()方法，区分 Android 10 以上和 Android 9 以下的版本给出不同的保存文件的编程实现。

运行结果：在 Android Studio 支持的模拟器上，运行 Activity_ APICamera2 项目。首次运行 App 时会有权限请求对话框出现，请选择允许授权。

在授权通过之后，运行结果如图 11-17 所示。进入初始运行界面，如果此手机在指定的保存目录中没有任何图片，会在屏幕中出现一个默认的图片，如图 11-17(a)所示。单击“拍照”按钮，如果是首次进入相机预览，会出现一个选择对话框，如图 11-17(b)所示。选择“GOT IT”即可进入相机拍照界面，如图 11-17(c)所示。单击“拍照”按钮即进行拍照，并保存照片到指定的目录 sdcard/DCIM/CameraV2 中，返回首页面，如图 11-17(d)所示。

注意，开发有关拍照应用时，使用不同的模拟器或不同的设备来执行程序，其执行效果可能会有所区别。主要是因为不同的硬件设备操作系统，对其图像处理的底层设计会有所不同。所以在应用项目测试时，要使用多个版本、多种型号机型来进行测试，才能保障应用项目的质量。

(a) 进入拍照初始界面

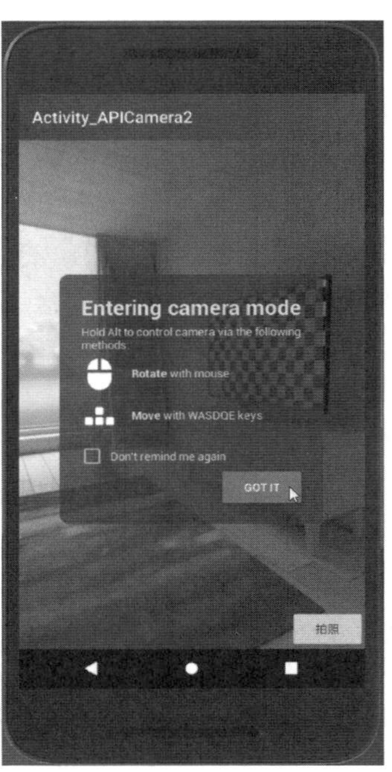

(b) 首次进入拍照预览界面

(c) 再次进入拍照预览界面

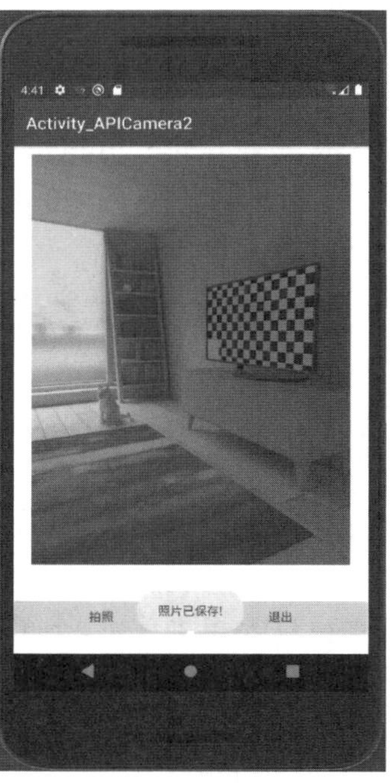

(d) 拍照完成返回首界面

图 11-17　使用 Camera2 类自定义拍照应用

小结

本章对运行于 Android 平台上的相关多媒体技术作了集中介绍。主要以案例的形式说明关于音频、视频的播放应用，录制声音和拍摄图像等应用的编程实现方法。由于 Android 的版本升级较快，对多媒体技术的支持度也随着版本的升级而更新，所以，在实践中，大家还需不断学习，不断地丰富对多媒体应用的编程技巧，掌握其更精湛的技术。

对 Android 平台的多媒体应用开发，大部分会与运行设备的软、硬件配置相关。多数应用的运行设备以手机为主。在第 12 章，将学习涉及手机基本功能应用的开发，请大家继续学习。

练习

1. 设计一个音乐播放应用，对第 10 章练习进行再次开发。要求音乐文件存储在 SD 卡的 Music 目录下，能对音乐进行播放控制和音量调节。在播放页面添加以下内容：一个歌曲播放进度条，一组插入/暂停、停止按钮，能调节音量。

2. 设计一个图片加载应用，从手机相册或使用拍照功能获得图像，并显示在固定的框线内。要求至少设计两个页面：一个是请求加载图片页面，另一个是图片预览、选择页面。

第 12 章 手机基本功能

本章将对手机自身固有的一些功能开发进行介绍。例如手机的配置、状态、通信信息、属性设置、通信、传感器等基本功能。虽然直接用手机就可以独立完成上述应用操作，但是有时在特定的应用系统中也需要涉及相关功能的开发。

12.1 手机基本特性

手机基本特性包括响应系统设置更改事件，手机的来电提醒设置，音量调节，SIM 卡和电信网络信息，手机电池电量，等等。在 Android 系统中，有些手机基本特性设置的变化会影响到应用项目的执行。

12.1.1 更改手机配置

Android 程序在运行时，手机中的一些配置可能会改变，例如：可以设置横竖屏的切换、键盘的可用性、语言切换、字体大小变换、屏幕分辨率改变，等等。这些关于手机配置的状态信息，封装在 Android 的 Configuration 类中。Configuration 类位于 android. content. res. Configuration 包中，主要用来描述和设置手机设备的配置信息，这些配置信息与应用项目能够获取的资源相关。

1. Configuration 相关属性

在 Activity 的逻辑代码中，通过 Configuration 类对象的属性，可以监听到手机的配置信息。常见的配置属性见表 12-1。

表 12-1 Configuration 类常用的属性及说明

属　性	说　明
densityDpi	屏幕密度
fontScale	当前用户设置的字体的缩放因子
hardKeyboardHidden	判断硬键盘是否可见。有两个可选值：HARDKEYBOARDHIDDEN_NO 为十六进制的 0，HARDKEYBOARDHIDDEN_YES 为十六进制的 1
keyboard	获取当前关联的键盘类型。该属性的返回值：KEYBOARD_12KEY（只有 12 个键的小键盘）、KEYBOARD_NOKEYS、KEYBOARD_QWERTY（普通键盘）

续表

属　　性	说　　明
keyboardHidden	该属性返回一个 Boolean 值用于标识当前键盘是否可用。该属性不仅会判断系统的硬件键盘,也会判断系统的软键盘(位于屏幕)
locale	用户当前的语言环境,语言环境与区域、语言种类有关
mcc	获取移动信号的国家码
mnc	获取移动信号的网络码
orientation	获取系统屏幕的方向。该属性的返回值: ORIENTATION_LANDSCAPE(横向屏幕),为十六进制的 2 ORIENTATION_PORTRAIT(竖向屏幕),为十六进制的 1
screenHeightDp	屏幕可用高度
screenWidthDp	屏幕可用宽度
screenLayout	屏幕的布局。用二进制码来标识屏幕布局
screenSize	屏幕大小
touchscreen	获取系统触摸屏的触摸方式。该属性的返回值: TOUCHSCREEN_NOTOUCH(无触摸屏)、TOUCHSCREEN_STYLUS(触摸笔式触摸屏)、TOUCHSCREEN_FINGER(接收手指的触摸屏)
uiMode	用二进制码来标识 UI 显示模式。UI 模式有:桌面模式、汽车模式、手表模式、夜间模式,等等

随着 Android 版本的不断更新发展,有些属性是高阶版本新加入的,例如表中的 densityDpi,该属性是 Android 4.2(Android API 17)新增的;还有些属性已经被高版本弃用,例如表中的 locale,该属性被 Android 7.0(Android API 24)弃用,之前的 locale 属性值可以通过 getLocales().get(0)替代。getLocales()获得的是一个 LocaleList,它包含了"语言""描述""国家(或地区)"等一系列的 Locale 信息。关于常用的语言与国家标识等相关常量见表 12-2。

表 12-2　Locale 类常用的标志和常量及说明

语言及地区	Locale 常量	说　　明
zh	Locale.CHINESE	中文
zh_CN	Locale.CHINA, Locale.SIMPLIFIED_CHINESE	中文(中国),中文简体
zh_TW	Locale.TRADITIONAL_CHINESE	中文(台湾),中文繁体
en	Locale.ENGLISH	英文
en_GB	Locale.UK	英文(英国)
en_US	Locale.US	英文(美国)
fr	Locale.FRENCH	法文
fr_FR	Locale.FRANCE	法文(法国)
de	Locale.GERMAN	德文
de_DE	Locale.GERMANY	德文(德国)
ja	Locale.JAPENESE	日文
ja_JP	Locale.JAPAN	日文(日本)
ko	Locale.KOREAN	韩文
ko_KR	Locale.KOREA	韩文(韩国)

2. 权限及属性设置

为实现应用项目对设备配置进行改变，需要在清单文件 AndroidManifest.xml 中添加允许配置改变权限 CHANGE_CONFIGURATION。即在< application >节点之前，添加修改配置权限，代码片段如下。

```
<uses-permission android:name="android.permission.CHANGE_CONFIGURATION" />
```

在 Android 6.0 之后加强了权限的安全机制管理，对权限进行分类，分为普通权限和危险权限。在清单文件中对于一些危险权限的声明会给出警告，即在代码下方出现红色波浪线，甚至有些权限需要在 Java 代码中动态授权。修改配置权限属于危险权限，如果确认需要该权限，可以在权限声明元素中添加 tools:ignore="ProtectedPermissions"，告诉系统，在该 App 中允许进行此类操作，避免出现警告提示。声明权限代码如下。

```
<uses-permission android:name="android.permission.CHANGE_CONFIGURATION"
    tools:ignore="ProtectedPermissions" />
```

一般情况下，对手机配置进行改变之后，Activity 会重启。例如：横竖屏的切换、键盘的可用性等。这些事件一旦发生，当前活动的 Activity 会重新启动，重启过程是：在销毁之前会先调用 onSaveInstanceState()方法去保存 Activity 中的一些数据，然后调用 onDestroy()方法，最后依次调用 onCreate()、onStart()、onResume()等方法启动一个新的 Activity。

有时，重启 Activity 给用户的体验不是很好。如果不想让重启发生，需要在清单文件 AndroidManifest.xml 中的 activity 元素内添加 android:configChanges 属性，该属性可以设置多个值，用"|"隔开，例如：locale|navigation|orientation。设置了 android:configChanges 属性后，当指定的属性发生变化时，不会去重新启动 Activity，而是通知程序去调用 Activity 的 onConfigurationChanged()方法。

下面通过一个实例来说明 Configuration 类的应用。

3. 应用案例

【案例 12.1】 当手机发生切换语言或切换屏幕显示方向时，显示手机的配置信息，并且能记住前次运行中的语言设置。

说明：设计一个按钮实现中、英文切换，再设计一个按钮实现横屏、竖屏切换。这两种配置一发生，就显示常用的配置信息。使用 SharedPreferences 存储技术保存语言设置信息，以便重启 Activity 时使用同一种语言。

在本案例中，要求创建项目的最低兼容版本在 Android 4.2(API 17)或以上。这是因为使用的配置属性中最新增加的是在 API 级别 17。

开发步骤及解析：过程如下。

1）创建项目

在 Android Studio 中创建一个名为 Activity_CfgChange 的项目。其包名为 ee.example.activity_cfgchange，最低兼容版本为 17。

2）声明权限添加属性

在 AndroidManifest.xml 中的< application >元素前添加权限声明，代码如下。

```
<uses-permission android:name="android.permission.CHANGE_CONFIGURATION"
    tools:ignore="ProtectedPermissions" />
```

在< activity >元素中，添加 android:configChanges 属性，代码片段如下。

```
<activity
    android:name=".MainActivity"
    android:configChanges="locale|orientation|screenSize"
>    <!-- 在 API 12 之后，如果要监测方向改变，screenSize 属性是必须的 -->
```

3）准备图片素材

将准备好的图片 bg.jpg 用作背景图，复制到 res/drawable 目录中。

4）准备文本资源

中、英文的内容由资源目录中的两个 strings.xml 文件提供，其中一个文件在 values 目录中，另一个是在 values-en 目录中。

编写 res/values 目录下的 strings.xml 文件，提供中文资源内容，其代码片段如下。

```
<resources>
    <string name="app_name">设备配置状态信息</string>
    <string name="btn">单击更改屏幕朝向</string>
    <string name="lbtn">英文</string>
    ...
</resources>
```

创建 res 目录下的 values-en 子目录，编写 strings.xml 文件，提供英文资源内容，其代码片段如下。

```
<resources>
    <string name="app_name">Configuration information</string>
    <string name="btn">Click to Change Orientation</string>
    <string name="lbtn">Chinese</string>
    ...
</resources>
```

要实现多语言切换功能，需要这两个文本资源文件的 string 元素名称要一致，不同的是 string 元素的内容一个是全中文的，另一个是对应的英文内容。

5）设计布局

编写 res/layout 目录下的 activity_main.xml 文件。在布局中添加两个按钮，一个按钮的 ID 为 lbtn，另一个按钮的 ID 为 obtn；下面再添加一个 TextView 控件，ID 为 et，用于显示手机相关的配置信息。

6）开发逻辑代码

在 java/ee.example.activity_cfgchange 包下编写 MainActivity.java 程序文件，代码如下。

```
package ee.example.activity_cfgchange;

import androidx.appcompat.app.AppCompatActivity;

import android.annotation.SuppressLint;
import android.content.Context;
import android.content.SharedPreferences;
```

```java
import android.content.pm.ActivityInfo;
import android.content.res.Configuration;
import android.content.res.Resources;
import android.os.Build;
import android.os.Bundle;
import android.util.DisplayMetrics;
import android.view.View;
import android.widget.Button;
import android.widget.EditText;
import android.widget.Toast;

import java.util.Locale;

public class MainActivity extends AppCompatActivity {

    private static final String SP_INF = "SP_Files";
    private static final String LANGUAGE = "Current_Locale";
    EditText et;
    @Override
    protected void onCreate(Bundle savedInstanceState) {
        super.onCreate(savedInstanceState);
        Languagerefresh();
        setContentView(R.layout.activity_main);
        Button lbtn = (Button) findViewById(R.id.lbtn);
        Button obtn = (Button)findViewById(R.id.obtn);
        et = (EditText)findViewById(R.id.et);

        lbtn.setOnClickListener(new View.OnClickListener() {
            @Override
            public void onClick(View v) {
                Configuration config = getResources().getConfiguration();
                if (Build.VERSION.SDK_INT >= Build.VERSION_CODES.N) {
                    if( config.getLocales().get(0).equals(Locale.SIMPLIFIED_CHINESE))
                        changeAppLanguage(MainActivity.this,"en_US");     //切换为英文
                    else
                        changeAppLanguage(MainActivity.this,"zh_CN");     //切换为中文
                }else{
                    if( config.locale.equals(Locale.SIMPLIFIED_CHINESE))
                        changeAppLanguage(MainActivity.this,"en_US");     //切换为英文
                    else
                        changeAppLanguage(MainActivity.this,"zh_CN");     //切换为中文
                }
            }
        });

        obtn.setOnClickListener(new View.OnClickListener() {
            @SuppressLint("SourceLockedOrientationActivity")
            @Override
            public void onClick(View v) {
                Configuration config = getResources().getConfiguration();
                //如果是横屏则切换成竖屏
                if(config.orientation == Configuration.ORIENTATION_LANDSCAPE)
                {
                    MainActivity.this.setRequestedOrientation(
```

```
                                    ActivityInfo.SCREEN_ORIENTATION_PORTRAIT);
                    }
                    //如果是竖屏则切换成横屏
                    if(config.orientation == Configuration.ORIENTATION_PORTRAIT)
                    {
                        MainActivity.this.setRequestedOrientation(
                                    ActivityInfo.SCREEN_ORIENTATION_LANDSCAPE);
                    }
                }
            });
        }

        private void changeAppLanguage(Context context,String localestr) {
            Resources resources = context.getResources();             //获得 res 资源对象
            //获得屏幕参数:主要是分辨率、像素等
            DisplayMetrics metrics = resources.getDisplayMetrics();
            Configuration cfg = resources.getConfiguration();          //获得配置对象
            //在这里设置需要转换成的语言,也就是选择用哪个 values 目录下的 strings.
            //xml 文件
            switch(localestr){
                case "en_US":                                          //设置美式英文
                  if (Build.VERSION.SDK_INT >= Build.VERSION_CODES.N) {
                      cfg.setLocale(Locale.US);
                  } else {
                      cfg.locale = Locale.US;
                  }
                  break;
                case "zh_CN":                                          //设置简体中文
                    if (Build.VERSION.SDK_INT >= Build.VERSION_CODES.N) {
                        cfg.setLocale(Locale.SIMPLIFIED_CHINESE);
                    } else {
                        cfg.locale = Locale.SIMPLIFIED_CHINESE;
                    }
                    break;
            }
            resources.updateConfiguration(cfg, metrics);           //更新配置文件
            refresh(cfg);
        }

        private void refresh(Configuration newconfig) {
            SharedPreferences sp = getSharedPreferences(SP_INF,MODE_PRIVATE);
            SharedPreferences.Editor editor = sp.edit();
            editor.clear();
            if (Build.VERSION.SDK_INT >= Build.VERSION_CODES.N) {
                editor.putString(LANGUAGE,newconfig.getLocales().get(0).toString());
            }else{
                editor.putString(LANGUAGE,newconfig.locale.toString());
            }
            editor.commit();
            this.recreate();                                       //刷新 Activity
            et.setText(getstatus(newconfig));
        }

        private void Languagerefresh() {
```

```
        DisplayMetrics mt = getResources().getDisplayMetrics();
        Configuration config = getResources().getConfiguration();
        SharedPreferences ssp = getSharedPreferences(SP_INF,MODE_PRIVATE);
        String slocale = ssp.getString(LANGUAGE,null);
        if(slocale!= null)
        {
            switch(slocale){
                case "zh_CN":                                    //设置简体中文
                    if (Build.VERSION.SD K_INT >= Build.VERSION_CODES.N) {
                        config.setLocale(Locale.SIMPLIFIED_CHINESE);
                    } else {
                        config.locale = Locale.SIMPLIFIED_CHINESE;
                    }
                    break;
                case "en_US":                                    //设置美式英文
                    if (Build.VERSION.SDK_INT >= Build.VERSION_CODES.N) {
                        config.setLocale(Locale.US);
                    } else {
                        config.locale = Locale.US;
                    }
                    break;
            }
        }
        getResources().updateConfiguration(config, mt);
    }

    @Override
    public void onConfigurationChanged(Configuration newconfig) {
        super.onConfigurationChanged(newconfig);
        Languagerefresh();
        Toast.makeText(this, getResources().getText(R.string.msgstr), Toast.LENGTH_LONG).show();
        et.setText(getstatus(newconfig));
    }

    private String getstatus(Configuration newconfig){
        StringBuffer statustr = new StringBuffer();
        statustr.append(getResources().getString(R.string.t_orientation));
        if(newconfig.orientation == 1) {
            statustr.append(getResources().getText(R.string.v_orient_P) + "\n");
        }else if(newconfig.orientation == 2) {
            statustr.append(getResources().getText(R.string.v_orient_L) + "\n");
        }
        statustr.append(getResources().getText(R.string.t_locale));
        statustr.append(getResources().getText(R.string.v_local) + "\n");
        statustr.append(getResources().getText(R.string.t_mcc));
        statustr.append(newconfig.mcc + "\n");
        statustr.append(getResources().getText(R.string.t_mnc));
        statustr.append(newconfig.mnc + "\n");
        statustr.append(getResources().getText(R.string.t_uiMode));
        statustr.append(newconfig.uiMode + "\n");
        statustr.append(getResources().getText(R.string.t_fontScale));
        statustr.append(newconfig.fontScale + "\n");
        statustr.append(getResources().getText(R.string.t_densityDpi));
        statustr.append(newconfig.densityDpi + "\n"); //要求 Android SDK 17 以上版本
```

```
167             statustr.append(getResources().getText(R.string.t_screenHeightDp));
168             statustr.append(newconfig.screenHeightDp + "\n");
169             statustr.append(getResources().getText(R.string.t_screenWidthDp));
170             statustr.append( newconfig.screenWidthDp + "\n");
171             statustr.append(getResources().getText(R.string.t_screenLayout));
172             statustr.append(newconfig.screenLayout + "\n");
173             return statustr.toString();
174         }
175 
176 }
```

(1) 第 29 行,调用 Languagerefresh()自定义方法。该方法用于从 SharedPreferences 中读取语言设置信息,并配置语言资源。注意,要先调用该方法,再调用 setContentView()方法。

(2) 第 112～137 行,定义 Languagerefresh()的实现。该方法主要实现四方面功能,第一是创建屏幕资源对象和配置资源对象(见第 113、114 行);第二是从 SharedPreferences 中读取语言记录信息(见第 115、116 行);第三是根据语言记录设置程序读取的文本资源文件,是 value 下的 strings. xml,还是 value-en 下的 strings. xml。在设置文本资源配置时,VERSION_CODES. N(Android 7.0)之后使用 config. setLocale()方法设置配置,所以在编程时需要根据不同的版本使用不同的设置配置实现代码,以保证程序的向下兼容性。最后,第 136 行,更新配置文件。

(3) 第 35～51 行,实现中、英文语言切换设置。切换语言功能是通过调用 changeAppLanguage()自定义方法来实现的。

(4) 第 72～96 行,定义 changeAppLanguage()的实现。具体的实现代码与 Languagerefresh()方法的主体代码差不多,只是在实现更新配置文件之后,需要调用 refresh()方法刷新一次 UI 中的控件,使得语言环境发生改变。

(5) 第 98～110 行,定义自定义方法 refresh()的实现。在该方法中将新设置的语言环境保存至 SharedPreferences 中(见第 99～107 行),其对应的保存信息文件名由 SP_INF 常量指定,默认存放在/data/data/ee. example. activity_cfgchange/shared_prefs 下。然后调用 Activity 的 recreate()方法刷新 Activity,重新设置 et 的文本信息。

(6) 第 53～70 行,实现横屏、竖屏切换设置。在 Configuration 中,屏幕方向的配置信息是由十六进制数值来表示的,也可以用系统的常量来表示。ORIENTATION _PORTRAIT 表示竖屏,数值为 1;ORIENTATION _LANDSCAPE 表示横屏,数值为 2。注意,在 ActivityInfo 中,表示屏幕方向的常量与 Configuration 中表示屏幕方向的常量是有区别的,在 ActivityInfo 中,表示竖屏的常量是 SCREEN_ORIENTATION _PORTRAIT,数值为 0;表示横屏的常量是 SCREEN_ORIENTATION _LANDSCAPE,数值为 1。第 54 行,是为了屏蔽 Android lint 的提示警告,其作用是在编译器中不出现警告,使得切换屏幕方向操作正常进行。

(7) 在本案例中,AndroidManifest. xml 文件里的< activity >元素中添加了如下属性:android:configChanges ="locale|orientation|screenSize",在程序运行过程中如果对语言环境、屏幕方向、屏幕大小的配置信息进行修改,会自动调用 onConfigurationChanged()方法(见第 140～145 行)。

(8) 第 147～174 行,定义自定义方法 getstatus()的实现,并返回一个字符串信息。在该方法中逐一获取指定的 Configuration 对象的指定属性值,并添加到返回字符串中。

运行结果：在 Android Studio 支持的模拟器上，运行 Activity_CfgChange 项目。首次运行项目，系统默认英文语言环境、竖屏显示，如图 12-1(a)所示。如果是再次运行项目，系统会记住上一次的语言环境配置信息。单击 CLICK TO CHANGE ORIENTATION 按钮，切换为横屏显示，并且显示当前设备的配置信息，如图 12-1(b)所示。单击 CHINESE 按钮，即刻切换为中文语言环境，屏幕中的可显示控件显示中文内容，包括提示消息内容，如图 12-1(c)所示。单击“单击更改屏幕朝向”按钮，屏幕切换成竖屏显示，如图 12-1(d)所示。

(a) 初始界面

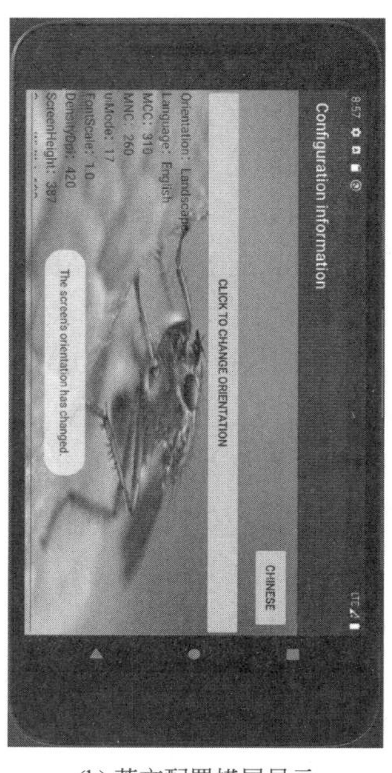

(b) 英文配置横屏显示

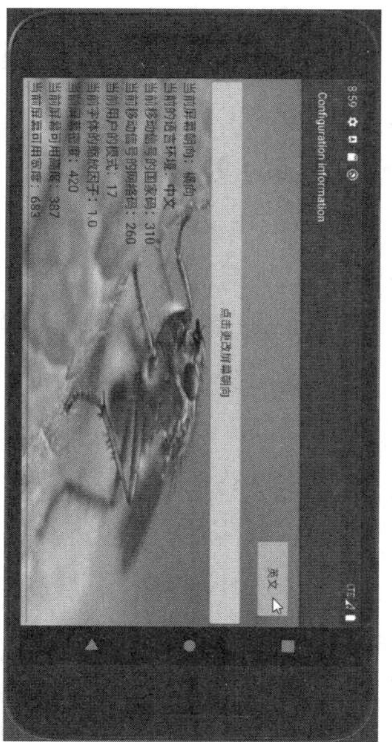

(c) 切换为中文显示

(d) 切换成竖屏显示

图 12-1　横屏/竖屏、中/英文切换应用项目运行效果

Configuration 类只能提供手机设备的配置信息，关于手机通信方面的信息要从另一个类中获取。

12.1.2 查看手机信息

Android 中的 TelephonyManager 类不仅可以获取手机卡及电信网络等信息，而且可以对手机通信进行管理。TelephonyManager 类位于 android. telephony. TelephonyManager 包中，该类提供了一系列用于访问与手机通信相关的状态和信息的 get...()方法，见表 12-3。

表 12-3　TelephonyManager 类中常用的 get...()方法及说明

get 方法	说　明
getDeviceId()	获取手机设备号，Android 8.0(Android API 26)以后弃用
getImei()	获取国际移动设备识别码 IMEI，Android 10(Android API 29)以后弃用
getDeviceSoftwareVersion()	获取设备软件版本
getLine1Number()	获取本机号码，Android 11(Android API 30)以后不能获取
getVoiceMailNumber()	获取语音信箱号码
getPhoneType()	获取手机制式
getNetworkCountryIso()	获取网络国别
getNetworkOperator()	获取网络运营商代号
getNetworkOperatorName()	获取网络运营商名称
getNetworkType()	获取网络制式
getSimCountryIso()	获取 SIM 卡国别
getSimOperator()	获取 SIM 卡运营商代号
getSimOperatorName()	获取 SIM 卡运营商名称
getSimSerialNumber()	获取 SIM 卡序列号，Android 10(Android API 29)以后不能获取
getSimState()	获取 SIM 卡状态
getSubscriberId()	获取国际移动用户识别码 IMSI，Android 10(Android API 29)以后不能获取

常用的手机通信方面信息如下。

(1) IMEI(International Mobile Equipment Identity)国际移动设备识别码，由 15 位数字组成，与每台手机一一对应，而且该码是全世界唯一的。

(2) IMSI(International Mobile Subscriber Identity)国际移动用户识别码，由 15 位码组成，其结构如下：MCC+MNC+MSIN。

(3) MCC 是移动国家码(3 位)。

(4) MNC 是移动网络码(2 位)。

(5) MSIN 是移动用户的识别号码(10 位)。

从表 12-3 中可以看到，随着 Android 版本的升级，越来越多的通信信息被列为危险级别信息，限制应用访问。

如果应用要获取手机的相关通信信息，首先要创建 TelephonyManager 对象实例，创建实例是通过 Context. getSystemService()方法实现的，代码如下。

```
Context.getSystemService(Context.TELEPHONY_SERVICE)
```

另外，还需要在清单文件 AndroidManifest. xml 中添加读取手机状态信息的权限，代码如下。

```
<uses-permission android:name="android.permission.READ_PHONE_STATE"/>
```

注意：由于获取手机信息涉及相关隐私，属于危险权限，因此在 Android 高版本的应用项目编程中，不仅要在清单文件中声明，而且还要在代码中动态申请 READ_PHONE_STATE 权限。

12.1.3 查看电池电量

现在的智能手机应用功能越来越强大，电量消耗的速度也越来越快。所以需要监测 App 的耗电情况，开发的 App 注重省电控制，是衡量 App 质量的重要指标之一。

在开发中，获取手机当前电池电量发生变化的信息是通过监听广播来实现的。具体的实现需要做以下两个方面的编程。

一是通过一个 Intent 广播，在创建该 Intent 时，需要附加获取电量相关的 action 信息，用 action 的常量来表示，表示电量信息的有 3 个常量：Intent. ACTION_BATTERY_CHANGED（电池电量发生改变时）、Intent. ACTION_BATTERY_LOW（电池电量达到下限时）和 Intent. ACTION_BATTERY_OKAY（电池电量从低恢复到高时）。

二是需要为应用项目注册 BroadcastReceiver 组件，以便在电池电量发生变化时，通过该 BroadcastReceiver 组件捕获 ACTION_BATTERY_CHANGED 动作，接收到系统发出相应的广播，来得到电池电量信息，并进行应用项目需要的应对处理。如果要停止监测手机电池电量，只需要注销该 BroadcastReceiver 组件即可。

下面通过一个案例，来介绍查看手机信息及电量信息的具体实现。

【案例 12.2】 开发一个随时可查看手机的通信信息和电量信息的应用，并且具有良好的 Android 版本兼容性。

说明：设计两个开关按钮，一个按钮控制显示或关闭手机通信信息，另一个按钮控制显示或关闭手机电量信息。

有些获取通信信息的方法已被弃用，在开发中需要根据 API 级别分别进行编程实现，以实现兼容性。

获取手机通信信息需要获得 READ_PHONE_STATE 权限，不仅要在清单文件中声明，而且要在代码中请求权限。

开发步骤及解析：过程如下。

1）创建项目

在 Android Studio 中创建一个名为 Activity_MobileInfo 的项目。其包名为 ee. example. activity_mobileinfo。

2）声明权限

在 AndroidManifest. xml 中的<application>元素前添加读取手机通信信息权限声明，

代码如下。

```
<uses-permission android:name="android.permission.READ_PHONE_STATE"/>
```

3）设计布局

编写 res/layout 目录下的 activity_main.xml 文件。在布局中添加两个开关按钮 ToggleButton，它们的 ID 分别为 tbtn1 和 tbtn2；考虑到显示的内容比较多，一屏可能放不下，所以在按钮下面添加一个 ScrollView 控件，其下添加两个 TextView，ID 分别为 title 和 info，分别用于显示标题和通信/电量信息。

4）开发逻辑代码

在 java/ee.example.activity_mobileinfo 包下编写 MainActivity.java 程序文件，代码如下。

```
package ee.example.activity_mobileinfo;

import androidx.annotation.RequiresApi;
import androidx.appcompat.app.AppCompatActivity;
import androidx.core.app.ActivityCompat;

import android.Manifest;
import android.content.BroadcastReceiver;
import android.content.Context;
import android.content.Intent;
import android.content.IntentFilter;
import android.content.pm.PackageManager;
import android.os.Build;
import android.os.Bundle;
import android.provider.Settings;
import android.widget.CompoundButton;
import android.widget.TextView;
import android.widget.ToggleButton;
import android.telephony.TelephonyManager;
import android.widget.CompoundButton.OnCheckedChangeListener;

import java.util.Objects;

public class MainActivity extends AppCompatActivity implements OnCheckedChangeListener {

    private ToggleButton tb1 = null, tb2 = null;
    private TextView title = null, info = null;
    private BatteryReceiver battreceiver = null;

    @Override
    protected void onCreate(Bundle savedInstanceState) {
        super.onCreate(savedInstanceState);
        setContentView(R.layout.activity_main);
        battreceiver = new BatteryReceiver();      //创建 BroadcastReceiver 组件对象

        title = (TextView) findViewById(R.id.title);
        info = (TextView) findViewById(R.id.info);

        tb1 = (ToggleButton) findViewById(R.id.tbtn1);
        tb2 = (ToggleButton) findViewById(R.id.tbtn2);
```

```
        tb1.setOnCheckedChangeListener(this);
        tb2.setOnCheckedChangeListener(this);
    }

    @RequiresApi(api = Build.VERSION_CODES.KITKAT)
    @Override
    public void onCheckedChanged(CompoundButton cb, boolean isChecked) {
        switch (cb.getId()) {
            case R.id.tbtn1: {
                title.setText("手机设备信息：");
                if (isChecked) {                    //获取手机信息
                    getSystemPhoneMessage();
                } else {                            //停止获取手机信息
                    title.setText(null);
                    info.setText(null);
                }
                break;
            }
            case R.id.tbtn2: {
                title.setText("电池电量信息：");
                if (isChecked) {                    //获取电池电量
                    IntentFilter battfilter =
                            new IntentFilter(Intent.ACTION_BATTERY_CHANGED);
                    registerReceiver(battreceiver, battfilter);
                                                    //注册 BroadcastReceiver
                } else {                            //停止获取电池电量
                    unregisterReceiver(battreceiver);   //注销 BroadcastReceiver
                    title.setText(null);
                    info.setText(null);
                }
                break;
            }
        }
    }

    @RequiresApi(api = Build.VERSION_CODES.KITKAT)
    private void getSystemPhoneMessage() {
        //声明 TelephonyManager 对象的引用
        TelephonyManager tm = (TelephonyManager)
                    MainActivity.this.getSystemService(Context.TELEPHONY_SERVICE);

        final int PERMISSION_ READ_PHONE_REQUEST_CODE = 1;   //读手机状态权限请求码
        int haspermission = ActivityCompat.checkSelfPermission
                                (this, Manifest.permission.READ_PHONE_STATE);
        if (haspermission != PackageManager.PERMISSION_GRANTED) {
            ActivityCompat.requestPermissions(this,
                    new String[]{Manifest.permission.READ_PHONE_STATE},
                    PERMISSION_ READ_PHONE_REQUEST_CODE);
        }

        //获取手机设备相关信息
        StringBuilder str = null;
        str.append("\n 手机设备相关信息:\n");
        String imei = null;                         //获取 IMEI:国际移动设备识别码
```

```
        if(Build.VERSION.SDK_INT < Build.VERSION_CODES.O){        //版本号< Android 8.0
            imei = tm.getDeviceId();
        }else if (Build.VERSION.SDK_INT < Build.VERSION_CODES.Q){    //版本号< Android 10
            imei = tm.getImei();
        }else {
            imei = Settings.System.getString(
                    this.getContentResolver(), Settings.Secure.ANDROID_ID);
        }
        str.append("设备编号(IMEI) = " + imei + "\n");      //IMEI:国际移动设备识别码
        str.append("设备软件版本:"
                + Objects.requireNonNull(tm).getDeviceSoftwareVersion() + "\n");
        if (Build.VERSION.SDK_INT < Build.VERSION_CODES.R) {   //版本号 < Android 11
            str.append("本机号码:" + tm.getLine1Number() + "\n");
        }
        str.append("语音信箱号码: " + tm.getVoiceMailNumber() + "\n");
        str.append("手机制式:" + tm.getPhoneType() + "\n");
        //获取网络相关信息
        str.append("\n 网络相关信息:\n");
        str.append("网络国别:" + tm.getNetworkCountryIso() + "\n");
        str.append("网络运营商代号:" + tm.getNetworkOperator() + "\n");
        str.append( "网络运营商名称:" + tm.getNetworkOperatorName() + "\n");
        str.append("网络制式:" + tm.getNetworkType() + "\n");
        //获取 SIM 卡相关信息
        str.append( "\nSIM 卡相关信息:\n");
        str.append("SIM 卡国别:" + tm.getSimCountryIso() + "\n");
        str.append( "SIM 卡运营商代号:" + tm.getSimOperator() + "\n");
        str.append("SIM 卡运营商名称: " + tm.getSimOperatorName() + "\n");
        if (Build.VERSION.SDK_INT < Build.VERSION_CODES.Q) { //版本号 < Android 10
            str.append( "SIM 卡序列号: " + tm.getSimSerialNumber() + "\n");
        }
        str.append( "SIM 卡状态: " + tm.getSimState() + "\n");
        //获取国际移动用户识别码 IMSI 的相关信息
        str.append("\n 国际移动用户识别码 IMSI 相关信息:\n");
        if (Build.VERSION.SDK_INT < Build.VERSION_CODES.Q) { //版本号 < Android 10
            str.append("SubscriberId(IMSI):" + tm.getSubscriberId() + "\n");
        }

          int mcc = getResources().getConfiguration().mcc;
          int mnc = getResources().getConfiguration().mnc;
          str.append("IMSI MCC (Mobile Country Code):" + String.valueOf(mcc) + "\n");
          str.append("IMSI MNC (Mobile Network Code):" + String.valueOf(mnc) + "\n");
          info.setText(str.toString());
    }

    private class BatteryReceiver extends BroadcastReceiver{
        @Override
        public void onReceive(Context context, Intent intent) {
            int current = intent.getExtras().getInt("level");    //获得当前电量
            int total = intent.getExtras().getInt("scale");      //获得总电量
            int percent = current * 100/total;
            info.setText("现在的剩余电量是:" + percent + " %.");
        }
    }
}
```

(1) 第 39～42 行，创建两个 ToggleButton 实例。ToggleButton 按钮是开关按钮，监听 onCheckedChanged 事件。

(2) 第 47～72 行，重写 onCheckedChanged()方法分别实现两个开关按钮的操作。由于调用获取手机信息方法 getSystemPhoneMessage()要求最低版本在 Android 4.4 (VERSION_CODES.KITKAT，API 19)以上，所以在方法前加上了@ RequiresApi()注解。

(3) 第 75～132 行，定义方法 getSystemPhoneMessage()的实现。在该方法中实现三大功能操作。一是创建 TelephonyManager 对象实例 tm(见第 77 行)，Context 对象就是 MainActivity.this，调用 MainActivity.this.getSystemService(Context.TELEPHONY_SERVICE)来创建 tm；二是动态请求读手机状态信息授权(见第 79～85 行)；三是依次通过调用 TelephonyManager 类的 get...()方法和 Configuration 的属性获取手机通信信息，并给 TextView 对象 info 赋值(见第 131 行)。

(4) 在 getSystemPhoneMessage()方法中，调用有版本适用性的 get...()方法时使用了 if((Build.VERSION.SDK_INT < Build.VERSION_CODES.XXX)语句来判断 Android 版本号，Build.VERSION_CODES.KITKAT 代表 Android 4.4，Build.VERSION_CODES.O 代表 Android 8.0，Build.VERSION_CODES.Q 代表 Android 10，Build.VERSION_CODES.R 代表 Android 11。第 91～98 行，获取设备的 IMEI 编号，在 Android 8.0 以前，可以使用 getDeviceId()方法获得；在 Android 8.0 及之后，到 Android 10 以前，可以使用 getImei()方法获得；Android 10 及之后，则使用 Settings.Secure.ANDROID_ID。

(5) 获取电量信息使用广播机制。第 62 行，创建广播的 IntentFilter 时，指定 action 为 Intent.ACTION_BATTERY_CHANGED，告诉系统只有当电池电量发生变化时进行广播。广播接收器 onReceive()方法在第 136～141 行中重写。在这里只给出了当前电量占比数值。

运行结果：在 Android Studio 支持的模拟器上，运行 Activity_MobileInfo 项目。运行初始界面，如图 12-2(a)所示。单击“获取手机信息”按钮，显示当前手机的通信信息，按钮显示“停止获取手机信息”，如图 12-2(b)所示。单击“停止获取手机信息”按钮，则关闭手机通信信息。单击“查看电量信息”按钮，显示手机的当前电量，按钮显示“关闭电量信息”，如图 12-2(c)所示。

12.1.4 振动设置

手机一般用响铃和振动来做提醒功能。在应用项目中也可以适当地运用振动特性。在 Android App 中不仅可以启动手机振动，还可以设置振动的周期、持续的时间等参数，当 App 退出时，所有的振动设置会随之停止。

1. 振动类 Vibrator

Vibrator 类是控制管理手机振动的类，该类位于 android.os.Vibrator 包中。如果想让手机启动振动，需要创建 Vibrator 对象，创建一个 Vibrator 对象需要调用 getSystemService()方法，如下所示。

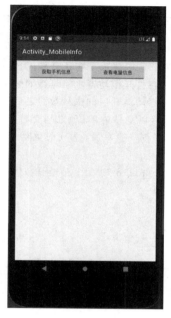

(a) 初始界面

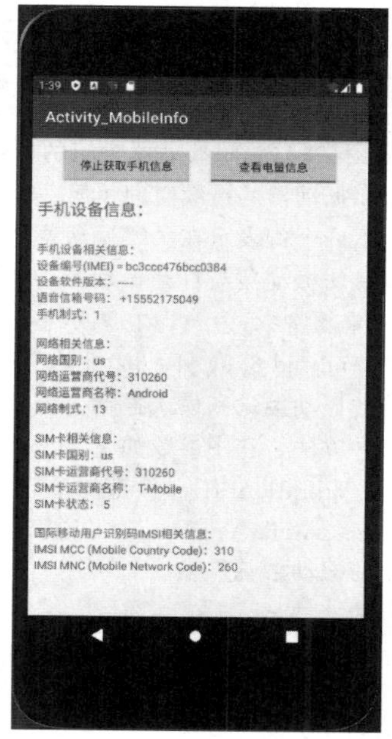

(b) 查看手机的通信信息和电池电量信息的运行

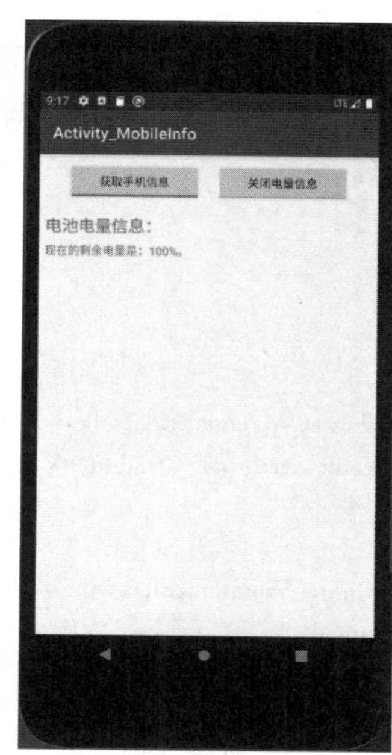

(c) 显示电池电量信息

图 12-2　显示手机通信信息及电池电量信息运行效果

```
Vibrator vibrator = (Vibrator)getSystemService(Service.VIBRATOR_SERVICE);
```

在创建语句中，使用了 Service. VIBRATOR_SERVICE 参数，这需要引入 android. app. Service 类。

启动或关闭手机振动，由 Vibrator 对象中的方法实现，Vibrator 对象常用的方法见表 12-4。

表 12-4　Vibrator 类常用方法及说明

方　法	说　明
hasAmplitudeControl()	检查振动器是否有振幅控制。如果硬件可以控制振动的幅度，则返回 true，否则返回 false
hasVibrator()	检查硬件是否有振动器。如果硬件有振动器，则返回 true，否则返回 false
vibrate(long milliseconds)	在指定的时间段内不间断振动。其中 milliseconds 为振动的毫秒数。 从 Android 8.0(API 26)开始弃用
vibrate(long milliseconds, AudioAttributes attributes)	在指定的时间段内进行指定振动模式振动。其中参数如下。 milliseconds：振动的毫秒数。 attributes：AudioAttributes 对应振动。示例如下。 AudioAttributes. USAGE_ALARM 表示报警振动。 AudioAttributes. USAGE_NOTIFICATION_RINGTONE 表示与来电相关的振动。 从 Android 8.0(API 26)开始弃用

续表

方　法	说　明
vibrate(long[] pattern, int repeat)	在指定的时间内间歇性地振动、停止。其中参数如下。 pattern：打开与关闭振动器时间的整数序列。这些整数是在几毫秒内打开与关闭振动器的持续时间。第一个值表示在打开振动器之前等待的毫秒数。下一个值表示在关闭振动器之前保持振动器开启的毫秒数。之后的值以关闭振动器与打开振动器的持续时间(以毫秒为单位)之间交替。 repeat：重复次数，为－1 时，表示不重复；为 0 时一直重复；为 n 时重复 n 次。从 Android 8.0(API 26)开始弃用
vibrate(VibrationEffect vibe, AudioAttributes attributes)	按一定的振动触觉和模式振动。该方法在 Android 8.0(API 26)新增，代替前三种方法。其中参数如下。 vibe：是 VibrationEffect 类型的参数。 attributes：AudioAttributes 对应振动
vibrate(VibrationEffect vibe)	按一定的振动触觉振动
cancel()	关闭振动

2. 振动模式类 VibrationEffect

从 Android 8.0(API 26)版本开始，增加了一个振动模式类 VibrationEffect，该类位于 android.os.VibrationEffect 包中。VibrationEffect 类描述由振动器执行的触觉效果，这些效果可以是单发振动，也可以是复杂的波形。

VibrationEffect 对象的常用方法如下。

1）创建一次性振动

一次性振动将以指定的振幅在指定的时间段内持续振动，然后停止。创建方法如下。

```
createOneShot(long milliseconds, int amplitude)
```

参数如下。

milliseconds：振动的毫秒数。

amplitude：振幅的强度，它必须是 1 和 255 之间的值，或 DEFAULT_AMPLITUDE。

返回值：VibrationEffect。

2）创建波形振动

波形振动是可能重复的一系列时序序列。对于每个序列，时序序列中的值确定振动的持续时间。并且将忽略时序值为 0 的任何序列。创建方法如下。

```
createWaveform(long[] timings, int repeat)
```

参数如下。

timings：交替开关时间的模式，从关闭开始。值为 0 将导致忽略时序序列。

repeat：重复次数，为－1 时，表示不重复；为 0 时一直重复；为 n 时重复 n 次。

返回值：VibrationEffect。

3）创建波形振幅

波形振动是可能重复的一系列时序序列和振幅序列。对于每个序列，振幅序列中的值确定振动的强度,时序序列中的值确定振动的时间长度。振幅为 0 意味着没有振动(即关闭),并且将忽略时序值为 0 的任何序列。创建方法如下。

```
createWaveform (long[ ] timings, int[ ] amplitudes, int repeat)
```

参数如下。

timings:交替开关时间的模式,从关闭开始。值为 0 将导致忽略时序/幅度序列。

amplitude:振幅的强度。它必须是 1 到 255 之间的值,或 DEFAULT_AMPLITUDE。

repeat:重复次数,为−1 时,表示不重复;为 0 时一直重复;为 n 时重复 n 次。

返回值:VibrationEffect。

例 1 定义一个单次振动,振动持续时间为 1 秒,代码片段如下。

```
Vibrator v = (Vibrator) getSystemService(Context.VIBRATOR_SERVICE);
if (Build.VERSION.SDK_INT < Build.VERSION_CODES.O) { // Android 8.0 版本之前
    if (v != null) {
        v.vibrate(1000);
    }
} else { // Android 8.0 及以上版本
    VibrationEffect effect =
            VibrationEffect.createOneShot(1000, VibrationEffect.DEFAULT_AMPLITUDE);
    v.vibrate(effect);
}
```

例 2 定义一个不重复的振动,振动方式是:等待 1 秒启动振动,持续振动 0.5 秒后停止,停止 0.5 秒后重新启动,持续振动 1 秒,代码片段如下。

```
Vibrator v = (Vibrator) getSystemService(Context.VIBRATOR_SERVICE);
if (Build.VERSION.SDK_INT < Build.VERSION_CODES.O) {
    if (v != null) {
        v.vibrate(new long[ ]{1000,500,500,1000}, -1);
    }
} else {
    VibrationEffect effect =
            VibrationEffect.createWaveform (new long[ ]{1000,500,500,1000}, -1);
    v.vibrate(effect);
}
```

最后,请大家记住,要使用振动应用,必须在清单文件 AndroidManifest.xml 中添加振动权限声明,代码如下。

```
<uses-permission android:name="android.permission.VIBRATE" />
```

12.2 手机即时通信

手机基本功能就是即时通信。比如发送短信、拨打电话等。有时候应用项目也需要执行发送、接收短信,拨打、接听电话等基本功能,可以通过相关的管理类来实现。

12.2.1 短信管理

手机中的短信服务(SMS,Short Message Service)使用概率比较大,可用 SmsManager

类提供的服务，该类主要负责实现发送短信的服务。接收短信服务使用的是广播接收机制。下面分别介绍短信的收发及管理编程技术。

1. 收发短信权限

在 Android 中的短信操作功能需要授权，其权限属于危险权限，不仅需要在清单文件 AndroidManifest.xml 中声明权限，而且需要在程序代码中动态申请授权。在开发中需要记住。

如果应用只是发送短信，那么要在< manifest >元素下添加发送权限，声明权限代码如下。

```
<uses-permission android:name="android.permission.SEND_SMS"/>
```

如果应用需要接收短信，那么要在< manifest >元素下添加接收权限，声明权限代码如下。

```
<uses-permission android:name="android.permission.RECEIVE_SMS "/>
```

2. 发送短信

短信一般是一段简短的文字信息，如果长度比较长，可以分割成若干段发送。在 Android 中 SmsManager 是发送短信的工具类，该类位于 android.telephony.SmsManager 包中，提供了相关方法来完成短信的处理和发送。常用于发送短信的方法见表 12-5。

表 12-5 SmsManager 类中常用方法及说明

get 方法	说明
divideMessage(String text)	当短信超过 SMS 消息的最大长度时，将短信分割为几段。其中，text 是初始的消息，不能为空。返回值为有序的 ArrayList < String >，可以重新组合为初始的消息
getDefault()	获取 SmsManager 的默认实例。返回值为 SmsManager 的默认实例
getSmscAddress()	从用户 SIM 卡中获得短信中心的地址
sendTextMessage(String destinationAddress, String scAddress, String text, PendingIntent sentIntent, PendingIntent deliveryIntent)	发送一个基于 SMS 的文本。其中，destinationAddress 是消息的目标地址；scAddress 是服务中心地址；text 是初始的消息；sentIntent 是当消息发送成功或失败时，这个对象就广播，这个参数最好不为空，否则会存在资源浪费的潜在问题；deliveryIntent 是当消息成功发送到接收者时，这个对象就广播
sendMultipartTextMessage(String destinationAddress, String scAddress, ArrayList < String > parts, ArrayList < PendingIntent > sentIntents, ArrayList < PendingIntent > deliveryIntents)	发送一个基于 SMS 大篇幅文本，调用者已经通过调用 divideMessage(String text)将消息分割成正确的大小。参数与 sendTextMessage 方法中一样，不过 sentIntents 和 deliveryIntents 是一组 PendingIntent

在上述方法中使用了 PendingIntent 参数，PendingIntent 是在将来的某个时刻发生的一个 Intent 广播。当成功发送出短信时，就发出 sentIntent 广播，广播发出的状态信息；当

接收者接收到短信时，就发送 deliveryIntent 广播，广播接收的状态信息。

3. 群发短信

群发短消息，实际上是设置一个循环，对需要发送的号码逐一发送即可。通常群发是通过 HashMap(链表散列)技术实现的，把需要发送的联系人地址信息记录在一个 HashMap 对象中，群发时，从 HashMap 中取出地址一一发送。

一般群发的地址是从手机通讯录中选取联系人，然后存放到 HashMap 中。在 Android 中有一个 ContentResolver 类，通过 ContentResolver 可以获取 Android 内部的数据集，比如联系人信息、系统的多媒体信息、短信信息等。通常是通过一个 Cursor 对象来获得这个数据集的。通讯录的 URI 为 ContactsContract. CommonDataKinds. Phone. CONTENT_URI。

假设 PhoneContact 是一个存储联系人的 ID 号、姓名和电话号码的数据结构，从通讯录中获得 PhoneContact 的列表，可以定义 getAllContactList()方法来实现，代码片段如下。

```
public static List<PhoneContact> getAllContactList(Context context) {
    List<PhoneContact> smsContactList = new ArrayList<>();
    try {
        Cursor cursor = context.getContentResolver().query(
                ContactsContract.CommonDataKinds.Phone.CONTENT_URI,
                new String[]{ContactsContract.CommonDataKinds.Phone.CONTACT_ID,
                        ContactsContract.CommonDataKinds.Phone.NUMBER,
                        ContactsContract.CommonDataKinds.Phone.DISPLAY_NAME},
                null, null, ContactsContract.CommonDataKinds.Phone.SORT_KEY_PRIMARY);
        if (cursor != null && cursor.getCount() > 0) {
            String lastPhone = null;
            while (cursor.moveToNext()) {
                PhoneContact smsContact = new PhoneContact();
                smsContact.contactId = cursor.getLong(cursor.getColumnIndex(
                    ContactsContract.CommonDataKinds.Phone.CONTACT_ID));   //联系人 ID
                smsContact.name = cursor.getString(cursor.getColumnIndex(
                    ContactsContract.CommonDataKinds.Phone.DISPLAY_NAME));   //联系人姓名
                String phone = cursor.getString(cursor.getColumnIndex(
                    ContactsContract.CommonDataKinds.Phone.NUMBER));   //联系人电话号码
                if (!TextUtils.isEmpty(phone)) {
                    //通讯录中的电话号码经过格式化,一般不是连续的数字,需要去掉格式
                    phone = formatNum(phone);
                    smsContact.number = phone;
                    if (TextUtils.isEmpty(phone) || TextUtils.equals(lastPhone, phone)) {
                        continue;
                    }
                    lastPhone = phone;
                }
                smsContactList.add(smsContact);
            }
        }

        if (cursor != null) {
            try {
                cursor.close();
```

```
                } catch (Throwable throwable) {
                }
            }
        } catch (Exception e) {
        }
        return smsContactList;
    }
    /* 去掉字符串中的空格和"-"符号 */
    private static String formatNum(String phone) {
        if (phone != null) {
            phone = phone.replace(" ", "");
            phone = phone.replace("-", "");
        }
        return phone;
    }
```

从通讯录中读取信息，需要有权限，不仅要在 AndroidMainfest. xml 中声明 android. permission. READ_CONTACTS 权限。自 Android 6.0(API 23)以后，还需要在代码中动态检查权限，如果没有授权则需要请求权限。

4. 接收短信

在 Android 中，接收短信由系统广播行为 android. provider. Telephony. SMS_RECEIVED 识别。系统广播接收器在后台进行监听，一旦接收到短信，就会触发广播。该广播接收器既可以在清单文件 AndroidManifest. xml 中静态注册，也可以在代码文件中动态注册。

在文件 AndroidManifest. xml 中注册，是在< application >元素下添加< receiver >元素属性定义，代码如下。

```
<receiver android:name = ".MyBroadcastReceiver">
    <intent-filter>
        <action android:name = "android.provider.Telephony.SMS_RECEIVED"/>
    </intent-filter>
</receiver>
```

这里，Android 设备收发 SMS 是以 PDU(Protocol Description Unit)编码形式来传输的。Android 的 SmsMessage 类对象表现形式是一组 pdus 串，包含了短信的详细信息。例如：发送者的电话号码，时间戳和短信息内容等。

接收短信的编程，需要实现一个接收 SMS 的广播接收器类的子类，在该子类中完成对短信内容的接收及相关应用的需求。为实现该子类，要完成以下四个关键性编程。

1）指定该广播接收器 Intent 的行为 action

接收短信的系统广播，其 action 为 android. provider. Telephony. SMS_RECEIVED。只有当系统接到的广播其行为是这个 action 值时，才能断定接收的是短信。

2）获得接收的一组短信内容

通过调用 Bundle. get("pdus")来获得接收到的短信消息。该方法返回了一个表示短信内容的数组。每一个数组元素表示一条短信。这就意味着通过 Bundle. get("pdus")可以返回多条系统接收到的短信内容。

3）通过 SmsMessage 对象获取单条短信

Bundle.get("pdus")返回的是一个短信数组，一般不能直接使用，需要通过调用 SmsMessage.createFromPdu()方法将这些数组元素转换成 SmsMessage 对象。每一个 SmsMessage 对象表示一条短信内容。

4）获取短信的具体内容

通过调用 SmsMessage 类的 getDisplayOriginatingAddress()方法可以获得发送短信的电话号码。通过调用 SmsMessage 类的 getDisplayMessageBody()方法可以获得短信的内容。

5. 查询发送状态

当发送短信的命令发出之后，会有多种状态，如发送成功，或发送暂停，或发送失败等。这些状态有时需要反馈给用户。查询短信发送状态需要注册广播接收器，并且其查询状态的代码要在广播接收器的 onReceive()方法中编写，根据得到的结果值判断短信发送的状态。

通常广播接收器接收到的 Intent 状态值有如下几种。

(1) Activity.RESULT_OK——发送成功。

(2) SmsManager.RESULT_ERROR_GENERIC_FAILURE——发送失败，值为 1。

(3) SmsManager.RESULT_ERROR_RADIO_OFF——无线连接异常，值为 2。

(4) SmsManager.RESULT_ERROR_NULL_PDU——PDU 为空值，值为 3。

(5) SmsManager.RESULT_ERROR_NO_SERVICE——服务不可用，值为 4。

下面通过一个案例来说明短信的收发应用。

【案例 12.3】 开发一个短信收发的应用，要求发出短信后能提示短信发送的状态信息，每次发出的短信长度不超过 70 个字；在界面中能看到发送的短信以及接收的短信内容。

说明：本案例将设计两个类，一个类是 App 主类，主要实现发送短信功能。另一个类继承 BroadcastReceiver 的子类，在该子类中实现对短信的监听，接收短信的 PDU 信息。

本案例的难点是将在子类中接收的短信信息传送到主类 UI 中显示。解决方案是在子类中定义一个接口，在主类中实现该接口。通过该接口传递文本信息。

开发步骤及解析：过程如下。

1）创建项目

在 Android Studio 中创建一个名为 Activity_SMSManager 的项目。其包名为 ee.example.activity_smsmanager。

2）声明权限

本案例既要发送短信，又要接收短信，所以要添加两个权限。在清单文件 AndroidManifest.xml 中的<application>元素前添加发送短信和接收短信的两个权限声明，代码如下。

```
<uses-permission android:name="android.permission.SEND_SMS"/> <!-- 发送短信权限 -->
<uses-permission android:name="android.permission.RECEIVE_SMS"/> <!-- 接收短信权限 -->
```

3）准备文本资源

在 res/values 目录下编辑 strings.xml 文件。在其中声明 totle、smscontent、sendbtn、message 等字符串内容。

4）设计布局

编写 res/layout 目录下的 activity_main.xml 文件。在布局中把屏幕分为三段区域，上段是一个 TextView，ID 为 tv01，一个 EditText，ID 为 smsTel，位于屏幕的顶端，用于输入对方的电话号；中段是一个 ScrollView，其内添加一个 TextView，ID 为 tv02，用于显示来往短信的内容；下段是一个 EditText，ID 为 smsContent，一个 Button，ID 为 smsSend，位于屏幕的底端，用于输入短信内容并发送。

5）开发逻辑代码

在 java/ee.example.activity_smsmanager 包下编写 MainActivity.java 文件，实现 App 主类的功能，包括发送短信，显示收、发短信记录内容；然后还要创建一个 BroadcastReceiver 子类 SMSReceiver.java 文件，实现接收短信功能。

MainActivity.java 程序文件代码如下。

```
package ee.example.activity_smsmanager;

import androidx.appcompat.app.AppCompatActivity;
import androidx.core.app.ActivityCompat;
import androidx.core.content.ContextCompat;

import android.Manifest;
import android.app.Activity;
import android.app.PendingIntent;
import android.content.BroadcastReceiver;
import android.content.Context;
import android.content.Intent;
import android.content.IntentFilter;
import android.content.pm.PackageManager;
import android.os.Bundle;
import android.telephony.SmsManager;
import android.view.View;
import android.widget.Button;
import android.widget.EditText;
import android.widget.TextView;
import android.widget.Toast;

import java.util.List;

public class MainActivity extends AppCompatActivity implements SMSReceiver.RecMsg{

    private static final int SEND_PERMISSION_REQUEST_CODE = 1 ;
    private static final int RECEIVE_PERMISSION_REQUEST_CODE = 2 ;
    private static final String SENT_SMS_ACTION = "SENT_SMS_ACTION";
    private static final String DELIVERED_SMS_ACTION = "DELIVERED_SMS_ACTION";
    private EditText telNoText,contentText;
    private TextView smstxt;
    private SMSReceiver smsReceiver;

```

```
    @Override
    protected void onCreate(Bundle savedInstanceState) {
        super.onCreate(savedInstanceState);
        setContentView(R.layout.activity_main);
        checkPermission();
        telNoText = (EditText)findViewById(R.id.smsTel);
        contentText = (EditText)findViewById(R.id.smsContent);
        smstxt = (TextView)findViewById(R.id.tv02);

        smsReceiver = new SMSReceiver();
        IntentFilter intentFilter = new IntentFilter("android.provider.Telephony.SMS_RECEIVED");
        registerReceiver(smsReceiver, intentFilter);
        smsReceiver.sendcontext(this);

        Button btn = (Button)findViewById(R.id.smsSend);
        btn.setOnClickListener(new View.OnClickListener() {
              public void onClick(View arg0) {
                    String telNo = telNoText.getText().toString();
                    String content = contentText.getText().toString();
                    if (validate(telNo, content))
                    {
                        sendSMS(telNo, content);
                    }
              }
        });
    }

    /* 发送 SMS 方法 */
    private void sendSMS(String phoneNumber, String message) {

        Intent sentIntent = new Intent(SENT_SMS_ACTION);      //生成 sentIntent 参数
        PendingIntent sentPI = PendingIntent.getBroadcast(this, 0, sentIntent,0);

        Intent deliverIntent = new Intent(DELIVERED_SMS_ACTION);   //生成 deilverIntent 参数
        PendingIntent deliverPI = PendingIntent.getBroadcast(this, 0, deliverIntent, 0);

        SmsManager sms = SmsManager.getDefault();
        if (message.length() > 70) {           //超出 70 个字的短信进行拆分发送
            List<String> msgs = sms.divideMessage(message);
            for (String msg : msgs) {
                sms.sendTextMessage(phoneNumber, null, msg, sentPI, deliverPI);
            }
        } else {
            sms.sendTextMessage(phoneNumber, null, message, sentPI, deliverPI);
        }
        Toast.makeText(MainActivity.this, R.string.message, Toast.LENGTH_LONG).show();

        //注册 BroadcastReceivers 广播短信发送状态
        registerReceiver(new BroadcastReceiver(){
            @Override
            public void onReceive(Context _context, Intent _intent)
            {
                  switch(getResultCode()){
                      case Activity.RESULT_OK:
```

```
                        Toast.makeText(getBaseContext(),
                            "SMS sent success actions",
                            Toast.LENGTH_SHORT).show();
                        break;
                    case SmsManager.RESULT_ERROR_GENERIC_FAILURE:
                        Toast.makeText(getBaseContext(),
                                "SMS generic failure actions",
                                Toast.LENGTH_SHORT).show();
                        break;
                    case SmsManager.RESULT_ERROR_RADIO_OFF:
                        Toast.makeText(getBaseContext(),
                                "SMS radio off failure actions",
                                Toast.LENGTH_SHORT).show();
                        break;
                    case SmsManager.RESULT_ERROR_NULL_PDU:
                        Toast.makeText(getBaseContext(),
                            "SMS null PDU failure actions",
                            Toast.LENGTH_SHORT).show();
                        break;
                }
            }
        }
        new IntentFilter(SENT_SMS_ACTION));

        registerReceiver(new BroadcastReceiver(){
            @Override
            public void onReceive(Context _context,Intent _intent)
            {
                Toast.makeText(getBaseContext(),
                    "SMS delivered actions",
                    Toast.LENGTH_SHORT).show();
            }
        },
        new IntentFilter(DELIVERED_SMS_ACTION));

        telNoText.setText("");
        contentText.setText("");
        smstxt.append("发出短信:" + message);
        smstxt.append("\n\n");
    }

    public void getSMSstr(String s){
        smstxt.append(s + "\n\n");
    }

    /* 判断输入信息的有效性 */
    public boolean validate(String telNo, String content){

        if((null == telNo)||("".equals(telNo.trim()))){
            Toast.makeText(this, "请输入电话号码!",Toast.LENGTH_LONG).show();
            return false;
        }
        if((null == content)||("".equals(content.trim()))){
            Toast.makeText(this, "请输入短信内容!",Toast.LENGTH_LONG).show();
```

```
133                 return false;
134             }
135             return true;
136         }
137
138         /* 检测用户是否同意权限 */
139         private void checkPermission() {
140             //判断所申请的权限是否已通过,没通过则返回 false; 通过则返回 true,提示出来并
                //拨打电话
141             if (ContextCompat.checkSelfPermission(this, Manifest.permission.SEND_SMS)
                                                  != PackageManager.PERMISSION_GRANTED) {
142                 //申请发送短信权限回调函数
143                 ActivityCompat.requestPermissions(
                                                  this,
                                                  new String[]{Manifest.permission.SEND_SMS},
                                                  SEND_PERMISSION_REQUEST_CODE);
144             } else {
145                 Toast.makeText(this, "发送短信权限已申请通过!",Toast.LENGTH_SHORT).show();
146             }
147             if (ContextCompat.checkSelfPermission(this, Manifest.permission.RECEIVE_SMS)
                                                  != PackageManager.PERMISSION_GRANTED) {
148             //申请接收短信权限回调函数
149             ActivityCompat.requestPermissions(
                                                  this,
                                                  new String[]{Manifest.permission.RECEIVE_SMS},
                                                  RECEIVE_PERMISSION_REQUEST_CODE);
150             } else {
151                 Toast.makeText(this, "接收短信权限已申请通过!", Toast.LENGTH_SHORT).show();
152             }
153         }
154
155     }
```

(1) 第 25 行,告诉系统,在创建 MainActivity 类的同时,还要实现 SMSReceiver 类的接口 RecMsg,在 MainActivity 类中完成 RecMsg 接口中的方法 getSMSstr()代码编程,即将字符串型参数值显示在文本控件 smstxt 中(见 120～122 行)。

(2) 第 39 行,调用 checkPermission()方法,用于检测权限,如果未授权则动态申请权限。该方法在第 139～153 行定义。

(3) 第 44～47 行,动态注册接收短信的广播接收器,第 47 行调用该接收器内的 sendcontext()方法将本类对象 this 作为上下文参数传递到广播接收器中。

(4) 第 63～118 行,定义 sendSMS()方法。在该方法中完成四个功能的实现。其中两个功能是创建两个广播的 PendingIntent 对象(见第 65～69 行),分别在第 83～103 行动态注册发出短信广播,发出短信广播接收器定义了各种发出短信的状态下的处理;在第 105～112 行动态注册短信发送到达广播接收器;第 71～80 行,执行发送短信功能,当短信字数超过 70 个字则使用 divideMessage()拆分信息,使用 sendTextMessage()方法发送信息;第 114～117 行,定义发送短信后 UI 界面的显示内容。

(5) 第 125～136 行,定义 validate()方法,主要功能是检查两个信息不能为空。一个是发送短信的接收方电话号码,另一个是发送的短信内容。

SMSReceiver.java 程序文件代码如下。

```
package ee.example.activity_smsmanager;

import android.content.BroadcastReceiver;
import android.content.Context;
import android.content.Intent;
import android.os.Bundle;
import android.telephony.SmsMessage;
import android.widget.Toast;

import java.text.SimpleDateFormat;
import java.util.Date;

public class SMSReceiver extends BroadcastReceiver {

    private static final String strRes = "android.provider.Telephony.SMS_RECEIVED";
                                                                    //收到的是短信
    private RecMsg recMsg;

    @Override
    public void onReceive(Context arg0, Intent arg1) {
    //判断接收到的广播是否为收到短信的 Broadcast Action
        if(strRes.equals(arg1.getAction())){
            StringBuilder sb = new StringBuilder();
            Bundle bundle = arg1.getExtras();          //接收由 SMS 传过来的数据
            //判断是否有数据
            if(bundle!= null){
                //通过 pdus 可以获得接收到的所有短信消息
                Object[] pdusArray = (Object[])bundle.get("pdus");
                //构建短信对象 array,并依据收到的对象长度来创建 array 的大小
                SmsMessage[] msg = new SmsMessage[pdusArray.length];
                for(int i = 0 ;i < pdusArray.length;i++){
                    msg[i] = SmsMessage.createFromPdu((byte[])pdusArray[i]);
                }
                //获取接收短信的时间
                SimpleDateFormat simpleDateFormat
                                = new SimpleDateFormat("yyyy/MM/dd HH:mm:ss");
                Date date = new Date(System.currentTimeMillis());    //HH:mm:ss 获取当
                                                                      //前时间
                //将送来的短信合并自定义信息于 StringBuilder 当中
                for(SmsMessage curMsg:msg){
                    sb.append(" [ ");
                    sb.append(curMsg.getDisplayOriginatingAddress());
                                                        //获取接收短信电话号码
                    sb.append("] (" + simpleDateFormat.format(date) + ")\n 发来短信:\n");
                    sb.append(curMsg.getDisplayMessageBody());      //获得短信的内容
                }
                Toast.makeText(arg0, "接到短信!" + sb.toString(),
                                 Toast.LENGTH_SHORT).show();

                recMsg.getSMSstr(sb.toString());    //将接收的短信内容写入接口方法
            }
        }
```

```
48          }
49
50          interface RecMsg{
51              void getSMSstr(String s);
52          }
53
54          public void sendcontext(RecMsg context){
55              recMsg = context;
56          }
57
58      }
```

(1) 第21行，判断接收到的广播是否为收到短信的广播。

(2) 第27～32行，获得接收到的一组短信，存储到pdus数组。

(3) 第37～42行，从每一条短信中获取电话号码和短信内容，拼成一个StringBuilder对象sb。

(4) 第50～52行，定义RecMsg()接口，接口中只有getSMSstr(String s)方法。

(5) 第55行，在该方法中指定接口对象的上下文。在MainActivity类中的第47行，smsReceiver.sendcontext(this)语句将MainActivity类对象的上下文传递到SMSReceiver类中，在SMSReceiver子类中的第45行，recMsg.getSMSstr(sb.toString())语句表示使用MainActivity类中定义的接口方法来执行，即将sb转换为字符串，显示在MainActivity类的文本控件smstxt中。

运行结果：在Android Studio中使用两个模拟器运行Activity_SMSManager项目，第一个启动的模拟器电话号码默认为(555)521-5554，在运行时也可以只用号码的后四位5554来代表整个号码。第二个启动的模拟器电话号码默认为(555)521-5556，在运行时也可以用5556来代表整个号码。本案例使用5554运行应用项目，使用5556作为接收方手机。

运行本项目，在初始界面中输入接收短信的电话号码5556，并在输入短信文本框中输入"Hello, This is 5554."，如图12-3(a)所示。单击"发送"按钮，这时可以在5556模拟器中听到接收短信的提示声，并且可以看到收到短信的通知，如图12-3(b)所示。点开接收短信项，可以看到与该号码的短信往来历史信息，也可以在输入框中回复短信，如图12-3(c)所示。在5554模拟器中可以看到应用项目的界面，显示了当前发送和接收短信的内容，如图12-3(d)所示。如果5554发送较长短信内容，如图12-3(e)所示；在5556中可以看到短信被拆分成两条来接收，如图12-3(f)所示。

12.2.2 电话管理

手机的电话功能是最基本的功能。在应用项目中，特别是通信类的应用项目，常做的事情是对电话进行控制、监听等服务，如能够拨打电话、监听电话的呼入，或对电话进行录音，或在应用项目中直接拨打电话等。

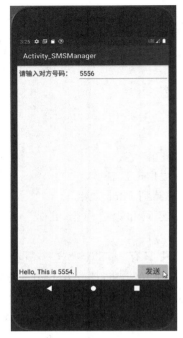

(a) 在5554上运行应用

(b) 在5556上接收短信

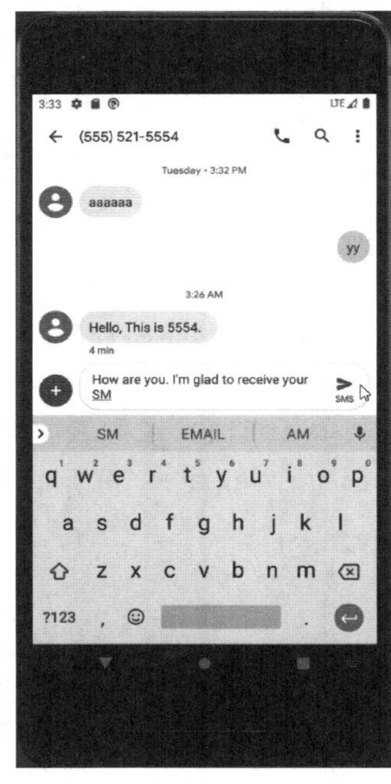

(c) 在5556上回复短信

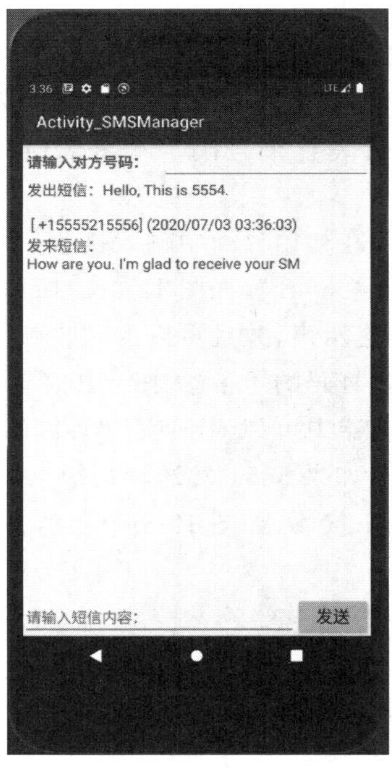

(d) 在5554上显示当前收、发短信内容

图 12-3　在 Android 应用中收、发短信的运行效果

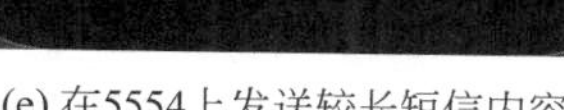
(e) 在5554上发送较长短信内容

(f) 在5556上分段接收较长短信内容

图 12-3　（续）

1. 电话应用权限

如果在应用中要拨打电话或监听电话的状态信息，需要在清单文件 AndroidManifest.xml 中声明权限。如拨打电话 CALL_PHONE 权限，读取电话状态 READ_PHONE_STATE 权限。注意，Android 9(API 28)以后，为了增强用户隐私，又进一步对若干行为权限进行了限制，该新增限制对手机通话的状态新增权限规则，即原来只需要有 READ_PHONE_STATE 权限就可以读取来电的号码，Android 9 以后，则需要添加 READ_CALL_LOG 权限才能获得来电的所有信息，包括来电的电话号码信息。

所以，对于一个对电话有来电、去电管理需求的应用，一般需要添加以下三个权限。在 AndroidManifest.xml 中声明权限，在代码文件中申请权限。声明权限代码如下。

```
<uses-permission android:name="android.permission.CALL_PHONE" />
<uses-permission android:name="android.permission.READ_PHONE_STATE" />
<uses-permission android:name="android.permission.READ_CALL_LOG" />
```

2. 拨打电话

开发拨打电话应用，有两种方式：一种是直接拨打指定号码，另一种是调用系统的 Dialer 拨号应用。这两种方式实际上都是通过启动 Intent 来实现的，不同的是 Intent 的 action。采用直接拨打方式，其 action 为 Intent.ACTION_CALL。而调用系统自带的拨号程序，其 action 为 Intent.ACTION_DIAL。

3. 监听电话状态

在 Android 应用中，所有与电话相关的应用主要由 TelephonyManager 类来管理。该类位于 android. telephony. TelephonyManager 包中。TelephonyManager 类作为一个 Service 接口提供给用户查询电话相关内容的功能，例如获取来电的 IMEI（国际移动设备识别码）、电话号码、来电状态等。在 TelephonyManager 类中使用系统常量来表示电话的状态。与来电有关的状态有三种。

（1）TelephonyManager. CALL_STATE_IDLE：表示待机状态，值为 0。

（2）TelephonyManager. CALL_STATE_RINGING：表示来电状态（振铃），值为 1。

（3）TelephonyManager. CALL_STATE_OFFHOOK：表示通话中（摘机），值为 2。

PhoneStateListener 类位于 android. telephony. PhoneStateListener 包中。该类对象实现对电话状态的监听，监听事件诸如：响铃、挂断、基站位置变化、语音邮件、呼叫转移等。该类常用的监听通话方法以及表示相关状态的常量见表 12-6。

表 12-6　PhoneStateListener 类常用监听方法及对应常量和说明

方　法	常　量	说　明
onCallForwardingIndicatorChanged (boolean cfi)	LISTEN_CALL_FORWARDING_INDICATOR	监听通话转移指示变化的回调方法和常量
onCallStateChanged (int state, String incomingNumber)	LISTEN_CALL_STATE	监听呼叫状态变化的回调方法和常量
onCellLocationChanged (CellLocation location)	LISTEN_CELL_LOCATION	监听设备单元位置变化的回调方法和常量
onDataActivity(int direction)	LISTEN_DATA_ACTIVITY	监听数据流量移动方向变化的回调方法和常量
onDataConnectionStateChanged (int state)	LISTEN_DATA_CONNECTION_STATE	监听数据连接状态的变化
onMessageWaitingIndicatorChanged (boolean mwi)	LISTEN_MESSAGE_WAITING_INDICATOR	监听消息等待指示变化的回调方法和常量
onServiceStateChanged (ServiceState serviceState)	LISTEN_SERVICE_STATE	监听网络服务状态变化的回调方法和常量
onSignalStrengthChanged(int asu)	LISTEN_SIGNAL_STRENGTH	监听网络信号强度变化的回调方法和常量

在 Android 中开发监听电话的拨打状态应用，需要联合使用 TelephonyManager 对象以及 PhoneStateListener 监听和回调方法来实现。实现步骤如下。

1）声明权限和申请权限

要获得电话的拨打状态，必须要在清单文件中添加权限声明，声明代码如下。

```
<uses-permission android:name="android.permission.READ_PHONE_STATE" />
<uses-permission android:name="android.permission.READ_CALL_LOG" />
```

2）创建 TelephonyManager 实例

使用 getSystemService(Context. TELEPHONY_SERVICE) 方法创建 TelephonyManager

对象。例如创建 tmr 对象,代码如下。

```
TelephonyManager tmr = (TelephonyManager) this.getSystemService(Context.TELEPHONY_SERVICE);
```

3) 创建对电话事件的监听

通过 TelephonyManager 设置对电话事件的监听。例如 TelephonyManager 的对象是 tmr,监听呼叫状态改变,代码如下。

```
tmr.listen(new TeleListener (),PhoneStateListener.LISTEN_CALL_STATE);
```

如果需要设置多个监听,可以在 listen()方法的第二个参数上多加几个常量,用“|”隔开。例如设置呼叫状态改变和网络服务状态改变监听,代码如下。

```
tmr.listen(new TeleListener (),
        PhoneStateListener.LISTEN_CALL_STATE | PhoneStateListener.LISTEN_SERVICE_STATE );
```

4) 创建 PhoneStateListener 子类,重写回调方法

实现对电话状态的监听,需要开发一个继承 PhoneStateListener 的子类,在其中重写相关回调方法。例如,PhoneStateListener.LISTEN_CALL_STATE 监听,需要重写回调方法 onCallStateChanged()。PhoneStateListener.LISTEN_SERVICE_STATE 监听,需要重写回调方法 onServiceStateChanged()。在这些方法中定制应用的需求操作。

5) 注销对电话的监听

当不需要对电话状态进行监听后,可以取消电话事件的监听,代码如下。

```
telephonyMgr.listen(TeleListener, PhoneStateListener.LISTEN_NONE);
```

下面通过一个案例来说明拨打电话以及监听电话状态的应用。

【案例 12.4】 开发一个拨打电话,显示来电是已接或未接状态记录的应用。要求拨打电话用两种方式。

说明:本案例设计两个按钮来分别实现两种拨打电话的方式。使用对来电状态的监听来确定已接或未接,从 TelephonyManager 对象获得来电的状态信息,从 PhoneStateListener 的回调方法来判断电话是已接还是未接。

开发步骤及解析:过程如下。

1) 创建项目

在 Android Studio 中创建一个名为 Activity_TelCall 的项目。其包名为 ee.example.activity_telcall。

2) 声明权限

本案例既要拨打电话,又要对来电的状态进行监听,所以要添加三个权限。在清单文件 AndroidManifest.xml 中的<application>元素前添加拨打电话、获取电话状态以及通话记录等三个权限声明,代码片段如下。

```
<uses-permission android:name = "android.permission.CALL_PHONE" />
<uses-permission android:name = "android.permission.READ_PHONE_STATE" />
<uses-permission android:name = "android.permission.READ_CALL_LOG" />
```

3）准备文本资源

在 res/values 目录下编辑 strings. xml 文件。在其中声明 inputmobile、buttoncall、buttondail、callintitle 等字符串内容。

4）设计布局

编写 res/layout 目录下的 activity_main. xml 文件。在布局中把屏幕分为两段区域，上段有四个控件，它们是一个 TextView，一个 EditText，ID 为 mobileno，两个按钮，ID 分别为 btncall 和 btndial，位于屏幕的顶端，用于输入要拨打的电话号和两种拨打电话方式的单击按钮；下段是两个 TextView，其中一个显示标题，另一个 ID 为 tvcall，显示来电接通/未接通信息。

5）开发逻辑代码

在 java/ee. example. activity_telcall 包下编写 MainActivity. java 文件，其代码如下。

```
package ee.example.activity_telcall;

import androidx.appcompat.app.AppCompatActivity;
import androidx.core.app.ActivityCompat;
import androidx.core.content.ContextCompat;

import android.Manifest;
import android.content.pm.PackageManager;
import android.os.Bundle;
import android.net.Uri;
import android.content.Intent;
import android.view.View;
import android.view.View.OnClickListener;
import android.widget.Button;
import android.widget.EditText;
import android.widget.Toast;
import android.telephony.PhoneStateListener;
import android.telephony.TelephonyManager;
import android.widget.TextView;

public class MainActivity extends AppCompatActivity implements OnClickListener{

    private static final int CALL_PERMISSION_REQUEST_CODE = 1 ;
    private static final int ListenCALL_PERMISSION_REQUEST_CODE = 2 ;
    private static final int CALLLOG_PERMISSION_REQUEST_CODE = 3 ;
    private EditText edt;
    private Button btc,btd;
    private TextView callstate ;
    private TelephonyManager telmanager ;
    private int lastState = TelephonyManager.CALL_STATE_IDLE;     //上次的状态,默认
                                                                  //为空闲状态

    @Override
    protected void onCreate(Bundle savedInstanceState) {
        super.onCreate(savedInstanceState);
        setContentView(R.layout.activity_main);
        //检测相关权限与请求授权
        checkPermission();

        edt = (EditText) findViewById(R.id.mobileno);
```

```
        btc = (Button) findViewById(R.id.btncall);
        btd = (Button) findViewById(R.id.btndial);
        callstate = (TextView)findViewById(R.id.tvcall);

        //获取电话服务对象
        telmanager = (TelephonyManager) this.getSystemService(TELEPHONY_SERVICE);
    }

    @Override
    protected void onStart() {
        super.onStart();
        btc.setOnClickListener(this);
        btd.setOnClickListener(this);
        telmanager.listen(new MyCallStateListener(), PhoneStateListener.LISTEN_CALL_STATE);
    }

    /* 检测用户是否同意权限 */
    private void checkPermission() {
        //判断所申请权限是否已经通过,没通过则返回 false; 通过则返回 true,提示出来并
        //拨打电话
        if (ContextCompat.checkSelfPermission(this, Manifest.permission.CALL_PHONE)
                                        != PackageManager.PERMISSION_GRANTED) {
            //申请拨打电话权限回调函数
            ActivityCompat.requestPermissions(this,
                                        new String[]{Manifest.permission.CALL_PHONE},
                                        CALL_PERMISSION_REQUEST_CODE);
        } else {
            Toast.makeText(this, "拨打电话权限已申请通过!", Toast.LENGTH_SHORT).show();
        }
        if (ContextCompat.checkSelfPermission(this, Manifest.permission.READ_PHONE_STATE)
                                        != PackageManager.PERMISSION_GRANTED) {
            //申请监听电话状态权限回调函数
            ActivityCompat.requestPermissions(this,
                                    new String[]{Manifest.permission.READ_PHONE_STATE},
                                    ListenCALL_PERMISSION_REQUEST_CODE);
        } else {
            Toast.makeText(this, "监听电话状态权限已通过!", Toast.LENGTH_SHORT).show();
        }
        if (ContextCompat.checkSelfPermission(this, Manifest.permission.READ_CALL_LOG)
                                        != PackageManager.PERMISSION_GRANTED) {
            //申请监听电话记录权限回调函数
            ActivityCompat.requestPermissions(this,
                                    new String[]{Manifest.permission.READ_CALL_LOG},
                                    CALLLOG_PERMISSION_REQUEST_CODE);
        } else {
            Toast.makeText(this, "监听电话记录权限已通过!", Toast.LENGTH_SHORT).show();
        }
    }

    @Override
    public void onClick(View v) {
        switch(v.getId()){
            case R.id.btncall:
                if (edt.getText().equals("")) {
```

```
84                      Toast.makeText(MainActivity.this, "号码不能为空!",
                                                    Toast.LENGTH_SHORT).show();
85                  } else {
86                      Intent intent = new Intent();
87                      intent.setAction(Intent.ACTION_CALL);
                                                        //指定其动作为直接拨打电话
88                      Uri data = Uri.parse("tel:" + edt.getText().toString());
                                                        //指定将要拨出的号码
89                      edt.setText("");
90                      intent.setData(data);
91                      startActivity(intent);
92                  }
93                  break;
94              case R.id.btndial:
95                  Intent dintent = new Intent();
96                  dintent.setAction(Intent.ACTION_DIAL);
                                                //指定动作为调出拨号界面来拨打电话
97                  Uri dt = Uri.parse("tel:" + edt.getText().toString());
98                  edt.setText("");
99                  dintent.setData(dt);
100                 startActivity(dintent);
101                 break;
102         }
103     }
104
105     /* 继承PhoneStateListener类,可以重新编写其内部的各种监听方法
106      * 然后通过手机状态改变时,系统自动触发这些方法来实现人们想要的功能
107      */
108     class MyCallStateListener extends PhoneStateListener{
109         @Override
110         public void onCallStateChanged(int state, String incomingNumber) {
111             super.onCallStateChanged(state, incomingNumber);
112             String result = "";
113             //如果当前状态为空闲、上次状态为响铃中,则认为是未接来电
114             if (lastState == TelephonyManager.CALL_STATE_RINGING
                 && state == TelephonyManager.CALL_STATE_IDLE) {
115                 result = "     未接来电:" + incomingNumber + ".\n";
116             }
117             //如果当前状态为接通、上次状态为响铃中,则认为是在接来电
118             if (lastState == TelephonyManager.CALL_STATE_RINGING
                  && state == TelephonyManager.CALL_STATE_OFFHOOK) {
119                 result = "接听来电:" + incomingNumber + ".\n";
120             }
121             callstate.append(result);        //来电接听状态显示在文本控件
122             lastState = state;               //最后改变当前值
123         }
124     }
125
126 }
```

(1) 第45行,创建一个TelephonyManager对象telmanager。

(2) 第80～103行,重写onClick()方法,分别实现两个按钮的单击操作,通过启动Intent实现拨打电话。第82～93行,实现的是直接拨打电话,所以必须保证要输入接听方

的电话号码；第 94～101 行，实现调用系统拨号应用，在拨号界面中还可以通过拨号输入接听方的电话号码，所以在代码中没有检查文本框是否输入号码。

(3) 第 53 行，对 telmanager 对象建立 PhoneStateListener. LISTEN_CALL_STATE 状态监听。当监听到有电话状态发生变化时，会自动回调 onCallStateChanged()方法。

(4) 第 108～124 行，定义 PhoneStateListener 类的监听子类，在子类中重写 onCallStateChanged()方法，在该方法内判断来电是否被接听。要判断来电是否被接听需要根据两次状态组合判定。如果当前一次状态为 TelephonyManager. CALL_STATE_RINGING(即响铃)、本次状态为 TelephonyManager. CALL_STATE_IDLE(即待机)时，可判断为未接听来电；如果当前一次状态为 TelephonyManager. CALL_STATE_RINGING (即响铃)、本次状态为 TelephonyManager. CALL_STATE_OFFHOOK(即摘机)时，则可判断为正在接听。所以，在方法中有一个记录上一次状态值的变量 lastState。

运行结果：在 Android Studio 上启动两个模拟器运行 Activity_TelCall 项目。本案例使用 5554 运行应用项目，使用 5556 接听本应用的电话，并向 5554 拨打电话。

运行本项目，在初始界面中输入要拨打的电话号码 5556，如图 12-4(a)所示。如果单击"拨打"按钮，即刻进行拨打操作，如图 12-4(b)所示；如果单击"拨号"按钮，则进入系统的拨号界面，如图 12-4(c)所示，在单击下方的电话图标按钮后，才进行拨打操作。执行拨打操作后，在 5556 模拟器中可以看到来电提醒界面，如图 12-4(d)所示。在 5556 模拟器上回拨电话给 5554，在 5554 上也能看到相同的来电提示界面。当按下 Answer 按钮后，则在 5554 上接听电话，这时可在应用页面看到接听了来电的状态信息，如图 12-4(e)所示；当再次来电时，按下 Decline 按钮，这时可在应用页面看到未接来电的状态信息，如图 12-4(f)所示。

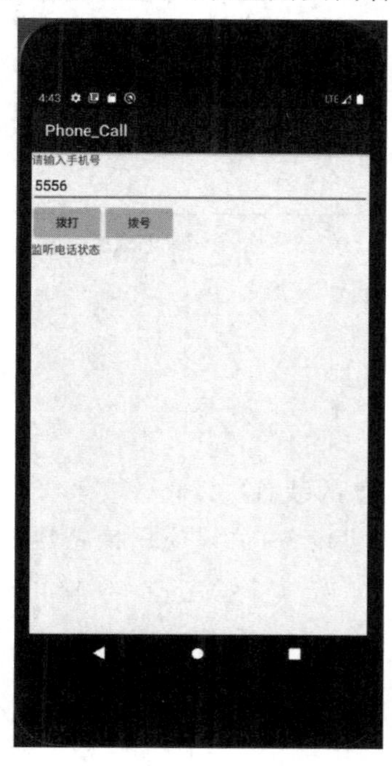

(a) 在5554上输入接听方电话号码

(b) 5554上的拨打电话界面

图 12-4　在应用中拨打电话及显示来电状态信息的运行效果

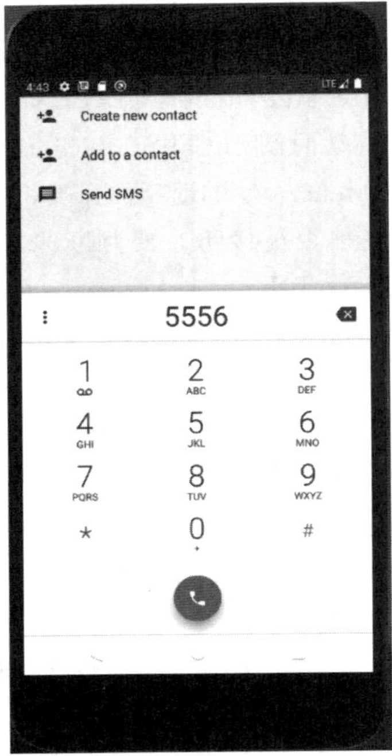

(c) 5554上的自带拨号应用界面

(d) 5556上的接收来电提醒界面

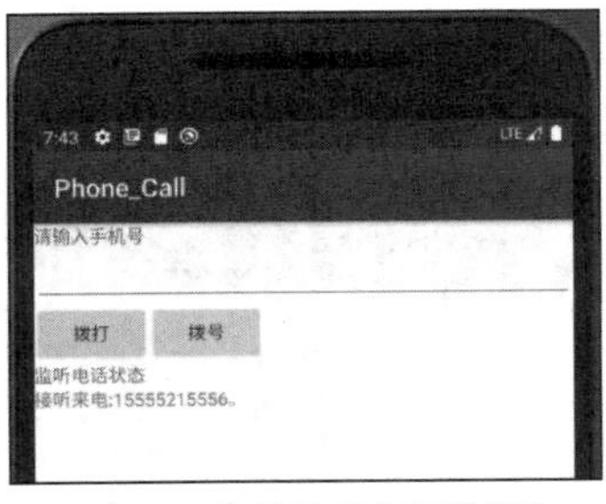

(e) 在5554上显示来电接听信息

(f) 在5554上显示来电未接信息

图 12-4 （续）

12.3 手机传感器

传感器是能够探测如光、热、重力、方向等信息的装置。Android 系统的一大亮点就是提供了对传感器的应用。以 Android 为系统的手机设备，可以应用传感器来感知外界条件的变化，开发出许多智能游戏和应用，比如“摇一摇”游戏，以及水平仪、指北针、计步器等。

12.3.1 Android 中的传感器

在 Android 的 android. hardware 包下提供了用于传感器编程的多个类，其中常用的是 Sensor 类和 SensorManager 类。

Sensor 类提供了 Android 系统支持的所有传感器，这些传感器是 Android 系统用来感知周围环境和运动信息的工具。但是，并不是每一部手机都能支持这么多感应器，越低端的手机支持的传感器就越少。在开发传感器应用时要注意这一点。Sensor 类中的传感器类型使用系统常量来表示。目前支持的传感器类型见表 12-7。

表 12-7　Sensor 类中传感器类型常量及说明

传感器类型	内部值	传感器名称	说　明
TYPE_ACCELEROMETER	1	加速度	指每个方向(3 轴)的加速度值(包括地球重力 9.8N/kg)。常用于“摇一摇”应用
TYPE_MAGNETIC_FIELD	2	磁场	用来感应磁场
TYPE_ORIENTATION	3	方向	从 API 15 以后被弃用，如果使用会出现警告。可使用 SensorManager 类中的 getOrientation()方法代替之
TYPE_GYROSCOPE	4	陀螺仪	指每个方向(3 轴)的角速度值，用来感应旋转和倾斜。 Pitch(前后倾斜)：绕 y 轴转动的角速度(x 轴在转，前后)。 Roll(左右倾斜)：绕 x 轴转动的角速度(y 轴在转，左右)。 Yaw(左右摇摆)：绕 z 轴转动的角速度
TYPE_GRAVITY	9	重力	用来感应重力
TYPE_PRESSURE	6	压力	用来感应气压，单位是 hPa(百帕斯卡)
TYPE_PROXIMITY	8	距离	用来感应距离
TYPE_LIGHT	5	光线	用来感应正面的光线强弱
TYPE_LINEAR_ACCELERATION	10	线性加速度	用来感应线性加速度
TYPE_TEMPERATURE	7	温度	从 API 15 以后被弃用，可使用下面的 TYPE_AMBIENT_TEMPERATURE 代替之
TYPE_AMBIENT_TEMPERATURE	13	环境温度	用来感应环境温度
TYPE_RELATIVE_HUMIDITY	12	相对湿度	用来感应相对湿度
TYPE_ROTATION_VECTOR	11	旋转矢量	融合了陀螺仪、加速度计和磁力计产生的原始数据，以产生四元数；主要用于 AR 等场景中。 三轴的分量与对应角度的计算值
TYPE_GAME_ROTATION_VECTOR	15	游戏旋转矢量	只融合了加速度计和陀螺仪数据，产生四元数
TYPE_GEOMAGNETIC_ROTATION_VECTOR	20	地磁旋转矢量	只融合了加速度计和地磁数据。该传感器的精度低于正常的旋转矢量传感器，但功耗降低
TYPE_ACCELEROMETER_UNCALIBRATED	35	未校准加速度	用来感应未校准加速度
TYPE_MAGNETIC_FIELD_UNCALIBRATED	14	未校准磁场	用来感应未校准磁场
TYPE_GYROSCOPE_UNCALIBRATED	16	未校准陀螺仪	用来感应未校准陀螺仪
TYPE_STEP_COUNTER	19	步行计数	记录激活后的步数

续表

传感器类型	内部值	传感器名称	说　明
TYPE_STEP_DETECTOR	18	步行检测	用户每走动一步值为 1，不走动值为 0
TYPE_DEVICE_PRIVATE_BASE	65536	设备私有传感器	设备附加的传感器，是 Android 能识别的最低传感器类型
TYPE_HEART_BEAT	31	心跳检测	可穿戴设备使用，如手环
TYPE_HEART_RATE	21	心跳速率	可穿戴设备使用，如手环
TYPE_HINGE_ANGLE	36	铰链角度	用来感应铰链角度
TYPE_LOW_LATENCY_OFFBODY_DETECT	34	低延迟身体检测	用来感应低延迟身体检测
TYPE_MOTION_DETECT	30	运动检测	用来感应运动检测
TYPE_POSE_6DOF	28	6 个自由度的姿势检测	用来感应 6 个自由度的姿势检测
TYPE_SIGNIFICANT_MOTION	17	特殊动作	用来感应特殊动作
TYPE_STATIONARY_DETECT	29	静止状态检测	用来感应静止状态检测

表中的传感器类型多数可在 Android 手机中使用，还有一些传感器是针对智能穿戴等其他设备的。对于某个传感器，它通过 get()方法获取具体信息。Sensor 类中常用的 get()方法见表 12-8。

表 12-8　Sensor 类中常用方法及说明

方　法	说　明
getMaximumRange()	获取最大取值范围。返回值为 float 类型
getName()	获取传感器名称。返回值为 String 类型
getPower()	获取该传感器使用时的功率。返回值为 float 类型，单位为 mA(毫安)
getResolution()	获取传感器的绝对单位。返回值为 float 类型
getType()	获取传感器类型。返回值为 int 类型，通常用表 12-7 中常量表示
getVendor()	获取设备供应商。返回值为 String 类型
getVersion()	获取设备版本号。返回值为 int 类型

下面对手机中常用的传感器进行介绍。

1. 加速度传感器

加速度传感器是最常见的感应器，大部分智能手机都内置该传感器。在注册了传感器监听器后，加速度传感器主要捕获 3 个参数：values[0]、values[1]和 values[2]，分别对应 x 轴、y 轴和 z 轴的加速度值，该数值包含地心引力的影响，单位为 m/s^2。如果将手机平放在水平的桌面上，x 轴默认为 0，y 轴默认为 0，z 轴默认为 9.81。该传感器广泛应用于“摇一摇”游戏，比如“摇一摇”寻找周边的朋友、投骰子等。

2. 磁场传感器

磁场传感器用于感应周围的磁感应强度，注册监听器后其主要捕获 3 个参数：values

[0]、values[1]、values[2]。3 个参数分别代表磁感应强度在空间坐标系中 3 个方向轴上的分量，单位为微特拉斯(μT)。通过该传感器可以开发出指南针、罗盘等磁场应用。

3. 方向传感器

方向传感器用于感应手机方位的变化。虽然该传感器已经被弃用，但是之前运用它开发的应用比较多，比如指南针、罗盘、水平仪等，所以在此还是有必要了解方向传感器，以及其替代方法。确定某个方向也需要一个三维坐标，坐标轴与手机的关系如图 12-5 所示。

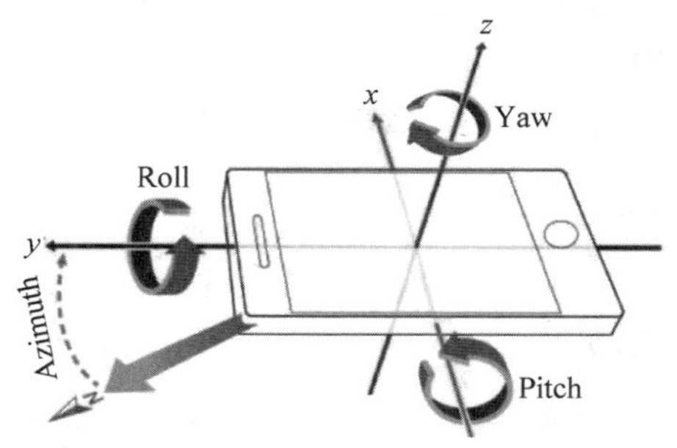

图 12-5　坐标轴与手机的位置关系

x 轴同 Pitch 轴，沿着屏幕水平方向从左到右；y 轴同 Roll 轴，从屏幕的底端开始沿着屏幕的垂直方向指向屏幕的顶端；z 轴同 Yaw 轴，当水平放置时，指向天空的方向。在注册了传感器监听器后方向传感器主要捕获以下 3 个参数。

values[0]：方位角，手机绕着 z 轴旋转的角度。0 表示正北(North)，90 表示正东(East)，180 表示正南(South)，270 表示正西(West)。

values[1]：倾斜角，手机翘起来的程度，当手机绕着 x 轴倾斜时该值会发生变化。取值范围是[－180，180]。

values[2]：滚动角，沿着 y 轴的滚动角度，取值范围为[－90，90]。

4. 计步传感器

计步器是通过手机前后摆动模拟步伐的监测。在 Android 中与计步器相关的传感器有步行检测(TYPE_STEP_DETECTOR)和步行计数(TYPE_STEP_COUNTER)。当步行检测返回 1 时表示检测到向前走了一步，步行计数返回的值是向前走的累计步数。

5. 光线传感器

光线传感器用于感应前方的光强，注册监听器后只捕获一个参数：values[0]。该参数代表前方的光照强度，单位为勒克斯(lux)。该传感器用于前置摄像头曝光应用。

6. 陀螺仪传感器

陀螺仪内部有一个陀螺，它的轴由于陀螺效应始终与初始方向平行，这样就可以通过与初始方向的偏差计算出实际方向。手机里陀螺仪实际上是一个结构非常精密的芯片，内部包含超微小的陀螺。陀螺仪对设备旋转角度的检测是瞬时的而且是非常精确的，能满足一些需要高分辨率和快速反应的应用，比如增强现实和虚拟现实应用。

12.3.2　传感器应用的开发

在 Android 系统中开发传感器应用，需要使用 SensorManager 类来处理。

1. SensorManager 类

SensorManager 类位于 android. hardware. SensorManager 包中，该类用于管理传感器。创建 SensorManager 对象通过调用 Context 的 getSystemService()方法实现，例如创建对象为 mySensorMgr，方法如下。

```
SensorManager mySensorMgr = (SensorManager) getSystemService(context.SENSOR_SERVICE);
```

其中 context 为当前的上下文对象。开发传感器应用，通常会用到 SensorManager 类中下列方法，见表 12-9。

表 12-9　SensorManager 类中常用方法及说明

方　法	说　明
getDefaultSensor(int type)	获取指定类型的默认传感器
getOrientation(float[] R, float[] values)	根据旋转矩阵获得设备的方向。其中参数 R 为旋转矩阵，values 为返回的方向向量
getRotationMatrix(float[] R, float[] I, float[] gravity, float[] geomagnetic)	计算出设备的旋转矩阵。其中参数 R 为设备与世界坐标系对齐时的单位矩阵，即设备的 X 轴指向东方、Y 轴指向北极、设备面向天空；I 为一个世界坐标空间的旋转矩阵，可以为 null；gravity 为加速度传感器的值；geomagnetic 为地磁传感器的值
getSensorList(int type)	获取指定类型的所有传感器。如果要获取所有类型的传感器，可使用常量 Sensor. TYPE_ALL
registerListener(SensorEventListener listener, Sensor sensor, int samplingPeriodUs)	为指定的传感器注册监听器。其中参数 listener 为监听传感器事件的监听器；sensor 是该监听器监听的传感器对象；samplingPeriodUs 指定获取传感器数据的频率
unregisterListener(SensorEventListener listener, Sensor sensor)	注销传感器的监听器

从 Android 4.0.3(API 15)以后，方向传感器类型被弃用，取而代之的是同时使用加速度传感器和地磁传感器来获取向量，然后通过 getRotationMatrix()方法计算出旋转矩阵，再通过 getOrientation()方法得到设备的方向。

加速度数值和磁力计数值均是向量，当手机水平放置时，加速度读数实际上就是重力向量，方向是竖直朝下的；磁力计表示本地的磁场，假设不考虑环境影响及磁偏角，认为磁场方向是水平南北朝向的。因此，首先对加速度和磁力计数据做一个差乘，得出一个水平东西方向的向量。经过这个运算，得到一个 3×3 的三维立体平面的矩阵，从而可以用来计算设备的方向。

registerListener()方法的第三个参数是获取传感器数据的速率，是延迟时间的精密度。传感器数据的获取速度值参数常见如下几种选择。

SensorManager. SENSOR_DELAY_NORMAL：(200 000μs)是默认的获取传感器数据的速度，标准延迟，对于一般的益智类游戏或者 EASY 级别的游戏可以使用，但过低的采样率可能对一些赛车类游戏有跳帧的现象。

SensorManager. SENSOR_DELAY_GAME：(20 000μs)适合开发游戏。一般实时性要求较高的游戏使用该速率。

SensorManager. SENSOR_DELAY_UI：(60 000μs)适合普通用户界面的速率。这种模式比较省电，系统开销比较小，但是延迟较大，常用于普通中小型应用项目。

SensorManager. SENSOR_DELAY_FASTEST：(0μs)延迟最小，一般不是特别灵敏的处理不推荐使用。该模式可能造成手机电量大量消耗，而且由于传递的是大量的原始数据，算法处理不好将会影响游戏逻辑和 UI 的性能。

2. SensorEventListener 接口

registerListener()方法对指定的传感器注册监听器，该监听器需要实现 SensorEventListener 接口。SensorEventListener 接口位于 android. hardware. SensorEventListener 包中，它是开发传感器应用最主要的接口。应用项目实现该接口来监视硬件中一个或多个可用传感器。

实现 SensorEventListener 接口主要需要实现以下两个方法。

onAccuracyChanged(int sensor, int accuracy)：该方法在传感器的精确度发生变化时调用，参数包括两个整数：一个表示传感器，另一个表示该传感器新的准确度值。传感器精度的变化通过四个状态常量表示，分别为 SENSOR_STATUS_ACCURACY_HIGH(高)、SENSOR_STATUS_ACCURACY_MEDIUM(中)、SENSOR_STATUS_ACCURACY_LOW(低)和 SENSOR_STATUS_UNRELIABLE(不可靠)。

onSensorChanged(SensorEvent event)：该方法在传感器的数据发生变化时调用，该方法只有一个 SensorEvent 类型的参数 event，其中 SensorEvent 类有一个 values 变量非常重要，该变量的类型是 float。但该变量最多只有 3 个元素，而且根据传感器的不同，values 变量中元素所代表的含义也不同。有些传感器只提供一个数据值，另一些则提供三个浮点值，例如加速度和磁场传感器都提供三个数据值。开发传感器应用的主要业务代码应该放在这里执行，例如读取数据并根据数据的变化进行相应的操作等。

3. 开发步骤

在 Android 系统中开发针对某一种或多种传感器的应用，是通过使用监听机制来实现的，开发步骤如下。

1) 创建 SensorManager 对象

调用 Context 的 getSystemService(Context. SENSOR_SERVICE)方法实现。

2) 获取指定传感器类型

调用 SensorManager 的 getDefaultSensor(int type)方法来得到指定类型的传感器。

3) 注册监听器

在 Activity 的 onResume()方法中调用 SensorManager 的 registerListener()方法来注册监听器。

4) 实现 SensorEventListener 接口

实现 SensorEventListener 接口就是重写 onAccuracyChanged(int sensor, int accuracy)和 onSensorChanged(SensorEvent event)两个方法。通常情况下，对传感器应用的编程只需要重写 onSensorChanged()方法中的代码。

5）注销监听器

在 Activity 的 onPause()方法中调用 SensorManager 的 unregisterListener()方法来注销监听器。一般来讲，在编程中，注册和注销的方法应该成对出现。

12.3.3 应用案例

对于传感器的应用是 Android 系统的特性之一，也是智能手机基本功能开发的特性之一。下面通过加速度传感器和磁场传感器来开发“摇一摇”和“水平仪”应用，以此介绍传感器应用的开发过程。

【案例 12.5】 开发一个简单的“摇一摇”应用，当手机捕捉到摇动发生时显示捕捉时刻的时间。

说明：本案例使用加速度传感器类型来开发，在代码中使用定义类的同时实现 SensorEventListener 接口方式。当捕捉到摇动时，辅助发出振动以提示用户手机感应到了“摇一摇”。

开发步骤及解析：过程如下。

1）创建项目

在 Android Studio 中创建一个名为 Activity_AccelerationSensor 的项目。其包名为 ee.example.activity_accelerationsensor。

2）声明权限

本案例需要使用振动功能，所以在清单文件 AndroidManifest.xml 中的<application>元素前要添加振动权限声明，代码片段如下。

```
<uses-permission android:name="android.permission.VIBRATE" />
```

3）设计布局

编写 res/layout 目录下的 activity_main.xml 文件。在布局上只添加一个 TextView，ID 为 tv_shakemsg。

4）开发逻辑代码

在 java/ee.example.activity_accelerationsensor 包下编写 MainActivity.java 文件，其代码如下。

```
package ee.example.activity_accelerationsensor;

import androidx.appcompat.app.AppCompatActivity;

import android.os.Bundle;
import android.content.Context;
import android.hardware.Sensor;
import android.hardware.SensorEvent;
import android.hardware.SensorEventListener;
import android.hardware.SensorManager;
import android.os.Vibrator;
import android.widget.TextView;

```

```
import java.text.SimpleDateFormat;
import java.util.Date;

public class MainActivity extends AppCompatActivity implements SensorEventListener{

    private TextView tv_msg;
    private SensorManager mySensorMngr;
    private Vibrator mVibrator;
    private static final int SENSOR_VALUE = 12;     //用于设置检测"摇一摇"的灵敏度
    private int number = 0;                         //用于统计捕捉"摇一摇"的次数

    @Override
    protected void onCreate(Bundle savedInstanceState) {
        super.onCreate(savedInstanceState);
        setContentView(R.layout.activity_main);
        tv_msg = (TextView) findViewById(R.id.tv_shakemsg);
        mySensorMngr = (SensorManager) getSystemService(Context.SENSOR_SERVICE);
        mVibrator = (Vibrator) getSystemService(Context.VIBRATOR_SERVICE);
    }

    @Override
    protected void onResume() {
        super.onResume();
        mySensorMngr.registerListener(this,
                mySensorMngr.getDefaultSensor(Sensor.TYPE_ACCELEROMETER),
                SensorManager.SENSOR_DELAY_NORMAL);
    }

    @Override
    protected void onPause() {
        super.onPause();
        mySensorMngr.unregisterListener(this);
    }

    @Override
    public void onSensorChanged(SensorEvent event) {
        if (event.sensor.getType() == Sensor.TYPE_ACCELEROMETER) {
            //values[0]:X 轴,values[1]:Y 轴,values[2]:Z 轴
            float[] values = event.values;
            //这里可以调节"摇一摇"的灵敏度
            if ((Math.abs(values[0]) > SENSOR_VALUE || Math.abs(values[1]) > SENSOR_VALUE
                    || Math.abs(values[2]) > SENSOR_VALUE)) {
                SimpleDateFormat s_format = new SimpleDateFormat("HH:mm:ss");
                tv_msg.setText(" 恭喜您在 " + s_format.format(new Date()) +
                        " 第 " + (++number) + " 次摇中啦!\n");
                //系统检测到"摇一摇"事件后,振动手机提示用户
                mVibrator.vibrate(500);
            }
        }
    }

    @Override
    public void onAccuracyChanged(Sensor sensor, int accuracy) {
        //当传感器精度改变时回调该方法,一般无须处理
```

```
67          }
68
69      }
```

① 第 30 行，创建一个 SensorManager 对象 mySensorMngr。

② 第 31 行，创建一个 Vibrator 对象 mVibrator。

③ 第 35～40 行，在 onResume()方法中注册传感器监听器，其中指定传感器是加速度类型，由于摇动手机对实时性要求不强，所以选择 SensorManager. SENSOR_DELAY_NORMAL 读取传感器中的数据速率。

④ 第 43～46 行，在 onPause()方法中注销监听器。

⑤ 第 49～62 行，重写 SensorEventListener 接口的 onSensorChanged()方法，这是本案例的重点。在捕捉到摇一摇后的操作都在该方法中实现。其中第 52 行，从传感器事件中获取 x、y、z 三轴的移动数据，第 54～60 行，通过判断，如果当手机在 x、y、z 三轴的位移有一个超过了 SENSOR_VALUE 值，就在屏幕上显示捕捉的摇动信息。本案例的灵敏度值为 12，该值可以随意调整；显示的内容为捕捉到摇动的时刻和次数，并且发出 0.5 秒的振动。

运行结果：在 Android Studio 支持的模拟器上，可以模拟运行大多数传感器应用项目。本案例可以选择使用模拟器运行，也可以选用真机运行。这里选择在模拟器上运行 Activity_AccelerationSensor 项目。

运行初始界面，如图 12-6(a)所示。在模拟器的右侧有一列竖立的菜单，该菜单列出了手机的常见操作项，如果要模拟摇动手机，需要选择“...”项(意思是“更多项”)，见图中标记。单击“...”，进入更多的对手机操作选项模拟界面，选择 Virtual sensors 项，然后选择右侧的 Move 单选选钮，即可模拟摇动手机操作，如图 12-6(b)所示。此时，用鼠标拖动屏幕中的手机来回摇晃，当手机在三个坐标轴方向移动，其中一个超过指定的灵敏度值 12 时，手机显示捕捉到摇一摇信息，如图 12-6(c)所示。

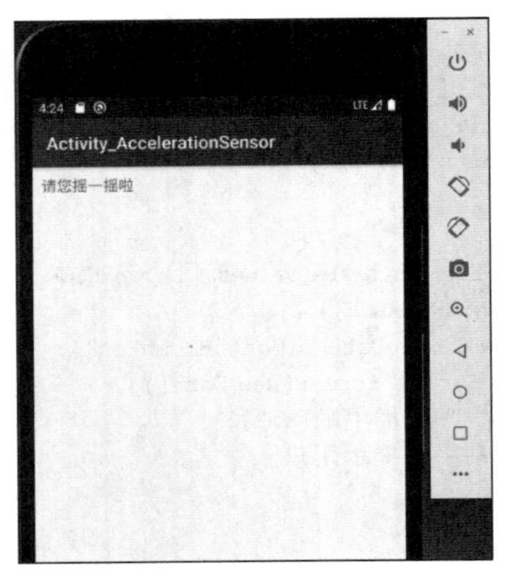

(a) 初始运行界面

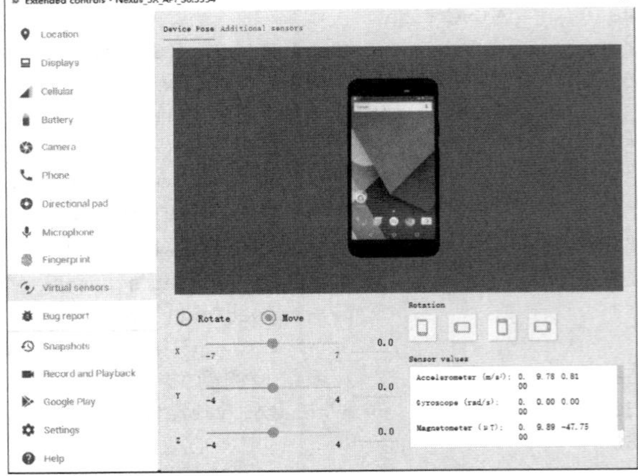

(b) 打开模拟器Virtual sensors项的Move界面

图 12-6　在模拟器中运行“摇一摇”应用的运行效果

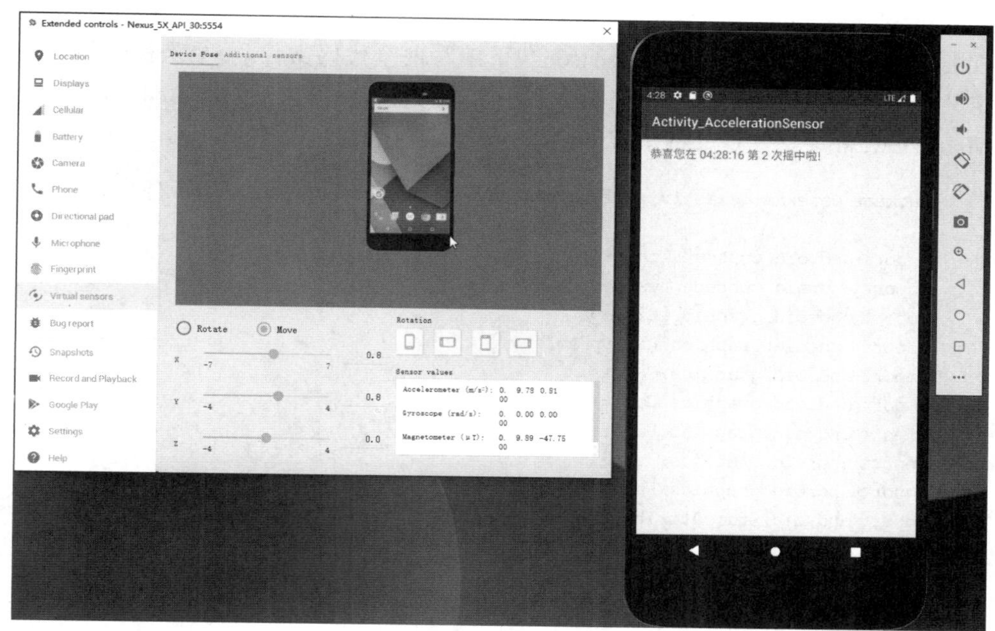

(c) 摇动模拟器中的手机，模拟“摇一摇”操作

图 12-6 （续）

【案例 12.6】 开发一个从横向、纵向、45°斜向以及圆形的全方位水平仪应用。

说明：案例中有横向、纵向、45°斜向、圆形四个水平仪，其中前三个是条形的，负责单一方向的水平测试，中间是一个圆形水平仪，负责全维度的水平测试。每个水平仪上有刻度和游动的气泡，包含元素比较多，绘制复杂。为此，专门定义一个 View 子类来初始化这四个水平仪。项目的布局设计由该类确定。

开发步骤及解析：过程如下。

1）创建项目

在 Android Studio 中创建一个名为 Activity_LevelSensor 的项目。其包名为 ee.example.activity_levelsensor。

2）准备图片资源

要事先准备好用于水平仪中的标尺背景图和位于其上的水泡图片，将它们复制到 res/drawable 目录中。

3）设计布局

编写 res/layout 目录下的 activity_main.xml 文件。在布局中添加一个自定义的视图控件类，由 MainView.java 定义。添加该控件的代码片段如下。

```
<ee.example.activity_levelsensor.MainView
    android:id="@+id/mainView"
    android:layout_width="match_parent"
    android:layout_height = "match_parent"
    />    <!-- 自定义 View -->
```

4）开发逻辑代码

在 java/ee. example. activity_levelsensor 包下编写两个代码文件，一个是初始化水平仪界面的定义类实现代码文件 MainView. java；另一个是主逻辑代码文件 MainActivity. java。

MainView. java 文件定义继承 View 类的子类，其代码如下。

```
package ee.example.activity_levelsensor;

import android.content.Context;
import android.content.res.Resources;
import android.graphics.Bitmap;
import android.graphics.BitmapFactory;
import android.graphics.Canvas;
import android.graphics.Color;
import android.graphics.Paint;
import android.graphics.RectF;
import android.graphics.Paint.Style;
import android.util.AttributeSet;
import android.util.DisplayMetrics;
import android.view.View;

public class MainView extends View{

Paint paint = new Paint();              //画笔
    Canvas canvas;
    int screenwidth = 0;                //屏幕宽度

    //图片资源的声明
    Bitmap CLB_Bitmap;                  //中间的大圆图
    Bitmap CLS_Bitmap;                  //中间的小气泡
    Bitmap HB_Bitmap;                   //上面的大矩形图
    Bitmap HS_Bitmap;                   //上面的气泡
    Bitmap VB_Bitmap;                   //左面的大矩形图
    Bitmap VS_Bitmap;                   //左面的气泡
    Bitmap HVB_Bitmap;                  //右下的矩形图
    Bitmap HVS_Bitmap;                  //右下的气泡

    //背景矩形的位置声明
    int circleB_X;                      //中间的大圆图坐标
    int circleB_Y;
    int horizontalB_X;                  //上面的大矩形图坐标
    int horizontalB_Y;
    int verticalB_X;                    //左面的大矩形图坐标
    int verticalB_Y;
    int HVB_X;                          //右下的矩形图坐标
    int HVB_Y;
    int horizontalS_X;                  //上面的气泡坐标
    int horizontalS_Y;
    int verticalS_X;                    //左面图的气泡坐标
    int verticalS_Y;
    int circleS_X;                      //中间的小气泡坐标
    int circleS_Y;
    int HVS_X;                          //右下的气泡坐标
    int HVS_Y;
```

```
    public MainView(Context context, AttributeSet attrs){
        super(context,attrs);
        initBitmap();                //初始化图片资源
        initLocation();              //初始化气泡的位置
    }

    private void initBitmap(){        //初始化图片的方法
        CLB_Bitmap = BitmapFactory.decodeResource(getResources(),R.drawable.cl1);
        CLS_Bitmap = BitmapFactory.decodeResource(getResources(),R.drawable.cl2);
        HB_Bitmap = BitmapFactory.decodeResource(getResources(),R.drawable.h1);
        HS_Bitmap = BitmapFactory.decodeResource(getResources(),R.drawable.h2);
        VB_Bitmap = BitmapFactory.decodeResource(getResources(),R.drawable.v1);
        VS_Bitmap = BitmapFactory.decodeResource(getResources(),R.drawable.v2);
        HVB_Bitmap = BitmapFactory.decodeResource(getResources(),R.drawable.hv1);
        HVS_Bitmap = BitmapFactory.decodeResource(getResources(),R.drawable.hv2);

        Resources resources = this.getResources();
        DisplayMetrics dm = resources.getDisplayMetrics();
        screenwidth = (int) (dm.widthPixels);
        circleB_X = screenwidth/2 - CLB_Bitmap.getWidth()/2;    //中间的大圆图坐标
        circleB_Y = circleB_X;
        horizontalB_X = screenwidth/2 - HB_Bitmap.getWidth()/2;     //上面的大矩形图坐标
        horizontalB_Y = circleB_Y - CLB_Bitmap.getWidth()/2;
        verticalB_X = horizontalB_Y;                                //左面的大矩形图坐标
        verticalB_Y = horizontalB_X;
        HVB_X = circleB_X + CLB_Bitmap.getWidth()/2 + 20;          //右下的矩形图坐标
        HVB_Y = HVB_X;
    }

    private void initLocation(){                                  //初始化气泡位置的方法
        circleS_X = circleB_X + CLB_Bitmap.getWidth()/2 - CLS_Bitmap.getWidth()/2;
        circleS_Y = circleB_Y + CLB_Bitmap.getHeight()/2 - CLS_Bitmap.getHeight()/2;
        horizontalS_X = horizontalB_X + HB_Bitmap.getWidth()/2 - HS_Bitmap.getWidth()/2;
        horizontalS_Y = horizontalB_Y + HB_Bitmap.getHeight()/2 - HS_Bitmap.getHeight()/2;
        verticalS_X = verticalB_X + VB_Bitmap.getWidth()/2 - VS_Bitmap.getWidth()/2;
        verticalS_Y = verticalB_Y + VB_Bitmap.getHeight()/2 - VS_Bitmap.getHeight()/2;
        HVS_X = HVB_X + HVB_Bitmap.getWidth()/2 - HVS_Bitmap.getWidth()/2;
        HVS_Y = HVB_Y + HVB_Bitmap.getHeight()/2 - HVS_Bitmap.getHeight()/2;
    }

    @Override
    protected void onDraw(Canvas canvas){                         //重写的绘制方法
        super.onDraw(canvas);
        canvas.drawColor(Color.WHITE);                            //设置背景色为白色
        paint.setColor(Color.BLUE);                               //设置画笔颜色
        paint.setStyle(Style.STROKE);                             //设置画笔为不填充
        canvas.drawRect(5, 5, screenwidth - 10, screenwidth - 10, paint);  //绘制外边框矩形

        //画背景图
        canvas.drawBitmap(CLB_Bitmap, circleB_X,circleB_Y, paint);              //中
        canvas.drawBitmap(HB_Bitmap, horizontalB_X,horizontalB_Y, paint);       //上
        canvas.drawBitmap(VB_Bitmap, verticalB_X,verticalB_Y, paint);           //左
        canvas.drawBitmap(HVB_Bitmap, HVB_X,HVB_Y, paint);                      //下
```

```
103
104            //开始绘制气泡
105            canvas.drawBitmap(CLS_Bitmap, circleS_X,circleS_Y, paint);               //中
106            canvas.drawBitmap(HS_Bitmap, horizontalS_X,horizontalS_Y, paint);  //上
107            canvas.drawBitmap(VS_Bitmap, verticalS_X,verticalS_Y, paint);         //左
108            canvas.drawBitmap(HVS_Bitmap, HVS_X, HVS_Y, paint);                        //下
109            paint.setColor(Color.GRAY);                 //设置画笔颜色用来绘制刻度
110
111            //绘制上面方框中的刻度
112            canvas.drawLine (horizontalB_X + HB_Bitmap.getWidth()/2 - 20,
                  horizontalB_Y, horizontalB_X + HB_Bitmap.getWidth()/2 - 20,
                  horizontalB_Y + HB_Bitmap.getHeight() - 2, paint);
113            canvas.drawLine (horizontalB_X + HB_Bitmap.getWidth()/2 + 20,
                  horizontalB_Y, horizontalB_X + HB_Bitmap.getWidth()/2 + 20,
                  horizontalB_Y + HB_Bitmap.getHeight() - 2, paint);
114
115            //绘制左面方框中的刻度
116            canvas.drawLine(verticalB_X,verticalB_Y + VB_Bitmap.getHeight()/2 - 20,
                  verticalB_X + VB_Bitmap.getWidth() - 2,
                  verticalB_Y + VB_Bitmap.getHeight()/2 - 20, paint);
117            canvas.drawLine(verticalB_X,verticalB_Y + VB_Bitmap.getHeight()/2 + 20,
                  verticalB_X + VB_Bitmap.getWidth() - 2,
                  verticalB_Y + VB_Bitmap.getHeight()/2 + 20, paint);
118
119            //绘制下面方框中的刻度
120            canvas.drawLine(HVB_X + HVB_Bitmap.getWidth()/2 - 30,
                  HVB_Y + HVB_Bitmap.getHeight()/2 - 60,
                  HVB_X + HVB_Bitmap.getWidth()/2 + 60,
                  HVB_Y + HVB_Bitmap.getHeight()/2 + 30, paint);
121            canvas.drawLine(HVB_X + HVB_Bitmap.getWidth()/2 - 60,
                  HVB_Y + HVB_Bitmap.getHeight()/2 - 30,
                  HVB_X + HVB_Bitmap.getWidth()/2 + 30,
                  HVB_Y + HVB_Bitmap.getHeight()/2 + 60, paint);
122
123            //中间圆圈中的刻度(小圆)
124            RectF oval  =  new RectF(circleB_X + CLB_Bitmap.getWidth()/2 - 18,
                  circleB_Y + CLB_Bitmap.getHeight()/2 - 18,
                  circleB_X + CLB_Bitmap.getWidth()/2 + 18,
                  circleB_Y + CLB_Bitmap.getHeight()/2 + 18);
125            canvas.drawOval(oval, paint);              //绘制基准线(圆)
126        }
127
128  }
```

(1) 第 50～54 行，是类的构造方法。其中第 52 行调用初始化图片的方法；第 53 行调用初始化气泡的方法。

(2) 第 56～77 行，实现图片初始化。其中第 57～64 行创建应用中的水平仪背景图片、气泡图片对象。第 66～68 行获取本视图的宽度。第 69～76 行计算出水平仪背景图的 X、Y 坐标。

(3) 第 79～88 行，计算气泡图片的位置坐标。

(4) 第 91～126 行，实现水平仪及刻度图形的绘制方法。其中第 93～96 行绘制一个矩形边框。第 99～102 行绘制四个水平仪的背景图。第 105～109 行绘制气泡。第 112、113

行绘制横向水平仪的中间刻度；第 116、117 行绘制纵向水平仪的中间刻度；第 120、121 行绘制斜向水平仪的中间刻度；第 124、125 行绘制中间圆形水平仪的中心圆圈刻度。

MainActivity.java 是本案例的主逻辑代码文件，实现水平仪的功能，其代码如下。

```
package ee.example.activity_levelsensor;

import androidx.appcompat.app.AppCompatActivity;

import android.hardware.Sensor;
import android.hardware.SensorEvent;
import android.hardware.SensorEventListener;
import android.hardware.SensorManager;
import android.os.Bundle;

public class MainActivity extends AppCompatActivity {

    private MainView mv;                    //主 View
    private int k = 3;                      //灵敏度

    private Sensor accelerometerSensor, magneticSensor;
    private SensorManager mySensorManager;
    private SensorEventListener myEventSensorListener;
    private float[] accelerometerValues = new float[3];      //加速度传感器的值
    private float[] magneticValues = new float[3];           //磁场传感器的值

    @Override
    protected void onCreate(Bundle savedInstanceState) {
        super.onCreate(savedInstanceState);
        setContentView(R.layout.activity_main);
        mv = (MainView) findViewById(R.id.mainView);        //获取水平仪视图实例

        //获取 SensorManager 对象
        mySensorManager = (SensorManager)getSystemService(SENSOR_SERVICE);

        //获取 Sensor 对象
        accelerometerSensor =
                    mySensorManager.getDefaultSensor(Sensor.TYPE_ACCELEROMETER);
        magneticSensor = mySensorManager.getDefaultSensor(Sensor.TYPE_MAGNETIC_FIELD);

        //实现传感器监听器
        myEventSensorListener = new SensorEventListener() {

            @Override
            public void onAccuracyChanged(Sensor sensor, int accuracy) {}

            @Override
            public void onSensorChanged(SensorEvent event) {    //重写 onSensorChanged()方法

                if (event.sensor.getType() == Sensor.TYPE_ACCELEROMETER){
                    accelerometerValues = event.values;     //获取加速度传感器的值
                };
                if (event.sensor.getType() == Sensor.TYPE_MAGNETIC_FIELD){
                    magneticValues = event.values;          //获取磁场传感器的值
```

```
        };
        float[] R = new float[9];                          //保存旋转的数组
        float[] values = new float[3];                     //保存方向数据的数组
        //获得一个包含旋转矩阵的函数
        SensorManager.getRotationMatrix(R,null,accelerometerValues,magneticValues);
        SensorManager.getOrientation(R,values);   //获取方向值
        float AngY = values[1];                            //y轴的旋转角度
        float AngZ = values[2];                            //z轴的旋转角度

        int x = 0; int y = 0;       //临时变量,计算中间水泡坐标时用
        int tempX = 0; int tempY = 0;                      //下面气泡的临时变量
        //开始调整x的值
        if(Math.abs(AngZ)<=k){
            mv.horizontalS_X = mv.horizontalB_X + (int)(((mv.HB_Bitmap.getWidth()
                - mv.HS_Bitmap.getWidth())/2.0) - (((mv.HB_Bitmap.getWidth()
                - mv.HS_Bitmap.getWidth())/2.0) * AngZ)/k);    //上面的
            x = mv.circleB_X + (int)(((mv.CLB_Bitmap.getWidth()
                - mv.CLS_Bitmap.getWidth())/2.0) - (((mv.CLB_Bitmap.getWidth()
                - mv.CLS_Bitmap.getWidth())/2.0) * AngZ)/k);   //中间的
        }else if(AngZ > k){
            mv.horizontalS_X = mv.horizontalB_X;
            x = mv.circleB_X;
        }else{
            mv.horizontalS_X = mv.horizontalB_X + mv.HB_Bitmap.getWidth()
                  - mv.HS_Bitmap.getWidth();
            x = mv.circleB_X + mv.CLB_Bitmap.getWidth() - mv.CLS_Bitmap.getWidth();
        }

        //开始调整y的值
        if(Math.abs(AngY)<=k){
            mv.verticalS_Y = mv.verticalB_Y + (int)(((mv.VB_Bitmap.getHeight()
                - mv.VS_Bitmap.getHeight())/2.0) + (((mv.VB_Bitmap.getHeight()
                - mv.VS_Bitmap.getHeight())/2.0) * AngY)/k);   //左面的
            y = mv.circleB_Y + (int)(((mv.CLB_Bitmap.getHeight()
                - mv.CLS_Bitmap.getHeight())/2.0)
                 + (((mv.CLB_Bitmap.getHeight()
                - mv.CLS_Bitmap.getHeight())/2.0) * AngY)/k);   //中间的
        }else if(AngY > k){
            mv.verticalS_Y = mv.verticalB_Y + mv.VB_Bitmap.getHeight()
                - mv.VS_Bitmap.getHeight();
            y = mv.circleB_Y + mv.CLB_Bitmap.getHeight() - mv.CLS_Bitmap.getHeight();
        }else{
            mv.verticalS_Y = mv.verticalB_Y;
            y = mv.circleB_Y;
        }

        //中间的水泡在圆内才改变坐标
        if(isContain(x, y)){
            mv.circleS_X = x;
            mv.circleS_Y = y;
        }

        //开始调整斜向的x、y的值
```

```
                tempX = -(int)(((mv.HVB_Bitmap.getWidth()/2 - mv.HVS_Bitmap.getWidth()) * AngZ
                    + (mv.HVB_Bitmap.getWidth()/2 - mv.HVS_Bitmap.getWidth()) * AngY)/k);
                tempY = (int)(((mv.HVB_Bitmap.getHeight()/2 - mv.HVS_Bitmap.getHeight()) * AngZ
                    + (mv.HVB_Bitmap.getHeight()/2 - mv.HVS_Bitmap.getHeight()) * AngY)/k);

                //限制斜向的气泡 x、y 值范围
                if(tempX > mv.HVB_Bitmap.getWidth()/2 - mv.HVS_Bitmap.getWidth()){
                    tempX = mv.HVB_Bitmap.getWidth()/2 - mv.HVS_Bitmap.getWidth();
                }
                if(tempX < -mv.HVB_Bitmap.getWidth()/2 + mv.HVS_Bitmap.getWidth()){
                    tempX = -mv.HVB_Bitmap.getWidth()/2 + mv.HVS_Bitmap.getWidth();
                }
                if(tempY > mv.HVB_Bitmap.getHeight()/2 - mv.HVS_Bitmap.getHeight()){
                    tempY = mv.HVB_Bitmap.getHeight()/2 - mv.HVS_Bitmap.getHeight();
                }
                if(tempY < -mv.HVB_Bitmap.getHeight()/2 + mv.HVS_Bitmap.getHeight()){
                    tempY = -mv.HVB_Bitmap.getHeight()/2 + mv.HVS_Bitmap.getHeight();
                }
                mv.HVS_X = tempX + mv.HVB_X + mv.HVB_Bitmap.getWidth()/2
                    - mv.HVS_Bitmap.getWidth()/2;
                mv.HVS_Y = tempY + mv.HVB_Y + mv.HVB_Bitmap.getHeight()/2
                    - mv.HVS_Bitmap.getWidth()/2;

                mv.postInvalidate();                          //重绘 MainView
            }

            public boolean isContain(int x, int y){           //判断点是否在圆内
                int tempx = (int)(x + mv.CLS_Bitmap.getWidth()/2.0);
                int tempy = (int)(y + mv.CLS_Bitmap.getWidth()/2.0);
                int ox = (int)(mv.circleB_X + mv.CLB_Bitmap.getWidth()/2.0);
                int oy = (int)(mv.circleB_X + mv.CLB_Bitmap.getWidth()/2.0);
                if(Math.sqrt((tempx - ox) * (tempx - ox) + (tempy - oy) * (tempy - oy))
                  >(mv.CLB_Bitmap.getWidth()/2.0 - mv.CLS_Bitmap.getWidth()/2.0)){
                                                              //不在圆内时
                    return false;
                }else{                                        //在圆内时
                    return true;
                }
            }
        };
    }

    @Override
    protected void onResume() {                               //重写 onResume()方法
        super.onResume();
        mySensorManager.registerListener(myEventSensorListener, accelerometerSensor,
                    SensorManager.SENSOR_DELAY_UI);
        mySensorManager.registerListener(myEventSensorListener, magneticSensor,
                    SensorManager.SENSOR_DELAY_UI);
    }

    @Override
    protected void onPause() {                                //重写 onPause()方法
```

```
137            mySensorManager.unregisterListener(myEventSensorListener);   //取消注册监听器
138            super.onPause();
139        }
140
141    }
```

(1) 第 26 行，获取自定义界面对象实例 mv。

(2) 第 32、33 行，创建加速度传感器和磁场传感器对象。水平仪的应用主要使用到加速度传感器和磁场传感器，从而确定手机的旋转和摆动的方向。

(3) 第 36～125 行，实现传感器的监听器。在其中重写了 onSensorChanged()方法，用于监听传感器方向的采样值变化，并根据其新的采样值重新绘制各部分气泡的位置，这些气泡都被限定在其各自的背景图范围内。实现该方法是本案例的重点。

其中第 44～49 行，获取传感器的 values 值。第 50～56 行，获取手机的 y 轴、z 轴的旋转角度。根据这两个旋转角度、水平仪背景图的坐标及范围，分别计算出每个气泡的新坐标(见第 58～109 行)。第 111 行，通过调用 postInvalidate()方法来重绘 MainView，该方法是 View 类中的方法，用于重绘视图。

(4) 第 129～133 行，在 onResume()方法内注册加速度、磁场两个传感器的监听器。

(5) 第 136～139 行，在 onPause()方法中注销传感器监听器。

运行结果：在 Android Studio 支持的模拟器上，运行 Activity_levelSensor 项目。

运行初始界面，如图 12-7(a)所示。在模拟器右侧的菜单中单击“...”，进入更多的操作选项模拟界面，选择 Virtual sensors 项，然后选择右侧的 Rotate 单选按钮，即可模拟旋转摇动手机操作，如图 12-7(b)所示。此时，用鼠标拖曳屏幕中的手机旋转，可以看到四个水平仪上的气泡随之跟着游动，如图 12-7(c)所示。当手机旋转到水平位置时，可以看到四个水平仪上的气泡都位于刻度中央，如图 12-7(d)所示。

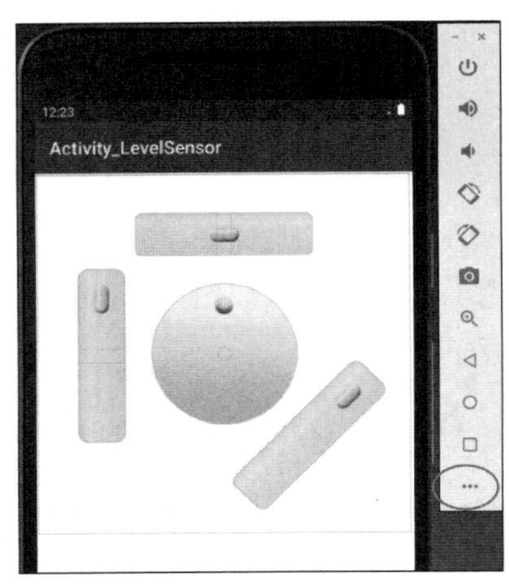

(a) 初始运行界面

(b) 打开模拟器Virtual sensors项的Rotate界面

图 12-7　在模拟器中运行水平仪应用的运行效果

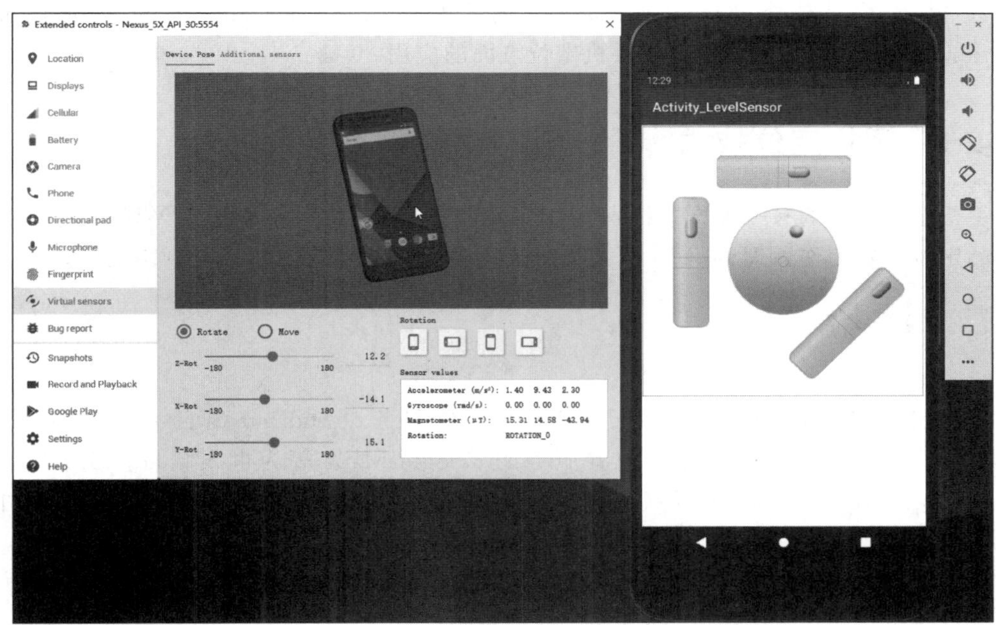

(c) 旋转摇动模拟器中手机，水平仪的气泡随之游动

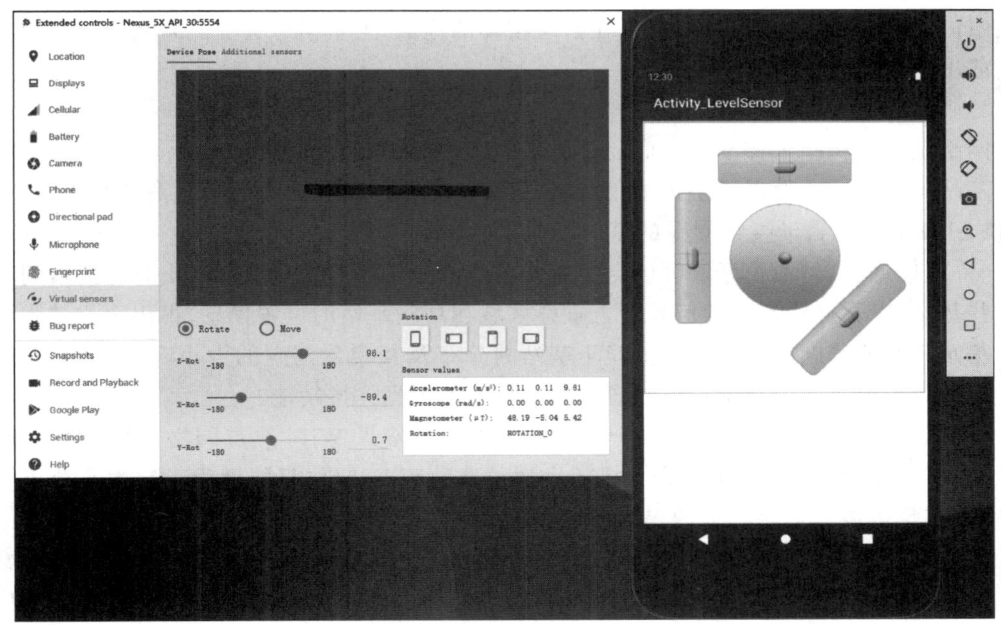

(d) 旋转模拟器中手机到水平位置，水平仪气泡复位

图 12-7　（续）

12.4　手机定位

手机定位功能是智能手机最普遍的应用之一。Android 平台支持多种定位方式，为手

机定位提供了定位条件器 Criteria、定位管理器 LocationManager 和定位监听器 LocationListener，可以方便地开发手机定位方面的应用。

12.4.1 手机定位技术

手机定位常见的有三种方式：基于 GPS 定位、基于基站定位和基于 WiFi 定位。根据不同的定位方式工作原理，可分为两大类定位技术：一是卫星定位，二是网络定位。

1. 定位分类简介

1）卫星定位

卫星定位最普遍使用的是基于 GPS 的定位服务。Android 出自 Google 之手，而 Google 公司又以搜索引擎和基于位置服务的 Google Maps 等为世人瞩目。GPS（Global Positioning System，全球定位系统）是较早的卫星定位系统，当然应用也较广。此外，中国的“北斗”、俄罗斯的“格洛纳斯”卫星定位服务也逐渐普及了。

卫星定位的工作原理，是根据多颗卫星与导航芯片的通信结果得到手机与卫星的距离，然后计算手机当前位置的经度、纬度、海拔高度等。

卫星定位的优点在于全球覆盖，在地球上的任何室外地方都可以对手机进行定位。但是在室内、地下车库、交通隧道、甚至在云层密集的地域，卫星信号会受到限制使得定位精度不高或出现定位盲区。

2）网络定位

网络定位包括基于移动运营商建立的基站定位和 WiFi 定位。

基站是移动运营商提供手机通信服务的必备设备。当手机插上运营商提供的 SIM 卡后，该 SIM 卡就会自动搜索周围的基站信号并接入通信服务。对于每一个基站，都携带有编号、位置信息、信号覆盖区域等信息。基站定位的工作原理，是监测 SIM 卡能搜索到周围的哪些基站，以及这些基站的信号覆盖重叠区域和这些基站的位置信息，得出手机的大致方位。基站定位的精度很大程度上依赖于基站的分布及其覆盖范围的大小，有些情况定位误差会超过一千米。使用基站定位需要开启手机上的数据连接功能。

WiFi 定位的工作原理，是在某个公共场所内手机接入 WiFi 热点网络，通过 WiFi 路由器的 MAC 地址和 IP 地址来确定手机的大致位置。WiFi 定位只能适用于室内小范围。使用 WiFi 定位需要开启手机上的 WLAN 功能。

无论是基站定位还是 WiFi 定位，都属于地表定位方式，在室内、隧道等地面建筑物中，只要在手机能接收到信号的地方都能实施定位，弥补了卫星定位的不足。但是网络定位都只能提供手机的大致位置，定位精度不高；其次，网络定位具有强制性，即使手机终端不请求位置服务，也进行主动定位，所以这类定位具有较强的监管能力。

2. 定位开启状态

Android 提供了 LocationManager 类来管理定位服务。LocationManager 类位于 android.location.LocationManager 包中。使用它可以获取定位信息的提供者，即定位服务是由哪一种定位方式提供的。LocationManager 类使用下列系统常量来识别提供者。

GPS_PROVIDER：表示卫星定位方式，要求开启 GPS 功能。

NETWORK_PROVIDER：表示网络定位方式，要求开启数据连接或 WLAN 功能。

PASSIVE_PROVIDER：表示无法定位，在未开启定位相关功能时返回定位提供者。

LocationManager 类的对象通过 Context.getSystemService(Context.LOCATION_SERVICE)方法获得实例。下面给出一些代码片段来介绍如何获取定位开关状态。

1）获取定位开关状态

如果要判断 GPS 定位、网络定位的开关状态，可以定义一个方法来判断。在方法中创建 LocationManager 对象 mylocatm，然后调用 LocationManager 的 isProviderEnabled()方法来判断指定的定位方式是否已开启。在该方法中对两种定位方式进行判断，并返回两个返回值。只要其中有一个值为 true，就表示定位状态已开启，代码如下。

```
public static Boolean getLocationStatus(Context c){
    LocationManager mylocatm = (LocationManager) c.getSystemService(Context.LOCATION_SERVICE);
    boolean gps_enabled = mylocatm.isProviderEnabled(LocationManager.GPS_PROVIDER);
    boolean network_enabled = mylocatm.isProviderEnabled(LocationManager.NETWORK_PROVIDER);
    return gps_enabled || network_enabled;
}
```

2）获取 WiFi 的开关状态

获取 WiFi 的开关状态是通过调用 WifiManager(WiFi 管理器)类的 isWifiEnabled()方法来判断的。下面定义一个方法来判断 WiFi 的开头状态，代码如下。

```
public static Boolean getWlanStatus(Context c){
    WifiManager mywfm = (WifiManager) c.getSystemService(Context.WIFI_SERVICE);
    boolean wifi_enabled = mywfm.isWifirEnabled();
    return wifi _enabled;
}
```

3）获取数据连接的开关状态

获取数据连接的开关状态是通过调用 ConnectivityManager(连接管理器)类的 getActiveNetworkInfo()方法实现的。如果该方法返回值不为空，同时为连接状态时，数据连接为打开状态。下面定义一个方法来判断数据连接是否开关，代码如下。

```
public static Boolean getMobileDataConnStatus(Context c){
    ConnectivityManager mymdc =
                        (ConnectivityManager) c.getSystemService(Context.CONNECTIVITY_SERVICE);
    boolean isCD_enabled = false;
    NetworkInfo activeNetworkInfo = mymdc.getActiveNetworkInfo();
    isCD_enabled = activeNetworkInfo != null && activeNetworkInfo.isConnected();
    return isCD_enabled;
}
```

3. 相关权限

开发手机定位应用，除了要开启相关的定位外，还需要在清单文件 AndroidManifest.xml 中添加对应的权限。如果只使用 GPS 定位，必须添加以下三个权限声明。

```
<!-- 定位 -->
<uses-permission android:name="android.permission.ACCESS_FINE_LOCATION" />
<uses-permission android:name="android.permission.ACCESS_COARSE_LOCATION" />
<!-- 查看手机状态 -->
<uses-permission android:name="android.permission.READ_PHONE_STATE" />
```

但是在定位应用中，往往不只使用一种定位，通常卫星定位与网络定位同时使用，所以在添加权限声明时会加上访问网络需要的权限。因此大多数情况下，定位应用要添加五个权限声明。

```
<!-- 定位 -->
<uses-permission android:name="android.permission.ACCESS_FINE_LOCATION" />
<uses-permission android:name="android.permission.ACCESS_COARSE_LOCATION" />
<!-- 查看网络状态 -->
<uses-permission android:name="android.permission.ACCESS_NETWORK_STATE" />
<uses-permission android:name="android.permission.ACCESS_WIFI_STATE" />
<!-- 查看手机状态 -->
<uses-permission android:name="android.permission.READ_PHONE_STATE" />
```

其中，定位的两个权限 ACCESS_FINE_LOCATION 和 ACCESS_COARSE_LOCATION，分别表示精确定位权限和粗糙定位权限，在代码中还需要动态申请。

12.4.2 手机定位信息

前面介绍了网络定位的工作原理，它们都是根据基站或 WiFi 路由器的地址来估算手机的位置。手机本身是无法获取位置信息的，必须借助于第三方位置服务提供商，比如 Google 地图、百度地图、高德地图等。所以在这里，手机定位信息主要来自 GPS 定位。通常情况下，手机的 GPS 都是处于开启状态的。此外，要获取手机定位信息还需要一些工具类的支持。

1. 定位条件器 Criteria

Criteria 用于设置定位的前提条件类，位于 android.location.Criteria 包中。该类就像一个过滤器，对指定的定位信息进行条件设置。比如能否获得海拔高度、速度、方向选择，定位信息精准度选择，是否产生资费选择，以及对电量的使用，等等。常用的方法见表 12-10。

表 12-10 Criteria 类中常用方法及说明

方法	说明
setAccuracy(int accuracy)	设置经纬度的精准度。可选参数有 ACCURACY_FINE(准确)、ACCURACY_COARSE(粗略)
setAltitudeRequired(boolean altitudeRequired)	设置是否需要获取海拔数据。取值 true 表示需要、false 为不需要
setBearingAccuracy(int accuracy)	设置方向的精确度。可选参数有 ACCURACY_LOW(低)、ACCURACY_MEDIUM(中)、ACCURACY_HIGH(高)、NO_REQUIREMENT(没有要求)
setBearingRequired(boolean bearingRequired)	设置是否需要获得方向信息。取值 true 表示需要、false 为不需要

续表

方　　法	说　　明
setCostAllowed(boolean costAllowed)	设置是否允许定位过程中产生资费，例如流量等。取值true表示需要、false为不需要
setPowerRequirement(int powerRequirement)	设置耗电量的级别。可选参数有POWER_LOW(低)、POWERMEDIUM(中)、POWER_HIGH(高)、NO_REQUIREMENT(没有要求)
setSpeedAccuracy(int accuracy)	设置速度的精确度。取值有三种：ACCURACY_HIGH，精度高，误差<100米；ACCURACY_MEDIUM，精度中等，100米≤误差≤500米；ACCURACY_LOW，精度低，误差>500米
setSpeedRequired(boolean speedRequired)	设置是否提供速度的要求。取值true表示需要、false为不需要

2. 定位管理器LocationManager

LocationManager用于获取定位信息的提供者、设置监听器，并获取最近一次的位置信息。该类的对象通过Context.getSystemService(Context.LOCATION_SERVICE)方法获得实例。一旦获得LocationManager的实例，就可以实现获取设备的位置、周期性地更新位置等功能。该类常用的方法见表12-11。

表12-11　LocationManager类中常用方法及说明

方　　法	说　　明
getBestProvider(Criteria criteria, boolean enabledOnly)	传入Criteria对象，返回与Criteria对象设置条件最匹配的LocationProvider，作为getLastKnownLocation方法的传入参数
getProvider(String name)	获得指定名称的定位提供者LocationProvider
isProviderEnabled(String provider)	判断指定的Provider是否可用
getLastKnownLocation(String provider)	根据Provider获得位置信息，其默认的Provider是GPS。该方法返回一个封装了经纬度等信息的Location对象
requestLocationUpdates(String provider, long minTimeMs, float minDistanceM, LocationListener listener)	添加一个LocationListener监听器。其中provider为注册的provider名
requestLocationUpdates(String provider, long minTime, float minDistance, PendingIntent intent)	通过给定的Provider名称，周期性地通知当前的Activity。其中minTime和minDistance代表地理位置更新的最小时间间隔及位移变化的最短距离
removeUpdates(LocationListener listener)	移除指定的LocationListener监听器

调用LocationManager类的getLastKnownLocation()方法只是主动地查询地理位置信息，如果需要在地理位置信息发生变化后自动通知系统，可以为LocationManager添加一个LocationListener监听器。

3. 定位监听器 LocationListener

LocationListener 是用于监听定位信息的变化的接口，位于 android. location. LocationListener 包中，如果定位提供者的开关、位置信息发生变化等，该监听器可以监听到。该接口有四个方法，见表 12-12。

表 12-12 LocationListener 接口中的方法及说明

方法	说明
onLocationChanged(Location location)	当设备位置信息发生变化时调用该方法。在此可获取最新的位置信息
onProviderDisabled(String provider)	当设备的定位提供者被禁用时调用该方法。如果注册监听器时设备已经禁用了 Location Provider，则会立即调用该方法
onProviderEnabled(String provider)	当设备的定位提供者被启用时调用该方法
onStatusChanged(String provider, int status, Bundle extras)	当设备的定位提供者状态发生变化时触发该方法，可取的状态为 TEMPORARILY_UNAVAILABLE（暂时不可用）、OUT_OF_SERVICE(在服务范围外)和 AVAILABLE（可用状态）。该方法在 Android Q(API 29)后被弃用

手机是便携设备，手机定位信息随时跟随人的移动而变化。开发手机定位应用，需要建立 Criteria、LocationManager 类对象实例和 LocationListener 监听器，才能时刻跟踪手机的位置信息变化。

下面通过一个案例来介绍如何获取定位信息的编程实现。

【案例 12.7】 开发一个获取手机经纬度、方向、海拔及精度等信息的应用。要求显示定位信息是可控的。

说明：本案例使用一个开/关按钮来控制显示/关闭手机定位信息。为了保证对手机位置的持续监听，设计一个子线程，间歇性地监测定位状态。

开发步骤及解析：过程如下。

1）创建项目

在 Android Studio 中创建一个名为 Activity_Location 的项目。其包名为 ee. example. activity_location。

2）声明权限

本案例需要使用定位功能，所以在清单文件 AndroidManifest. xml 中的< application >元素前要添加至少三个权限声明，代码片段如下。

```
<uses-permission android:name="android.permission.ACCESS_FINE_LOCATION" />
<uses-permission android:name="android.permission.ACCESS_COARSE_LOCATION" />
<uses-permission android:name="android.permission.READ_PHONE_STATE" />
```

3）设计布局

编写 res/layout 目录下的 activity_main. xml 文件。在布局中添加一个 ToggleButton 开关按钮，ID 为 tbtn，用于控制定位开关；添加一个 TextView 控件，ID 为 tv_locationmsg，用于显示定位信息。

4）开发逻辑代码

在 java/ee. example. activity_location 包下编写 MainActivity. java 文件，其代码如下。

```
package ee.example.activity_location;

import androidx.appcompat.app.AppCompatActivity;
import androidx.core.app.ActivityCompat;

import android.Manifest;
import android.app.PendingIntent;
import android.content.Intent;
import android.content.pm.PackageManager;
import android.net.Uri;
import android.os.Bundle;
import android.annotation.SuppressLint;
import android.content.Context;
import android.location.Criteria;
import android.location.Location;
import android.location.LocationListener;
import android.location.LocationManager;
import android.os.Handler;
import android.util.Log;
import android.widget.CompoundButton;
import android.widget.TextView;
import android.widget.ToggleButton;

import java.text.SimpleDateFormat;
import java.util.Date;

public class MainActivity extends AppCompatActivity
                              implements CompoundButton.OnCheckedChangeListener {
    private final static String TAG = "LocationActivity";
    private TextView tv_location;
    private ToggleButton btnSS;
    private String mLocation = "";
    private String myProvider = "";
    private LocationManager mLocationMgr;
    private Criteria mCriteria = new Criteria();
    private Handler mHandler = new Handler();
    private boolean bLocationEnable = false;

    @Override
    protected void onCreate(Bundle savedInstanceState) {
        super.onCreate(savedInstanceState);
        setContentView(R.layout.activity_main);
        initWidget();
    }

    private void initWidget() {
        tv_location = (TextView) findViewById(R.id.tv_locationmsg);
        btnSS = (ToggleButton) findViewById(R.id.tbtn);        //结束定位/开始定位开关按钮
        btnSS.setOnCheckedChangeListener(this);

        mLocationMgr = (LocationManager) getSystemService(Context.LOCATION_SERVICE);
```

```

        mCriteria.setAccuracy(Criteria.ACCURACY_FINE);   //设置定位精确度为比较精细
        mCriteria.setAltitudeRequired(true);             //设置需要海拔信息
        mCriteria.setBearingRequired(true);              //设置需要方位信息
        mCriteria.setCostAllowed(false);                 //设置运营商不付费
        mCriteria.setPowerRequirement(Criteria.POWER_LOW); //设置耗电量为低功耗
    }

    @Override
    public void onCheckedChanged(CompoundButton buttonView, boolean isChecked) {
        if (isChecked) {                                 //显示定位信息
            initLocation() ;
            mHandler.postDelayed(mRefresh, 1000);        //间隔 1 秒
        } else {                                         //停止定位
            if (mLocationMgr != null) {
                mLocationMgr.removeUpdates(mLocationListener);
            }
            tv_location.setText(null);
        }
    }

    private void initLocation() {
        myProvider = mLocationMgr.getBestProvider(mCriteria, true); //获取 GPS 信息
        if (myProvider == null) {
            openGPSSettings();
            myProvider = LocationManager.GPS_PROVIDER;   //获取 GPS 提供者
        }
        if (mLocationMgr.isProviderEnabled(myProvider)) {
            tv_location.setText("正在获取" + myProvider + "定位对象");
            mLocation = String.format("定位类型：% s", myProvider.toUpperCase());
            beginLocation(myProvider);
            bLocationEnable = true;
        } else {
            tv_location.setText("\n" + myProvider + "定位不可用");
            bLocationEnable = false;
        }
    }

    //开启 GPS 定位
    private void openGPSSettings() {
        Intent gpsIntent = new Intent();
        gpsIntent.setClassName("com.android.settings",
                      "com.android.settings.widget.SettingsAppWidgetProvider");
        gpsIntent.addCategory("android.intent.category.ALTERNATIVE");
        gpsIntent.setData(Uri.parse("custom:3"));
        try {
            PendingIntent.getBroadcast(MainActivity.this, 0, gpsIntent, 0).send();
        } catch (PendingIntent.CanceledException e) {
            e.printStackTrace();
        }
    }

    private void beginLocation(String method) {
        //检测定位精度权限,请求授权
```

```
        final int LOCATION_PERMISSION_REQUEST_CODE = 1;
        if (ActivityCompat.checkSelfPermission(this,
                            Manifest.permission.ACCESS_FINE_LOCATION) !=
                                        PackageManager.PERMISSION_GRANTED
                && ActivityCompat.checkSelfPermission(this,
                            Manifest.permission.ACCESS_COARSE_LOCATION) !=
                                        PackageManager.PERMISSION_GRANTED) {
            ActivityCompat.requestPermissions(this,
                    new String[]{Manifest.permission.ACCESS_FINE_LOCATION,
                            Manifest.permission.ACCESS_COARSE_LOCATION},
                    LOCATION_PERMISSION_REQUEST_CODE);
        }

        mLocationMgr.requestLocationUpdates(method, 300, 0, mLocationListener);
        Location location = mLocationMgr.getLastKnownLocation(method);
        setLocationText(location);
    }

    @SuppressLint("DefaultLocale")
    private void setLocationText(Location location) {
        if (location != null) {
            String desc = String.format("%s\n时间:%s\n" +
                            "\n手机位置信息如下:" +
                            "\n\t经度:%f" + "\n\t纬度:%f" +
                            "\n\t高度:%d米" + "\n\t精度:%d米" +
                            "\n\t方向:%f",
                    mLocation,
                    getNowDateTimeFormat(),
                    location.getLongitude(), location.getLatitude(),
                    Math.round(location.getAltitude()), Math.round(location.getAccuracy()),
                    location.getBearing());
            tv_location.setText(desc);
            Log.d(TAG, desc);
        } else {
            tv_location.setText(mLocation + "\n暂未获取到定位对象");
        }
    }

    public static String getNowDateTimeFormat() {
        String format = "yyyy-MM-dd HH:mm:ss";
        SimpleDateFormat s_format = new SimpleDateFormat(format);
        Date d_date = new Date();
        String s_date = "";
        s_date = s_format.format(d_date);
        return s_date;
    }

    //位置监听器
    private LocationListener mLocationListener = new LocationListener() {
        @Override
        public void onLocationChanged(Location location) {
            setLocationText(location);
        }
        @Override
```

```
150         public void onProviderDisabled(String arg0) {
151         }
152         @Override
153         public void onProviderEnabled(String arg0) {
154         }
155         @Override
156         public void onStatusChanged(String arg0, int arg1, Bundle arg2) {
157         }
158     };
159
160     private Runnable mRefresh = new Runnable() {
161         @Override
162         public void run() {
163             if (bLocationEnable == false) {
164                 initLocation();
165                 mHandler.postDelayed(this, 1000);
166             }
167         }
168     };
169
170     @Override
171     protected void onDestroy() {
172         if (mLocationMgr != null) {
173             mLocationMgr.removeUpdates(mLocationListener);
174         }
175         super.onDestroy();
176     }
177 }
```

(1) 第 45～57 行，定义一个初始化控件的方法 initWidget()。在该方法中创建了 TextView 对象用于显示手机定位信息；创建开/关按钮对象并对该按钮单击进行监听；创建一个 LocationManager 对象实例 mLocationMgr。并且对 Criteria 对象进行设置，该对象在第 34 行创建。

(2) 第 60～70 行，实现开/关按钮的单击事件，如果在开状态下，调用 initLocation()方法，初始化定位，并调用 mHandler 的 postDelayed()方法进行间歇刷新定位；如果在关状态下，只要 mLocationMgr 存在，就移除 LocationListener 监听器。

(3) 第 72～87 行，实现初始化定位方法。在该方法中首先取得定位提供者对象 myProvider，如果 GPS 没有开启则调用 openGPSSettings()方法打开 GPS 定位；然后在 myProvider 可用状态下调用 beginLocation(myProvider)方法从该对象中获取定位信息。

(4) 第 90～100 行，实现 openGPSSettings()方法。打开 GPS 定位功能是通过发送 Intent 广播实现的。

(5) 第 102～112 行，实现 beginLocation()方法。在该方法中首先要检测定位精度权限，如果没有权限则需要请求授权。然后注册位置更新监听器，将最近一次的位置信息写入 Location 对象中，最后调用 setLocationText(location)方法将相关信息显示在屏幕上。

(6) 第 115～132 行，实现 setLocationText()方法。在该方法中使用带格式的方式构建输出字符串，并同步将输出字符串发送到运行日志。

(7) 第 144～158 行，实现位置监听器 LocationListener 接口的定义。在监听器内部只重写了 onLocationChanged()方法，当位置信息发生变化时调用 setLocationText(location)方法显示新的位置信息。

(8) 第 160～168 行，实现子线程 Runnable 接口的定义。重写 run()方法，当定位提供者不可用时，重新开启定位。

运行结果：在 Android Studio 支持的模拟器上，运行 Activity_Location 项目。运行初始界面，如图 12-8(a)所示。单击"开始定位"按钮，即可看到当前的定位类型、时间和手机位置信息，如图 12-8(b)所示。单击"停止定位"按钮，则可立即停止显示定位信息，返回到初始运行界面。

(a) 初始运行界面

(b) 显示手机当前的定位等信息

图 12-8　在模拟器中运行显示当前定位信息的运行效果

小结

本章主要介绍了常用的有关手机特性的 Android 应用的开发。对于手机系统的设置变化、手机电信网络信息、手机电池电量的监测，短信发送接收、电话监听控制，以及手机传感器的应用，手机 GPS 定位应用等，给出了案例，进行了详细地介绍。在开发手机特性的应用中，有较多的应用需要在清单文件中添加权限声明，有些权限还需要在代码中动态请求授权，请大家要留意。另外，手机传感器应用涉及手机硬件方面的支持，在开发中，必须要了解相应的硬件设备的配置情况。

在手机定位的提供者方面，GPS 定位属于卫星定位。在实现应用时，卫星定位与网络定位往往要联合使用，关于网络定位将涉及网络通信方面的编程技术，将在第 13 章中系统学习。

练习

1. 设计"我的音乐盒"的一个歌曲播放列表，在每个列表栏项添加一个播放按钮。要求：当单击播放按钮时检测手机的电量，如果电量不足 25%，给出振动提示，并弹出一个对话框，提醒用户：手机电量不足，是否继续执行播放操作。

2. 设计一个计步器，能够统计用户每天的行走步数。

第 13 章 网络通信技术

Google 公司以其强大的互联网搜索引擎业务，在互联网领域曾经独占鳌头。Android 是由 Google 公司推出的一款手机操作系统，其网络功能自然会非常强大。随着无线互联网的迅猛发展，人们可以不受时间、空间的限制，随时随地地进行数据交换、网页浏览、事务处理，例如手机聊天、手机购物、手机银行、手机炒股、移动办公、共享单车等与互联网相关的应用。

本章主要介绍在 Android 平台下进行网络通信技术的常见应用编程知识，并涉及 Web 应用服务器 Tomcat 端的简单应用。内容包括浏览网页、网络通信、获取网络中数据、用户登录、上传文件等应用。

13.1 网络访问权限

任何 Android 应用需要访问到网络，都要求被赋予访问网络的权限。如果当前的网络连接不可用，则无须执行 App 的相关操作，直接提示用户连通网络，然后才能开始网络的应用。所以，在开发网络应用时，首先要在 AndroidManifest.xml 中添加网络访问权限，有些应用还需要添加更多的权限，如网络状态测试权限、WiFi 状态测试权限等。

在 AndroidManifest.xml 文件中声明权限，一般放在< application >节点之前。声明网络访问权限代码如下。

```
< uses - permission android:name = "android.permission.INTERNET" />
```

记住！应用项目只有在 AndroidManifest.xml 文件中声明了访问网络的权限，才能连接网络。

随着 WiFi 和 5G 应用的推广，越来越多的 Android 应用项目需要调用网络资源，检测网络连接状态也就成为网络应用项目所必备的功能。在 Android 应用中，常见的与网络操作相关的权限见表 13-1。

表 13-1 与网络相关的常用权限及说明

权 限 名	说 明
android.permission.INTERNET	允许程序打开网络套接字
android.permission.ACCESS_NETWORK_STATE	允许程序访问有关 GSM 网络信息
android.permission.ACCESS_COARSE_LOCATION	允许一个程序访问 CellID 或 WiFi 热点来获取粗略的位置

续表

权　限　名	说　　明
android. permission. ACCESS_WIFI_STATE	允许程序访问 WiFi 网络状态信息
android. permission. BLUETOOTH	允许程序连接到已配对的蓝牙设备
android. permission. BLUETOOTH_ADMIN	允许程序发现和配对蓝牙设备
android. permission. CHANGE_NETWORK_STATE	允许程序改变网络连接状态
android. permission. CHANGE_WIFI_STATE	允许程序改变 WiFi 连接状态

关于网络连接信息可以由 ConnectivityManager 对象提供。ConnectivityManager 是专门负责网络连接检测及管理的类，可以获取详细的网络连接信息，被称为网络连接管理器。该类位于 android. net. ConnectivityManager 包中。该类对象从系统服务 Context. CONNECTIVITY_SERVICE 中获得。本书在此不做详细介绍，需要用到相关技术时可以查阅 Android API 开发手册（https://developer. android. google.cn/reference/android/net/ConnectivityManager?hl=en）。

13.2　浏览网页

在 Android 中进行浏览网页的开发非常简单，可使用两种方式，一是通过用隐式 Intent 来启动系统默认的浏览器，二是使用 WebView 控件。只要传入一个网页的链接地址 URI 即可浏览网页。

注意：*在 android 9（API 28）以后，Android 默认禁止使用 HTTP 协议，必须使用 HTTPS 协议，否则会报错误。如果要使用 HTTP 协议，那么在 AndroidManifest. xml 的 <application>中必须添加属性：android:usesCleartextTraffic="true"。*

13.2.1　通过 Intent 启动浏览器

Intent 是 Android 平台上的一个重要的组件，它可以实现页面与页面之间的通信。从应用项目的页面到网站的页面，也是页面与页面之间的关系。只要传入一个网页的链接地址 URI，通过发送隐式的 Intent，就可以调用 Android 系统自带的浏览器打开该指定的网页。具体实现只需要三行代码。例如从应用项目中要打开百度搜索首页，代码片段如下。

```
Uri uri = Uri.parse("https://www.baidu.com");
Intent intent = new Intent(Intent.ACTION_VIEW, uri);
startActivity(intent);
```

如果移动设备上有多个浏览器，而且应用项目有对特定浏览器的要求，可以使用 Intent 的打开外部应用方法 setClassName(Context packageContext, String className)，来显式地启动指定浏览器打开网页。例如使用 UC 浏览器打开百度网页，代码如下。

```
Uri uri = Uri.parse("https://www.baidu.com");
Intent intent = new Intent(Intent.ACTION_VIEW,uri);
intent.setClassName("com.UCMobile","com.uc.browser.InnerUCMobile");
startActivity(intent);
```

使用 QQ 浏览器打开百度网页，代码如下。

```
Uri uri = Uri.parse("https://www.baidu.com");
Intent intent = new Intent(Intent.ACTION_VIEW,uri);
intent.setClassName("com.tencent.mtt","com.tencent.mtt.MainActivity");
startActivity(intent);
```

一般情况推荐使用隐式 Intent 方式。因为使用显式 Intent 指定浏览器，有时候会出现错误，原因是指定的浏览器可能存在版本升级导致包名改变，或其他兼容问题，导致网页无法正常打开。

下面简单地用一个案例介绍其用法。

【案例 13.1】 使用 Intent 组件浏览指定网址的网页。

说明： 设计一个文本输入框，由用户输入网址，然后通过隐式 Intent 启动系统默认的浏览器打开指定的网页。在本案例中将模拟用户的操作习惯，提供输入后按 Enter 键即打开网页和输入后单击按钮打开网页两种方式。

开发步骤及解析： 过程如下。

1）创建项目

在 Android Studio 中创建一个名为 BrowserInternet 的项目。其包名为 ee.example.browserinternet。

2）声明访问网络权限

在 AndroidManifest.xml 中的< application >元素前添加权限声明。该权限声明代码如下。

```
<uses-permission android:name="android.permission.INTERNET" />
```

3）设计布局

编写 res/layout 目录下的 activity_main.xml 文件。在布局中添加一个 EditText，ID 为 url_field，用于输入网址，添加一个按钮，ID 为 go_button，单击它进入浏览器。

4）开发逻辑代码

在 java/ee.example.browserinternet 包下编写 MainActivity.java 程序文件，代码如下。

```
package ee.example.browserinternet;

import androidx.appcompat.app.AppCompatActivity;
import android.os.Bundle;
import android.content.Intent;
import android.net.Uri;
import android.view.KeyEvent;
import android.view.View;
import android.view.View.OnClickListener;
import android.view.View.OnKeyListener;
import android.widget.Button;
import android.widget.EditText;

public class MainActivity extends AppCompatActivity {

```

```
16      private EditText urlText;
17      private Button goButton;
18
19      @Override
20      protected void onCreate(Bundle savedInstanceState) {
21          super.onCreate(savedInstanceState);
22          setContentView(R.layout.activity_main);
23
24          urlText = (EditText) findViewById(R.id.url_field);
25          goButton = (Button) findViewById(R.id.go_button);
26
27          goButton.setOnClickListener(new OnClickListener() {
28              public void onClick(View view) {
29                  openBrowser();
30              }
31          });
32
33          urlText.setOnKeyListener(new OnKeyListener() {
34              public boolean onKey(View view, int keyCode, KeyEvent event) {
35                  if (keyCode == KeyEvent.KEYCODE_ENTER) {
36                      openBrowser();
37                      return true;
38                  }
39                  return false;
40              }
41          });
42      }
43
44      //打开 EditView 中输入网址的网页
45      private void openBrowser() {
46          Uri uri = Uri.parse(urlText.getText().toString());
47          Intent intent = new Intent(Intent.ACTION_VIEW,uri);
48          startActivity(intent);
49      }
50  }
```

(1) 第 27～31 行，为按钮添加一个 OnClickListener()监听，第 33～41 行，为 EditView 的控件 urlText 添加一个 OnKeyListener()监听，在两个监听的回调方法中都调用了 openBrowser()方法，在该方法中编写打开网页的代码。

(2) 第 35 行，if (keyCode == KeyEvent.KEYCODE_ENTER)判断按下的键是否为 Enter 键，只有在按下 Enter 键时才能调用 openBrowser()方法。

运行结果：在 Android Studio 支持的模拟器上，运行 BrowserInternet 项目。在项目初始界面中输入百度网站的网址：https://www.baidu.com，如图 13-1(a)所示。单击 GO 按钮，启动系统的自带浏览器 Chrome。注意，首次启动 Chrome 浏览器，会出现浏览器欢迎页面，如图 13-1(b)所示。单击 Accept & continue 进入 Chrome 浏览器，会提示登入 Chrome 账户，如图 13-1(c)所示。选择左下角的 No Thanks，进入网页，浏览器会自动检测到所用设备是手机，会推送手机版网页对话框，如图 13-1(d)所示。在此可选择 Allow(允许)，接下来便进入网页浏览状态，如图 13-1(e)所示。再次输入网址即可立即进入网页浏览状态。

(a) 初始界面

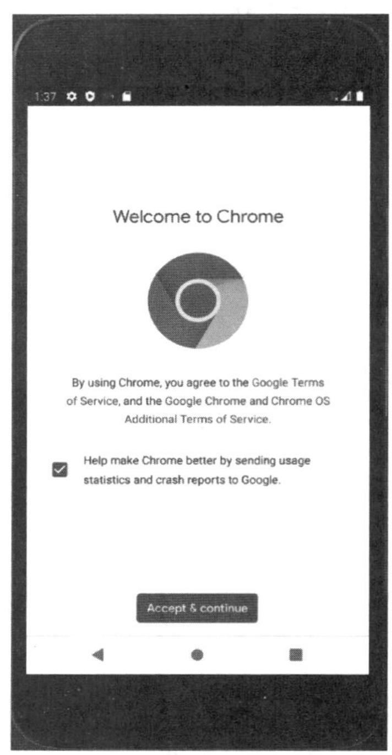

(b) Chrome浏览器欢迎页面

(c) 首次使用Chrome提示注册

(d) 提示转入手机版浏览器

(e) 进入页面界面

图 13-1　通过 Intent 使用系统自带浏览器打开网页

从运行结果看，此时已经完全脱离了原来的应用，就好像直接从手机上打开浏览器浏览网页一样。

13.2.2 使用 WebView 控件浏览网页

WebView 在 Android 平台上专门用来显示网页的 View，这个 WebView 类位于 android. webkit. WebView 包中，其内部实现是采用渲染引擎(WebKit)来展示 View 的内容，可以提供网页前进后退、网页放大、缩小、搜索等功能。在 Android 4.4(API 19)引入了一个基于 Chromium 的新版本，使得 WebView 能够支持 HTML5、CSS3 以及 JavaScript。

1. WebView 单独使用

WebView 可以单独使用，其用法与 ImageView 组件的用法基本相似，实现将网页显示在界面中。在应用中常用的方法见表 13-2。

表 13-2 WebView 常用的方法及说明

方 法	说 明
loadUrl (url: String)	打开 url 指定的网页
getSettings()	获得一个 WebSettings 子类对象
canGoBack()	设置网页是否可以后退
goBack()	后退网页，向后显示浏览过的页面
canGoForward()	是否可以前进
goForward()	前进网页，向前显示浏览过的页面
goBackOrForward(int steps)	以当前页为起始点前进或者后退到历史记录中指定的 steps 页面。其中，steps 为负数则为后退、为正数则为前进
clearCache(true)	清除网页访问留下的缓存。该方法不仅仅针对 WebView 对象，并且针对整个应用项目
clearHistory()	清除当前 WebView 访问的历史记录，当前访问记录除外
clearFormData()	清除自动完成填充的表单数据，但不会清除 WebView 存储到本地的数据
loadDataWithBaseURL((String baseUrl, String data, String mimeType, String encoding, String historyUrl)	从 WebView 的 baseUrl 中加载数据。在加载时，它会把 baseUrl 和 historyUrl 传到 List 列表中，当作历史记录来使用。当前进或后退时，它会通过 baseUrl 来寻找 historyUrl 的路径从而加载历史页面
destroy()	销毁 WebView。在关闭 Activity 时，如果 WebView 中正在播放音乐或视频，必须先从父容器中移除 WebView，然后再销毁 WebView

例如，在 Activity 中销毁 WebView 对象的代码片段如下。

```
@Override
protected void onDestroy() {
    if (webView != null) {
        webView.loadDataWithBaseURL(null, "", "text/html", "utf - 8", null);
        webView.clearHistory();
        ((ViewGroup)webView.getParent()).removeView(mWebView);
        webView.destroy();
        webView = null;
```

```
        }
        super.onDestroy();
    }
```

要销毁 WebView，首先要让 WebView 加载 null 内容，然后清除 WebView 访问的历史记录，从其父对象中移除 WebView，再销毁 WebView，最后把 WebView 设置为 null。

2. WebView 与子类联合使用

如果为了更好地管理网页浏览，获取浏览过程中的一些状态信息，往往需要联合使用其子类，设置与 JavaScript 交互等。常用的子类主要有三个，分别是 WebSettings 类、WebViewClient 类和 WebChromeClient 类。

1）WebSettings 类

WebSettings 类对 WebView 进行配置和管理。常用的设置方法见表 13-3。

表 13-3　WebSettings 类常用的方法及说明

方　法	说　明
setJavaScriptEnabled(boolean flag)	设置 WebView 是否支持 Javascript
setUseWideViewPort(boolean use)	设置是否将网页调整到适合 WebView 的大小
setLoadWithOverviewMode(boolean overview)	设置是否将网页缩放至屏幕的大小
setSupportZoom(boolean support)	设置是否支持缩放
setDisplayZoomControls(boolean enabled)	设置是否支持原生的缩放控件
setJavaScriptCanOpenWindowsAutomatically (boolean flag)	设置是否支持通过 JS 打开新窗口
setLoadsImagesAutomatically(boolean flag)	设置是否支持自动加载图片
setDefaultTextEncodingName(String encoding)	设置编码格式，例如 utf-8
setCacheMode(int mode)	是否关闭 WebView 中缓存。缓存模式常量如下。 LOAD_CACHE_ONLY：不使用网络，只读取本地缓存数据。 LOAD_DEFAULT：（默认）根据 cache-control 决定是否从网络上获取数据。 LOAD_NO_CACHE：不使用缓存，只从网络获取数据。 LOAD_CACHE_ELSE_NETWORK：只要本地有，无论是否过期，或者 no-cache，都使用缓存中的数据。 例如不使用缓存，方法如下。 setCacheMode(WebSettings.LOAD_NO_CACHE)

2）WebViewClient 类

WebViewClient 类用于处理各种通知和请求事件。常用的设置方法见表 13-4。

表 13-4　WebViewClient 类常用的方法及说明

方　法	说　明
shouldOverrideUrlLoading()	打开网页时，不调用系统浏览器打开，而是在本 WebView 中直接显示
onPageStarted()	开始载入页面时调用此方法，在这里可以设定一个 loading 的页面，告诉用户程序正在等待网络响应

续表

方法	说明
onPageFinished()	在页面加载结束时调用。可以关闭 loading 条,切换程序动作
onLoadResource()	在加载页面资源时会调用,每一个资源(比如图片)的加载都会调用一次

例如,打开网页 https://www.baidu.com 不调用系统浏览器,而是在本 WebView 中显示,代码片段如下。

```
webView.loadUrl("https://www.baidu.com/");
webView.setWebViewClient(new WebViewClient(){
    @Override
    public boolean shouldOverrideUrlLoading(WebView view, String url) {
        view.loadUrl(url);
        return true;
    }
});
```

例如,在网页加载前显示“开始加载”,在网页加载完成后显示“结束加载”,代码片段如下。

```
mWebview.setWebViewClient(new WebViewClient() {
    //设置加载前的函数
    @Override
    public void onPageStarted(WebView view, String url, Bitmap favicon) {
        beginLoading.setText("开始加载了");
    }
    //设置结束加载函数
    @Override
    public void onPageFinished(WebView view, String url) {
        endLoading.setText("结束加载了");
    }
});
```

3) WebChromeClient 类

WebChromeClient 类辅助 WebView 处理 Javascript 的对话框,如网站图标、网站标题等。

例如,获得网页的加载进度并显示,代码片段如下。

```
webview.setWebChromeClient(new WebChromeClient(){
    @Override
    public void onProgressChanged(WebView view, int newProgress) {
        if (newProgress < 100) {
            String progress = newProgress + "%";
            progress.setText(progress);
        } else {
            progress.setText("100%");
        }
    }
});
```

每个网页的页面都有一个标题,比如 www.baidu.com 这个页面的标题即“百度一下,你就知道”。例如,获取 Web 页中的标题,代码片段如下。

```
webview.setWebChromeClient(new WebChromeClient(){
```

```
        @Override
        public void onReceivedTitle(WebView view, String title) {
            titleview.setText(title);
        }
    });
```

下面使用一个简单案例来说明 WebView 的用法。

【案例 13.2】 使用 WebView 控件开发一个浏览网页应用。

说明： 设计一个文本输入框，由用户输入网址，两个文本框显示加载网页标题和加载进度，然后使用 WebView 打开指定的网页。在本案例中将屏蔽系统自带的浏览器。

开发步骤及解析： 过程如下。

1）创建项目

在 Android Studio 中创建一个名为 BrowserView 的项目。其包名为 ee.example.browserview。

2）声明访问网络权限

在 AndroidManifest.xml 中的< application >元素前添加权限声明。

```
<uses-permission android:name="android.permission.INTERNET" />
```

另外，从 Android 9(API 级别 28)开始，默认情况下限制了明文流量的网络请求。所以在清单文件的< application >元素中，要添加一个开关属性，用于允许明文流量的网络请求，使得未加密流量请求允许被使用。例如 HTTP 的 URL 能被 WebView 加载。设置允许明文流量网络请求的属性，代码片段如下。

```
<application
    ...
    android:usesCleartextTraffic="true"
    ...>
</application>
```

3）准备颜色资源

编写 res/values 目录下的 colors.xml 文件，添加 greenbg 等颜色。

4）设计布局

编写 res/layout 目录下的 activity_main.xml 文件。在布局中将屏幕分为三段区域，第一段为网址输入区，位于屏幕顶端，其内添加一个 EditText 控件，ID 为 url_field，用于输入网址，添加一个按钮，ID 为 btngo，单击它进入浏览器。第二段位于网址输入区下面，添加两个 TextView 控件，ID 为 title 的显示网站的标题，ID 为 Loadingprg 的显示网页加载的进度。第三段取屏幕的其余空间，使用 WebView 控件显示网页。声明 WebView 控件的代码片段如下。

```
<WebView
    android:id="@+id/mywebview"
    android:layout_below="@id/Loadingprg"
    android:layout_width="match_parent"
    android:layout_height="match_parent"
    android:layout_marginTop="6dp"/>
```

5）开发逻辑代码

在 java/ee. example. browserview 包下编写 MainActivity. java 程序文件，代码如下。

```
package ee.example.browserview;

import androidx.appcompat.app.AppCompatActivity;
import android.os.Bundle;
import android.view.KeyEvent;
import android.view.View;
import android.view.View.OnClickListener;
import android.view.View.OnKeyListener;
import android.view.ViewGroup;
import android.webkit.WebSettings;
import android.webkit.WebView;
import android.widget.Button;
import android.widget.EditText;
import android.widget.TextView;
import android.webkit.WebChromeClient;
import android.webkit.WebViewClient;

public class MainActivity extends AppCompatActivity {
    private EditText urlText;
    private Button goButton;
    private WebView mywebview;
    private TextView mytitle, loadingprg;
    private WebSettings mWebSettings;

    @Override
    protected void onCreate(Bundle savedInstanceState) {
        super.onCreate(savedInstanceState);
        setContentView(R.layout.activity_main);
        urlText = (EditText) findViewById(R.id.url_field);

        goButton = (Button) findViewById(R.id.btngo);
        mywebview = (WebView) findViewById(R.id.mywebview);
        mytitle = findViewById(R.id.title);
        loadingprg = findViewById(R.id.Loadingprg);

        mWebSettings = mywebview.getSettings();

        urlText.setOnKeyListener(new OnKeyListener() {
        public boolean onKey(View view, int keyCode, KeyEvent event) {
                if (keyCode == KeyEvent.KEYCODE_ENTER) {
                    openBrowser();
                    return true;
                }
                return false;
            }
        });
        goButton.setOnClickListener(new OnClickListener() {
            public void onClick(View view) {
                openBrowser();
            }
        });
```

```
        //设置不用系统浏览器打开,直接显示在当前 WebView
        mywebview.setWebViewClient(new WebViewClient() {
            @Override
            public boolean shouldOverrideUrlLoading(WebView view, String url) {
                view.loadUrl(url);
                return true;
            }
        });

        //设置 WebChromeClient 类
        mywebview.setWebChromeClient(new WebChromeClient() {
            //获取网站标题
            @Override
            public void onReceivedTitle (WebView view, String title){
                System.out.println("标题在这里");
                mytitle.setText(title);
            }

            //获取加载进度
            @Override
            public void onProgressChanged(WebView view, int newProgress) {
                if (newProgress < 100) {
                    String progress = newProgress + "%";
                    loadingprg.setText("下载进度: " + progress);
                } else if (newProgress == 100) {
                    String progress = newProgress + "%";
                    loadingprg.setText("下载进度: " + progress);
                }
            }
        });

    }

    //打开 EditView 中输入网址的网页
    private void openBrowser() {
        mWebSettings.setJavaScriptEnabled(true);
        mywebview.loadUrl(urlText.getText().toString());
    }

    //单击返回上一页面而不是退出浏览器
    @Override
    public boolean onKeyDown(int keyCode, KeyEvent event) {
        if (keyCode == KeyEvent.KEYCODE_BACK && mywebview.canGoBack()) {
            mywebview.goBack();
            return true;
        }
        return super.onKeyDown(keyCode, event);
    }

    //销毁 WebView
    @Override
    protected void onDestroy() {
        if (mywebview != null) {
```

```
106                 mywebview.loadDataWithBaseURL(null, "", "text/html", "utf-8", null);
107                 mywebview.clearHistory();
108                 ((ViewGroup) mywebview.getParent()).removeView(mywebview);
109                 mywebview.destroy();
110                 mywebview = null;
111             }
112             super.onDestroy();
113         }
114 
115 }
```

(1) 第 36 行，声明一个 WebView 的 WebSettings 对象。

(2) 第 54～60 行，重写 WebviewClient 类中的 shouldOverriderUrlLoading()方法，设置屏蔽系统的自带浏览器，用 WebView 对象 view 打开网页。

(3) 第 63～82 行，设置 WebChromeClient 类，获取网页标题，并用 mytitle 文本框显示标题内容；获取网页加载的进度数 newProgress，并传入 loadingprg 文本框中显示。

(4) 第 88 行，设置 WebView 支持与 Javascript 互动。

(5) 第 104～113 行，设置销毁 WebView 对象。其中第 108 行从其父对象中移除 WebView 对象。

运行结果：在 Android Studio 支持的模拟器上，运行 BrowserView 项目。在项目初始界面中输入 360 手机网站的网址：https://m.360.cn，如图 13-2(a)所示。单击“浏览”按钮，开始加载，加载完成后进入 360 网站的首页，如图 13-2(b)所示。单击“360 商城”，进入“360 商城”页面，如图 13-2(c)所示。单击“360 热卖榜单”，进入“360 热卖榜单”页面，如图 13-2(d)所示。接下来单击左下角的返回按钮(见标记处)，便可以逐页后退页面，回到 360 首页；如果再单击返回按钮，则退出项目。

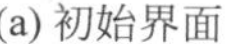
(a) 初始界面

(b) 加载360网页完毕

图 13-2　使用 WebView 打开网页

(c) 进入360商城页面

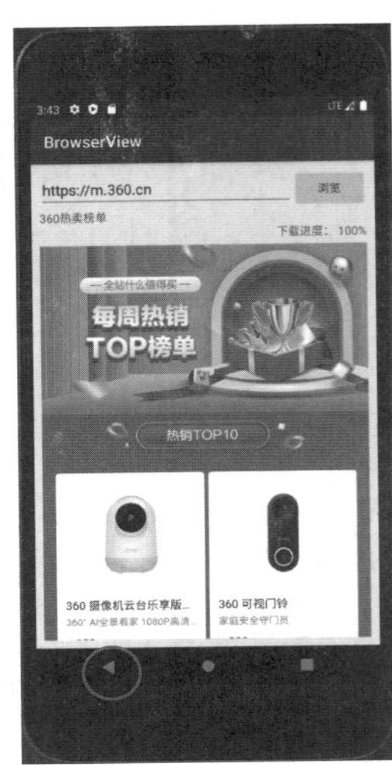

(d) 进入360热卖榜单页面

图 13-2 （续）

注意：使用 WebView 加载网页，有时会遇到加载失败的情况，会出现 net::ERR_UNKNOWN_URL_SCHEME 这样的错误。导致失败的原因是多方面的，有可能是网络连接问题，或是网络连接超时，或是网址不正确等。此时可以重写 WebViewClient 里面的 onReceivedError()方法，给用户良好的交互提示。

13.3 基于 HTTP 协议的接口通信

Android 网络通信，即 Android 应用客户端与服务器端之间的通信，在 Android 应用中占据重要地位。实现网络通信，从宏观上可分为调用原生类和使用第三方框架两种方式。而 Android 原生类网络通信，也分为两种方式，一是基于 HTTP 协议的接口调用方式；二是基于 TCP 协议的 Socket 网络请求方式。本节主要介绍基于 HTTP 协议的接口调用网络通信。

13.3.1 HTTP 协议

HTTP(HyperText Transfer Protocol，超文本传输协议)是互联网上应用最为广泛的一种网络协议。HTTP 定义了浏览器与服务器之间的文件交换通道协议，是一种“请求-响应”的方式，即在客户端向服务器端发送请求时建立连接通道，然后服务器端才能向客户端返回数据。HTTP 是互联网可靠的交换文件的基础，例如交换文本、图像、声音以及视频等。

HTTP 协议永远是客户端发起、服务器端响应，而且 HTTP 在每次请求结束后都会主动释放连接，所以 HTTP 是一种“短连接”“无状态”的连接模式。如果要保持客户端程序的在线状态，需要不断地向服务器端发起连接请求，服务器端在收到该请求后对客户端进行回复，表明知道客户端“在线”；若服务器端长时间无法收到客户端的请求，则认为客户端“下线”。反之，若客户端长时间没有收到服务器端的回复，则认为网络已经断开。

1. URL

URL称为统一资源定位符（Uniform Resource Locator），是互联网的标准资源地址，包含了用于查找某个资源的足够信息。URL格式主要分为三个部分。

（1）协议名，如http、https。

（2）存有该资源的IP地址，有时候也包括端口号。如IP地址加端口号192.168.1.112:8080，或使用域名网址www.baidu.com。

（3）主机资源的具体地址。如目录名、文件名等，如：index.jsp或aaa.jpg。

在URL中，第(1)、(2)部分之间用“://”隔开，不可缺少；IP地址与端口号之间用“:”隔开；在第(2)、(3)部分之间用“/”分开。第(3)部分可以缺省。

2. 请求方法

客户端向服务器端发送的请求包括请求方法和路径。常用的请求方法见表13-5。

表13-5 HTTP常用的请求方法及说明

请求方法	说明
GET	向服务器传送数据，请求获取Request-URI所标识的资源
POST	向服务器传送数据，在Request-URI所标识的资源后附加新的数据
HEAD	请求获取由Request-URI所标识的资源的响应消息报头
PUT	请求服务器存储一个资源，并用Request-URI作为其标识
TRACE	请求服务器回送收到的请求信息，主要用于测试或诊断
OPTIONS	请求查询服务器的性能，或者查询与资源相关的选项和需求

其中，GET和POST两种请求的使用最为频繁。并且从效果上看，都可以向服务器传送数据并请求获取Request-URI所标识的资源。

释疑：

GET和POST请求有什么区别呢？

通过表13-6对比了解GET和POST的区别吧。

表13-6 GET和POST的区别

POST	GET
数据不在URL里传递，而放在HTML HEADER内向服务器传送	数据以参数队列形式加在URL里传递。URL与参数之间以“?”隔开，参数各字段之间以“&”号隔开，参数按name=value的形式。例如：http://www.baidu.com/s?w=6&T=20
传送的数据量较大，一般被默认为不受限制	传送的数据量较小，不能大于2KB
可以提交表单信息	只用于提交简单的数据信息
数据对用户不可见	数据对用户可见
安全性较高	安全性非常低

3. 传输数据类型

HTTP允许传输任意类型的数据对象。在HTTP协议的消息头中，使用Content-Type来表示具体请求中传输数据的媒体类型信息。常见的互联网媒体类型如下。

(1) text/html：HTML格式。

(2) text/plain：纯文本格式。

(3) text/xml：XML格式。

(4) image/gif：GIF图片格式。

(5) image/jpeg：JPG图片格式。

(6) image/png：PNG图片格式。

(7) application/xhtml+xml：XHTML格式。

(8) application/xml：XML数据格式。

(9) application/json：JSON数据格式。

(10) application/pdf：PDF格式。

(11) application/msword：Word文档格式。

(12) application/octet-stream：二进制流数据(如常见的文件下载)。

(13) multipart/form-data：当在表单中进行文件上传时，使用该格式。

13.3.2 HTTP访问网络

1. HTTP网络请求方式

在Android中发送HTTP网络请求有多种方式，常用的有四种，分别为HttpURLConnection、HttpClient、AndroidHttpClient和OKHTTP。

HttpURLConnection继承自URLConnection，是java.net.*提供的与网络操作相关的标准Java接口，可用于指定URL并发送GET请求、POST请求。

HttpClient是Apache提供的HTTP网络访问接口，也可以完成HTTP的GET请求和POST请求。但在Android 6.0版本后摒弃了该类库，如果仍然想要使用该接口，需要在build.gradle文件中进行配置。

AndroidHttpClient是Android.net.*提供的网络接口，继承自HttpClient，常常进行Android特有的网络编程，使用较少。

OkHttp由移动支付公司Square推出，是一款开源的处理HTTP请求的轻量级框架，提供了一套相对成熟的支持包。OkHttp提供了替代HttpURLConnection和Apache HttpClient的解决方案。在Android端，近来使用OkHttp的人气迅速攀升。

本书主要介绍HttpURLConnection接口的编程，并简单地介绍OkHttp的用法。

开发网络访问的多数应用是一个客户端-服务器端的请求-响应应用，开发的Android应用是客户端的应用程序。当它需要访问网络时，会发送访问请求给网络服务器，在网络服务器端有一个应用程序，它负责接收并处理由客户端传来的请求，这个应用程序被称为Web应用管理程序。最后由Web应用管理程序将处理结果传回给客户端，这个过程被称为响应。所以网络应用是一个客户端-服务器端的请求-响应通信应用。

管理这个服务器端程序的系统被称为 Web 服务器。Web 服务器实际上是存储在网络中某台计算机上的被动程序，只有当用户通过客户端程序向服务器发送 HTTP 请求，服务器程序才开始解析请求，处理请求中相应操作。在应用中通常使用 Tomcat 服务器作为 Web 服务器。

2. Tomcat 服务器

Tomcat 是一种最常用的开源 Web 服务器，它原本是由 Apache 组织提供的一个 Servlet 容器，它实现了对 Servlet 和 JSP 的支持，并提供了作为 Web 服务器的一些特有功能，Tomcat 本身内置了一个 HTTP 服务器，是为 Java EE 的 Web 应用提供运行的轻量级容器。所以它被视作一个独立的 Web 服务器。在中小型系统和并发量小的场合下普遍使用 Tomcat。

1）Tomcat 的重要目录

在 Tomcat 的目录（在 Windows 中称为文件夹）下，有若干子目录，如下所示。

（1）bin：Tomcat 脚本文件存放目录，如启动、关闭等脚本文件。其中 *.sh 文件用于 Unix 系统；*.bat 文件用于 Windows 系统。

（2）conf：Tomcat 配置文件目录。

（3）lib：Tomcat 的 JAR 包文件目录。

（4）logs：Tomcat 默认日志目录。

（5）webapps：webapp 运行的目录。Web 端应用管理项目在此目录中发布，该目录被称为项目发布目录，其内的每个 WAR 包就是 webapp 的压缩包。发布目录下的主要子目录及其文件归档如下。

① META-INF：META-INF 目录用于存放项目自身相关的一些信息，包括配置清单文件。通常由开发工具和开发环境自动生成。

② WEB-INF：Java Web 应用的安全目录。所谓安全就是客户端无法访问，只有服务器端可以访问的目录。

③ /WEB-INF/classes：存放应用项目所需要的所有 Java Class 文件。

④ /WEB-INF/lib：存放应用项目所需要的所有 JAR 文件。

⑤ /WEB-INF/web.xml：Web 应用的部署配置文件。它是应用项目中最重要的配置文件，它描述了 servlet 和组成应用的其他组件，以及应用初始化参数、安全管理约束等。

2）Tomcat 的下载安装

Tomcat 服务器的安装配置十分方便。注意，在选择下载 Tomcat 服务器的版本时一定要与计算机上的 JDK 版本相匹配。Tomcat 下载安装配置步骤如下。

步骤 1：确定 JDK 版本号。

因为 Tomcat 的版本要与 JDK 的版本相匹配，所以在下载 Tomcat 之前，首先要检查一下本机的 JDK 版本号，方法是在 Windows 命令行窗口输入 java-version，按 Enter 键后如图 13-3 所示。

步骤 2：下载 Tomcat。

Apache Tomcat 的官网 http://tomcat.apache.org/上有多个 Tomcat 版本，找到需要

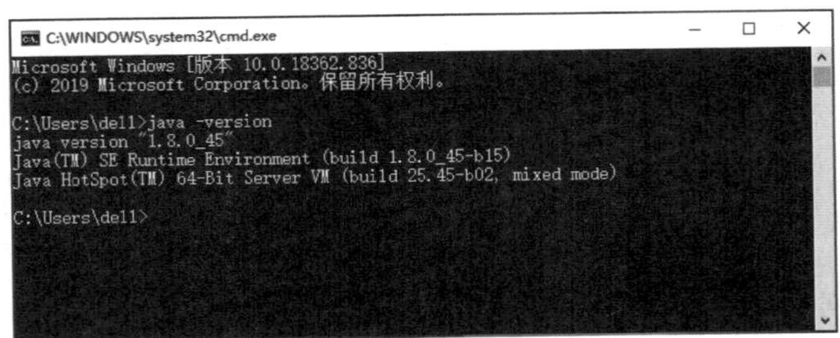

图 13-3　查看 JDK 版本信息

的 Tomcat 包下载。本书所用计算机安装的是 JDK 1.8.0_45，那么应该选择 Tomcat 8，下载 zip (pgp, sha512)，如图 13-4 所示。

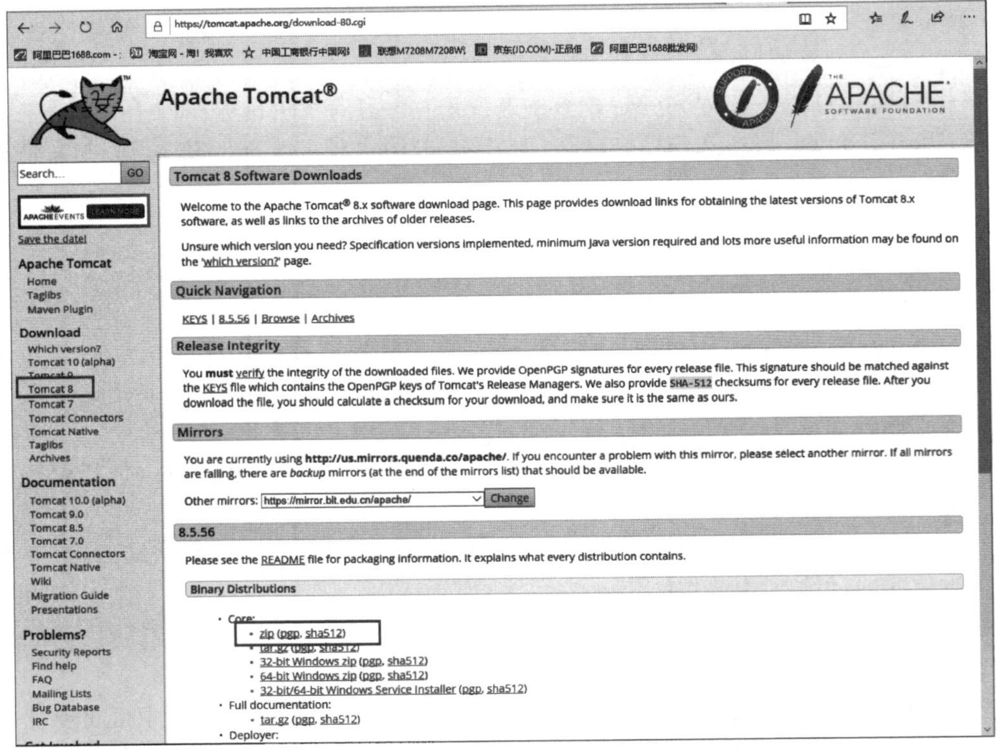

图 13-4　Tomcat 官网的下载页面

步骤 3：安装配置。

直接将下载包解压到计算机上的规划位置。原则上对安装解压位置没有约束限制，可以选择任何盘符任何文件夹路径。接下来是配置 Tomcat 的环境变量，操作方法与配置 JDK 环境变量相似。本书所用计算机将 Tomcat 解压在 D 盘中，并在 Windows 系统的环境变量中添加用户变量 CATALINA_HOME，指向 D:\apache-tomcat-8.5.55，如图 13-5 所示。

步骤 4：测试服务器。

配置完成后需要进行测试。测试方法分为两步，先开启服务器，然后在浏览器上测试。

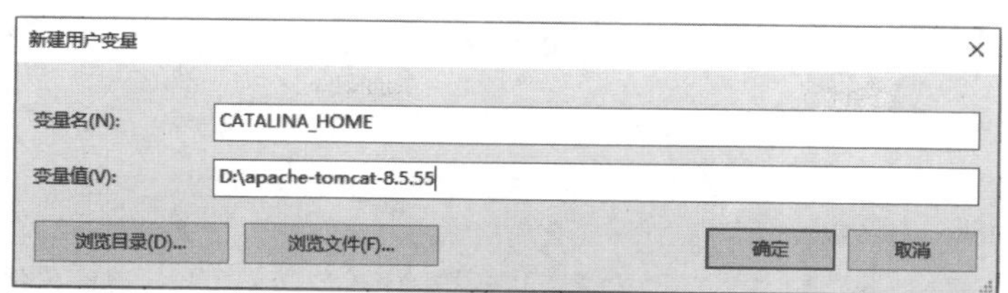

图 13-5 配置环境变量

开启服务器操作是在 D:\apache-tomcat-8.5.55\bin 文件夹下，打开 startup.bat 批处理文件，在命令行窗口显示运行后的信息，如图 13-6 所示。如果最后一行信息是 org.apache.catalina.startup.Catalina.start Server startup in 1095 ms，说明 Tomcat 服务器在 1095 毫秒内启动成功。

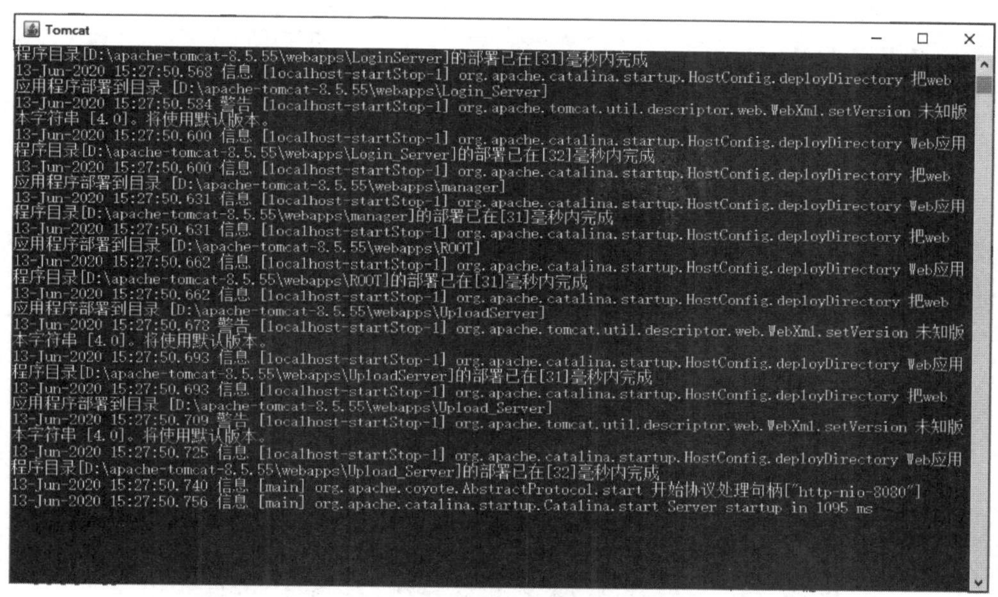

图 13-6 Tomcat 服务器启动信息

Tomcat 在本机的地址可以写成 localhost，端口号默认为 8080。在浏览器的地址栏输入 http://localhost:8080/，如果能顺利进入 Tomcat 网页，如图 13-7 所示，说明前面的配置成功。

这里，localhost 是一个特殊的 DNS 主机名，意思是本地主机，可以代表本机的网络地址，与 IP 地址的作用相同。如果要知道本机的 IP 地址，在命令行窗口输入 ipconfig 命令即可查阅，如图 13-8 所示。图中的 IPv4 地址，就是本机的 IP 地址。

运行在 Tomcat 服务器上的程序是 Servlet。Servlet 是用 Java 语言编写的服务器端程序，主要功能在于与客户端进行交互，处理请求信息，生成动态 Web 内容等。Servlet 是一种用来处理网络请求的一套规范。通常开发 Tomcat 服务器端程序，就是编写 Servlet 项目中的相关类代码和相关配置文件。只有 Servlet 项目开发完成，把项目目录部署到 D:\apache-tomcat-8.5.55\webapps 目录下，才能使得 Web 应用服务器开始工作。

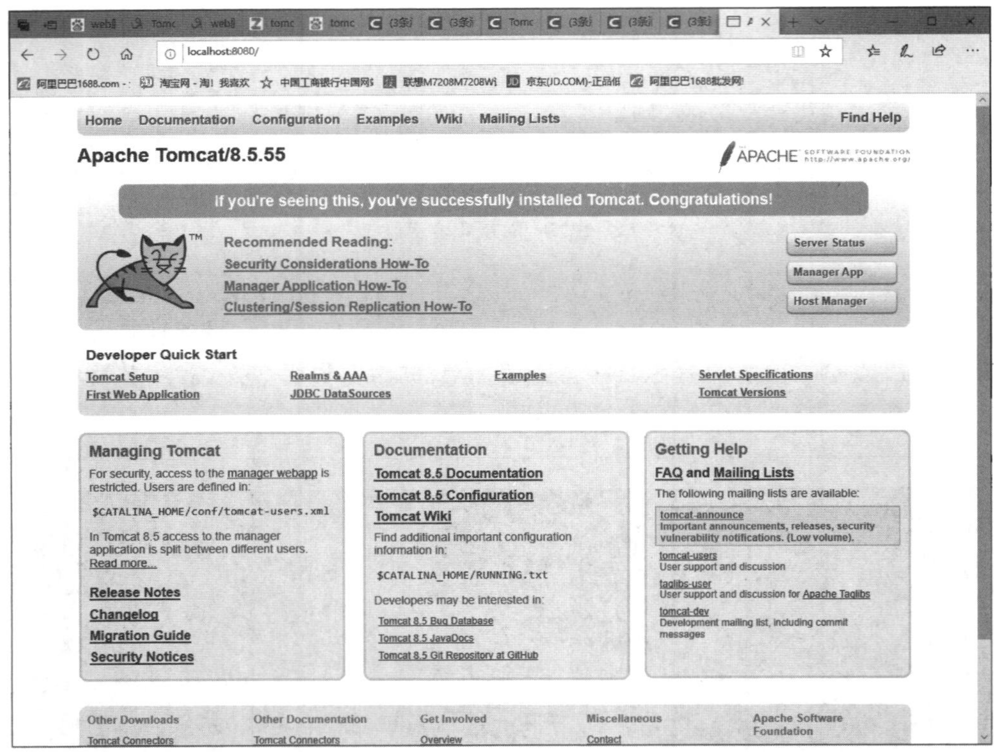

图 13-7　进入 Tomcat 网页

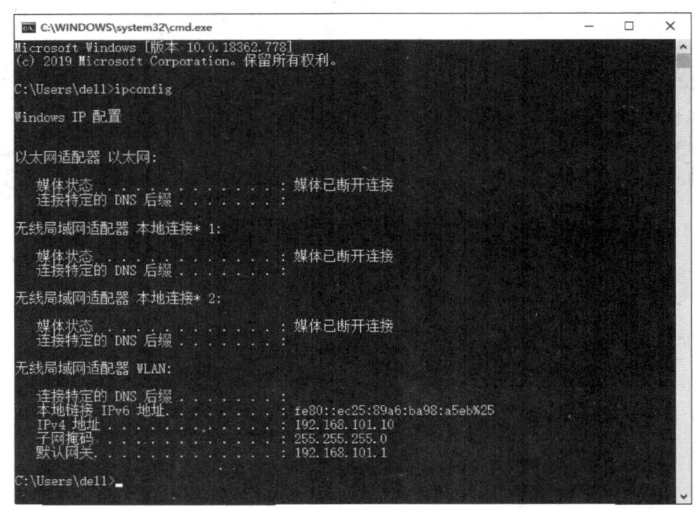

图 13-8　查阅本机的 IP 配置信息

13.3.3　HttpURLConnection 接口应用

HttpURLConnection 是移动客户端应用经常采用的轻量级网络通信接口，几乎可以实现所有 HTTP 网络访问操作。HttpURLConnection 对象由 URL 的 openConnection()方法创建,该对象表示应用项目和 URL 之间的通信连接,通过该对象的方法实现网络访问应用。常用的方法见表 13-7。

表 13-7　HttpURLConnection 类常用的方法及说明

方　法	说　明
setRequestMethod(String method)	设置请求类型。GET 表示 get 请求,POST 表示 post 请求
setConnectTimeout(int timeout)	设置连接的超时时间
setReadTimeout(int timeout)	设置读取的超时时间
setRequestProperty(String key, String value)	设置请求包头的属性信息
setDoOutput(boolean dooutput)	设置是否允许发送数据。采用 POST 请求方式时必须设置该方法
getOutputStream()	获取 HTTP 输出流。该方法返回一个 OutputStream 对象,调用该对象的 write()方法和 flush()方法可写入要发送的数据
connect()	建立 HTTP 连接。该方法在 getOutputStream()之后、getInputStream()之前调用
setDoInput(boolean doinput)	设置是否允许接收数据
getInputStream()	获取 HTTP 输入流。该方法返回一个 InputStream 对象,调用该对象的 read()方法可读取接收的数据
getResponseCode()	获取 HTTP 返回码
getHeaderField(int n)	获取应答数据包头的第 n 个属性值
getHeaderFields()	获取应答数据包头的所有属性列表
disconnect()	断开 HTTP 连接

注意:如果既要使用输入流读取 HttpURLConnection 响应的内容,也要使用输出流发送请求参数,一定要先使用输出流,再使用输入流。

实现一个 HttpURLConnection 接口对象的访问网络应用,通常需要如下几个步骤。

(1) 通过调用 URL 对象的 openConnection()方法来创建连接 URLConnection 对象。

(2) 设置 URLConnection 的参数和普通请求属性。

(3) 如果发送 GET 方式请求,使用 connect()方法建立和远程资源之间的实际连接即可;如果发送 POST 方式请求,需要 URLConnection 实例对应的输出流来发送请求参数。

(4) 远程资源可用时,程序可以访问远程资源的头字段,或通过输入流读取远程资源的数据。

记住!从 Android 4.0(API Level 14)版本之后,所有与网络通信相关的操作都必须放在子线程中完成。因为 HTTP 协议是一种不断地“请求-响应”的方式,以及远程传输数据需要占用的时间和资源也相当多,所以放在子线程中执行,不会造成主线程的阻塞。

下面通过一些案例来介绍如何使用 HttpURLConnection 实现访问网络应用。

1. 读取网络资源

在 Android 中,子线程是不能直接访问主线程中的 UI 控件的。由于对网络访问的操作都要在子线程中执行,那么从网络上获取了资源后,怎样能够把信息更新到主线程的 UI 控件中呢?有两种方式可以实现:一是使用 Handler,二是使用 Activity 中的

runOnUiThread()方法。第一种是采用传递消息方式，即调用 Handler 中的方法来处理消息，从而更新视图。如果需要频繁高效地更新 UI 信息，应该使用第二种方式，即将需要执行的代码放到 Runnable 的 run()方法中。例如在子线程中要更新主线程的 TextView 控件内容，代码片段如下。

```
getActivity().runOnUiThread(new Runnable() {
    @Override
    public void run() {
        // TODO Auto - generated method stub
        myTextView.setText("更新");              //UI 更新
    }
});
```

下面通过案例来说明获取网络资源的用法。

【案例 13.3】 使用 URL 读取指定网页的源代码资源。

说明：通过输入一个 URL 网址，向互联网发送请求访问指定网页，并读取该网页的源代码。该源代码信息是纯文本。

开发步骤及解析：过程如下。

1）创建项目

在 Android Studio 中创建一个名为 Application_UrlConn 的项目。其包名为 ee.example.application_urlconn。

2）声明访问网络权限

在 AndroidManifest.xml 中的< application >元素前添加访问网络权限声明，代码如下。

```
<uses - permission android:name = "android.permission.INTERNET" />
```

另外，在< application >元素内，添加允许明文流量网络请求的属性，代码如下。

```
android:usesCleartextTraffic = "true"
```

3）设计布局

编写 res/layout 目录下的 activity_main.xml 文件。在布局中添加两个区域，第一个区域位于屏幕的顶端，添加一个输入框用于输入网址，ID 为 url_field，添加一个按钮，ID 为 go_button，用于启动网络访问。第二个区域是一个文本框，ID 为 textView，设置 android:scrollbars 为 vertical，用于显示访问网页的文本内容。当内容超过屏幕显示时出现垂直方向的滚动条实现上、下滚动浏览。

4）开发逻辑代码

在 java/ee.example.application_urlconn 包下编写 MainActivity.java 程序文件，代码如下。

```
package ee.example.application_urlconn;

import androidx.appcompat.app.AppCompatActivity;
import android.os.Bundle;
import android.text.method.ScrollingMovementMethod;
import android.util.Log;
```

```
import android.view.View;
import android.widget.Button;
import android.widget.EditText;
import android.widget.TextView;

import java.io.BufferedReader;
import java.io.IOException;
import java.io.InputStream;
import java.io.InputStreamReader;
import java.io.Reader;
import java.net.HttpURLConnection;
import java.net.URL;

public class MainActivity extends AppCompatActivity {
    TextView mTextView;
    EditText urlStr;
    @Override
    protected void onCreate(Bundle savedInstanceState) {
        super.onCreate(savedInstanceState);
        setContentView(R.layout.activity_main);
        Button button = (Button) findViewById(R.id.go_button);
        urlStr = (EditText) findViewById(R.id.url_field);
        mTextView = (TextView) findViewById(R.id.textView);
        mTextView.setMovementMethod(ScrollingMovementMethod.getInstance());

        button.setOnClickListener(new View.OnClickListener() {
            @Override
            public void onClick(View v) {
                new Thread(){
                    @Override
                    public void run() {
                        super.run();
                        getNetwork();
                    }
                }.start();
            }
        });

    }

    private void getNetwork() {
        InputStream inputStream = null;
        Reader reader = null;
        BufferedReader bufferedReader = null;
        HttpURLConnection urlConnection = null;
        try {
            URL url = new URL(urlStr.getText().toString());
            urlConnection = (HttpURLConnection) url.openConnection();
            urlConnection.setRequestMethod("GET");
            urlConnection.setConnectTimeout(8000);
            urlConnection.setRequestProperty("key","value");

            inputStream = urlConnection.getInputStream();      //获取输入字节流
            reader = new InputStreamReader(inputStream);       //实例化字符流
```

```
61                  bufferedReader = new BufferedReader(reader);        //实例化缓冲流
62
63                  final StringBuilder result = new StringBuilder();;
64                  String temp;
65                  while ((temp = bufferedReader.readLine()) != null) {
66                      result.append(temp);
67                  }
68                  Log.i("MainActivity", result.toString());
69                  MainActivity.this.runOnUiThread(new Runnable() {
70                      @Override
71                      public void run() {
72                          mTextView. TextView.setText(result);
73                      }
74                  });
75                  inputStream.close();
76                  reader.close();
77                  bufferedReader.close();
78
79              } catch (Exception e) {
80                  Log.i("MainActivity", e.getMessage());
81                  e.printStackTrace();
82              }finally {
83                  if (reader!= null){
84                      try {
85                          reader.close();
86                      } catch (IOException e) {
87                          e.printStackTrace();
88                      }
89                  }
90                  if (inputStream!= null){
91                      try {
92                          inputStream.close();
93                      } catch (IOException e) {
94                          e.printStackTrace();
95                      }
96                  }
97                  if (bufferedReader!= null){
98                      try {
99                          bufferedReader.close();
100                     } catch (IOException e) {
101                         e.printStackTrace();
102                     }
103                 }
104                 if (urlConnection != null){
105                     urlConnection.disconnect();
106                 }
107             }
108         }
109     }
```

(1) 第32～43行，在按钮单击监听事件的onClick()中，创建子线程。在重写的run()方法中调用getNetwork()自定义方法，在该自定义方法中完成相关网络通信操作。

(2) 第47～108行，实现自定义方法getNetwork()。可以看出，本案例的主要代码工作都集中在此方法内。

(3) 第 52～79 行,由于网络通信应用可能会受到多方面的因素干扰,所以在编程中需要使用 try...catch 语句,在 try 语句块内编写与网络通信相关的操作。如下所示。

① 第 54 行,由 url 对象创建一个 HttpURLConnection 接口对象。因为只需要向服务器传输网址,信息量不多,所以第 55 行使用 GET 请求方式。

② 第 60 行,reader 是一个 Reader 对象,Reader 是一个抽象类,其对象可以从输入流中读取字符流。

③ 第 61 行,bufferedReader 是一个 BufferedReader 对象,BufferedReader 是一个包装类,它可以包装字符流,可以从字符输入流中读取文本,起到缓冲的作用。

④ 第 65～67 行,通过一个循环,从缓冲流 bufferedReader 对象中不断读取一个文本行内容,将读取的文本行添加到 StringBuilder 对象 result 中,直到读完缓冲流 bufferedReader。这里,bufferedReader. readLine()是从缓冲流 bufferedReader 对象中读取一行文本。StringBuilder 是一个可变长的字符串类。

⑤ 第 69～74 行,将从网页中读入的源代码内容显示在 UI 界面的 TextView 控件中。一般地,UI 界面的控件只能在主线程中更新,在子线程中要更新 UI 控件,进而调用 runOnUiThread()方法,重写 run()方法更新该 TextView 控件。

(4) 第 83～106 行,对前面生成的各对象实例一一进行关闭。

运行结果:在 Android Studio 支持的模拟器上,运行 Application_UrlConn 项目。在项目初始界面中输入百度网站的网址:https://www. baidu. com,如图 13-9(a)所示。单击“访问网络数据”按钮,开始读取百度首页的源代码,并显示在下面的文本框内,如图 13-9(b)所示。

(a) 初始界面

(b) 显示网页源代码

图 13-9　读取网页源代码资源

2. 用户登录

在大多数网络应用中，少不了用户登录模块，其操作是用户在客户端输入用户名和密码提交到服务器，然后由服务器处理并返回登录是否成功的结果。这是典型的客户端-网络服务器端的请求-响应应用。

下面以用户登录模块为例，来介绍如何配置 Web 服务器端程序和开发客户端应用项目。

【案例 13.4】 开发一个在移动设备上进行用户登录的应用。

说明：本案例需要开发两个项目，一是使用 Android 开发客户端应用项目，二是服务器端的 Servlet 项目。即客户端-服务器端程序（或称为前端-后端程序）。

Android 客户端应用项目主要实现用户输入用户名和密码，提交给服务器，接收从服务器返回的登录信息。Servlet 项目用于接收用户的输入，判断输入信息的正确性，并返回判断结果等逻辑处理操作。

本案例使用本地机作为服务器主机，其 IP 地址为 192.168.101.10，端口号为 9090。

注意，192.168.101.10 是本机的 IP 地址。网络上不同的主机，IP 地址都不相同。读者调试网络应用项目时，应将 IP 地址改为自己主机的 IP 地址，端口号不需修改。

本案例客户端分别使用 GET、POST 两种方式发送请求。

客户端开发步骤及解析：过程如下。

1）创建项目

在 Android Studio 中创建一个名为 Login_Client 的项目。其包名为 ee.example.login_client。

2）声明访问网络权限

在 AndroidManifest.xml 中的< application >元素前添加访问网络权限声明，代码如下。

```
<uses-permission android:name="android.permission.INTERNET" />
```

在< application >元素内，需要添加允许明文流量网络请求的属性，代码如下。

```
android:usesCleartextTraffic="true"
```

3）设计布局

编写 res\layout 目录下的 activity_main.xml 文件。本书已多次设计过登录界面，布局代码在此不赘述。本案例使用 ID 为 et_name 的输入框输入用户名，使用 ID 为 et_pwd 的输入框输入密码，记住，这些输入框需要添加 android:inputType="textPassword"属性。在输入框的下面添加两个按钮，ID 为 bt_get 的按钮启动以 GET 方式发送请求，ID 为 bt_post 的按钮启动以 POST 方式发送请求。

4）开发逻辑代码

在 java/ee.example.login_client 包下编写 MainActivity.java 程序文件，代码如下。

```
package ee.example.login_client;

import android.os.Bundle;
```

```
import android.os.Handler;
import android.os.Message;
import android.view.View;
import android.widget.Button;
import android.widget.EditText;
import android.widget.Toast;

import androidx.appcompat.app.AppCompatActivity;

import java.io.BufferedReader;
import java.io.BufferedWriter;
import java.io.IOException;
import java.io.InputStreamReader;
import java.io.OutputStream;
import java.io.OutputStreamWriter;
import java.io.UnsupportedEncodingException;
import java.net.HttpURLConnection;
import java.net.MalformedURLException;
import java.net.URL;
import java.net.URLEncoder;
import java.util.HashMap;
import java.util.Map;

public class MainActivity extends AppCompatActivity implements View.OnClickListener {

    private EditText etName;
    private EditText etPwd;
    private Button btgetLogin, btpostLogin;
    private String LOGIN_URL = "http://192.168.101.10:9090/Login_Server/login";

    @Override
    protected void onCreate(Bundle savedInstanceState) {
        super.onCreate(savedInstanceState);
        setContentView(R.layout.activity_main);

        etName = findViewById(R.id.et_name);
        etPwd = findViewById(R.id.et_pwd);
        btgetLogin = findViewById(R.id.bt_get);
        btpostLogin = findViewById(R.id.bt_post);

        btgetLogin.setOnClickListener(this);
        btpostLogin.setOnClickListener(this);
    }

    @Override
    public void onClick(View v) {
        final View lv = v;
        final Map<String, String> paramsmap = new HashMap<>();
        paramsmap.put("username", etName.getText().toString());
        paramsmap.put("password", etPwd.getText().toString());

        new Thread() {
        @Override
        public void run() {
```

```
                String loginresult = "";
                try {
                    switch (lv.getId()) {
                        case R.id.bt_get:
                            loginresult = LoginByGet(LOGIN_URL, paramsmap);
                            break;
                        case R.id.bt_post:
                            loginresult = LoginByPost(LOGIN_URL, paramsmap);
                            break;
                    }
                } catch (Exception e) {
                    e.printStackTrace();
                    threadRunToToast("登录时程序发生异常");
                }
                //返回消息
                Message msg = new Message();
                msg.what = 0x11;
                msg.obj = loginresult;
                handler.sendMessage(msg);
            };
            Handler handler = new Handler(getMainLooper()) {
                @Override
                public void handleMessage(Message msg) {
                    if (msg.what == 0x11) {
                        if ("success".equals(msg.obj.toString())) {
                            threadRunToToast("登录成功!");
//                              Intent intent = new Intent();
//                              intent.setClass(MainActivity.this,studentList.class);
//                              startActivity(intent);
                        } else if("failed".equals(msg.obj.toString())) {
                            threadRunToToast("用户名或密码错误!");
                        }
                    }
                }

            };
        }.start();
    }

    /**
     * HttpURLConnection GET 方式请求
     * 拼接后字符串:http://192.168.101.10:9090/Login_Server/login?name=admin&pwd=123456
     **/
    private String LoginByGet(String urlStr, Map<String,String> map) {

        StringBuilder result = new StringBuilder();   //用于单线程多字符串拼接,返回参数

        //拼接路径地址:http://192.168.101.10:9090/Login_Server/login?name=admin&pwd=123456
        StringBuilder pathString = new StringBuilder(urlStr);
        pathString.append("?");
        pathString.append(getStringFromEntry(map));

        //以下是 HttpURLConnection GET 访问代码
        try{
```

```
                //第一步 包装网络地址
                URL url = new URL(pathString.toString());
                //第二步 创建连接对象
                HttpURLConnection httpURLConnection = (HttpURLConnection) url.openConnection();
                //第三步 设置请求方式 GET
                httpURLConnection.setRequestMethod("GET");
                //第四步 设置读取和连接超时时长
                httpURLConnection.setReadTimeout(5000);
                httpURLConnection.setConnectTimeout(5000);
                //第五步 发出请求:只有 httpURLConnection.getResponseCode();非 - 1 时,才向
                //服务器发请求
                int responseCode = httpURLConnection.getResponseCode();
                //第六步 判断请求码是否成功
                if(responseCode == HttpURLConnection.HTTP_OK) {
                    //第七步 获取服务器响应的流
                    BufferedReader reader = new BufferedReader(
                        new InputStreamReader(httpURLConnection.getInputStream()));
                    String temp;
                    while((temp = reader.readLine()) != null) {
                        result.append(temp);
                    }
                }else{
                    return "failed";
                }
                httpURLConnection.disconnect();
            } catch (MalformedURLException e) {
                e.printStackTrace();
                threadRunToToast("登录失败,请检查网络!");
            } catch (IOException e) {
                e.printStackTrace();
                threadRunToToast("IO 发生异常");
            }
            return result.toString();
        }

        /**
        HttpURLConnection POST 方式请求
            路径:http://192.168.101.10:9090/Login_Server/login
            参数:
            name = admin
            pwd = 123456
         **/
        private String LoginByPost(String urlStr, Map<String,String> map) {

            StringBuilder result = new StringBuilder();   //用于单线程多字符串拼接,返回参数
            String paramsString = getStringFromEntry(map);   //拼接参数:name = admin&pwd = 123456

            //以下是 HttpURLConnection POST 方式访问代码
            try{
                //第一步 包装网络地址
                URL url = new URL(urlStr);
                //第二步 创建连接对象
                HttpURLConnection conn = (HttpURLConnection) url.openConnection();
                //第三步 设置请求方式 POST
```

```
            conn.setRequestMethod("POST");
            //第四步 设置读取和连接超时时长
            conn.setReadTimeout(5000);
            conn.setConnectTimeout(5000);
            //第五步 允许对外输出
            conn.setDoOutput(true);
            conn.setInstanceFollowRedirects(true);
            //第六步 得到输出流,并把实体输出写出去
            OutputStream outputStream = conn.getOutputStream();
            BufferedWriter writer = new BufferedWriter(
                    new OutputStreamWriter(outputStream,"utf-8"));
            writer.write(paramsString);
            writer.flush();
            writer.close();
            outputStream.close();
            //第七步 判断请求码是否成功:只有在执行 conn.getResponseCode()时才向服务
            //器发送请求
            if(conn.getResponseCode() == HttpURLConnection.HTTP_OK){
                //第八步 获取服务器响应的流
                BufferedReader reader = new BufferedReader(
                   new InputStreamReader(conn.getInputStream()));
                String temp;
                while((temp = reader.readLine()) != null) {
                    result.append(temp);
                }
            }else{
                return "failed";
            }
            conn.disconnect();
        } catch (MalformedURLException e) {
            e.printStackTrace();
            threadRunToToast("登录失败,请检查网络!");
        } catch (IOException e) {
            e.printStackTrace();
            threadRunToToast("IO 发生异常");
        }
        return result.toString();
    }

    /**
        将 map 转换成 key1 = value1&key2 = value2 的形式
        @return
        @throws UnsupportedEncodingException
    **/
    private String getStringFromEntry(Map<String, String> map) {

        StringBuilder sb = new StringBuilder(); //StringBuilder 用于单线程多字符串拼接
        boolean isFirst = true;
        try {
            for (Map.Entry<String, String> entry : map.entrySet()) {
                if (isFirst)
                    isFirst = false;
```

```
213                 else
214                     sb.append("&");
215                 sb.append(URLEncoder.encode(entry.getKey(), "utf - 8"));
216                 sb.append(" = ");
217                 sb.append(URLEncoder.encode(entry.getValue(), "utf - 8"));
218             }
219         } catch (UnsupportedEncodingException e) {
220             e.printStackTrace();
221         }
222         return sb.toString();
223     }
224
225     /**
226       在子线程中提示,属于 UI 操作
227      **/
228     private void threadRunToToast(final String text) {
229         runOnUiThread(new Runnable() {
230             @Override
231             public void run() {
232                 Toast.makeText(getApplicationContext(), text, Toast.LENGTH_SHORT).show();
233             }
234         });
235     }
236
237 }
```

(1) 第 32 行,声明一个符号常量,用于表示服务器 IP 地址及端口号。

(2) 第 49～95 行,重写 onClick()方法,在该方法中创建一个子线程(第 55～94 行),在子线程中分别定义两个按钮的单击事件所调用的方法。LoginByGet()定义向服务器发送 GET 请求,LoginByPost()定义向服务器发送 POST 请求。这两个方法都会从服务器获得返回信息 loginresult,第 76 行,将 loginresult 作为消息发送给 Handler 对象。

(3) 第 78～93 行,定义 Handler 对传入消息的处理操作。如果获得的消息是 success,表示登录用户与密码输入正确,登录成功(第 82～87 行),否则登录失败。正常情况下,登录成功后会进入下一个 Activity 中进行后续操作,代码如第 84～86 行所示。但本案例只限于对登录成功与否进行判断,所以把这三行代码加了注释符。

(4) 第 101～143 行,定义 LoginByGet()方法。分七个步骤完成向服务器发送 GET 请求。注意,GET 请求的网址与参数是合并到一起发送到服务器的,这些内容在浏览器的地址栏都可以看到。第 122 行(第五步骤),执行 httpURLConnection.getResponseCode(),才是向服务器发出请求的关键代码。只有在发送请求成功后才对输入流进行读取操作,并转换提取到 result 这个 BufferedReader 对象中,在第 124 行中,HttpURLConnection.HTTP_OK 是发送请求成功的请求码。

(5) 第 152～198 行,定义 LoginByPost()方法。分八个步骤完成向服务器发送 POST 请求。注意,POST 请求的网址与参数是分开发送到服务器的,参数由输出流对象负责提交给服务器。第 173 行,创建一个 BufferedWriter 对象 writer,writer 是一个缓冲字符输出流,它将使用默认大小的输出缓冲区,保存参数。第 174～177 行,完成参数写入输出流。

（6）第 205～223 行，完成用户名与密码参数的拼接，返回拼接字符串。

（7）第 228～235 行，实现自定义方法 threadRunToToast()，定义在子线程中向主线程发送提示信息的操作代码。因为第 70、83、88、137、140、192、195 行都是在子线程中执行的操作。从子线程向 UI 界面发送提示信息，也就是要更新 UI 界面，需要调用 runOnUiThread()方法。

服务器端开发步骤及解析：过程如下。

使用 IntelliJ IDEA 作为集成开发工具，进行服务器 Servlet 项目的开发。由于该部分内容不是本书的重点，在此只对项目的创建过程进行介绍，给出必要的源代码。

1）创建 Servlet 项目

在 IntelliJ IDEA 中创建一个名为 Login_Server 的项目，步骤如下。

（1）依次选择菜单 File→New→Project...，在 New Project 对话框的左侧，选择 Java Enterprise，在右侧 Additional Libraries and Frameworks 选项框中勾选 Web Application (4.0)选项，如图 13-10 所示，然后单击 Next 按钮。

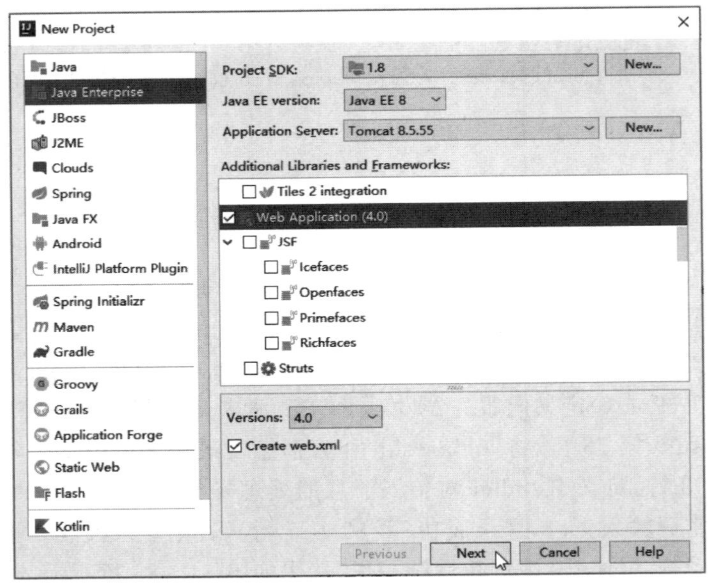

图 13-10 新建项目选项对话框

（2）为新建项目命名和指定存储位置。在此定义项目名称为 Login_Server。如图 13-11 所示。然后单击 Finish 按钮。

（3）创建子目录。在 IntelliJ IDEA 左侧目录栏，选择 web/WEB-INF 目录，分别创建 classes 和 lib 两个子目录。具体操作为右击鼠标弹出菜单，在弹出菜单中依次选择 New→Directory，然后在对话框中命名新建子目录名。完成后，可以在项目的目录结构中看到刚刚新创建的子目录，如图 13-12 所示。

2）配置 Servlet 项目

作为 Tomcat 的 Servlet 项目，还需要做一些必要的配置操作，步骤如下。

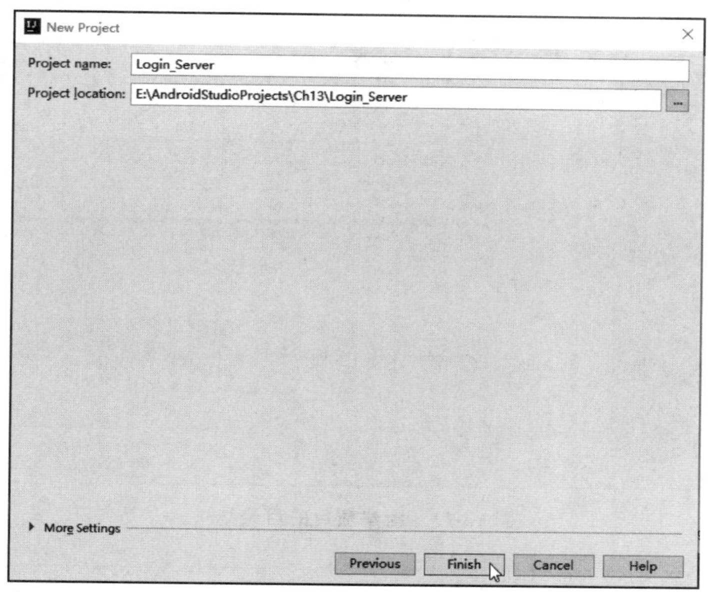

图 13-11　为新建项目命名

(1) 依次选择菜单 File→Project Structure...，在 Project Structure 对话框的左侧，选择 Modules，在中间，选择项目名称 Login_Server，如果只有一个项目，默认选择该项目，在右侧选项卡中选择 Paths 项，指定 Output path 和 Test output path 的目录，两个目录都指向本项目的 classes 目录所在路径，如图 13-13 所示。然后单击 Apply 按钮。

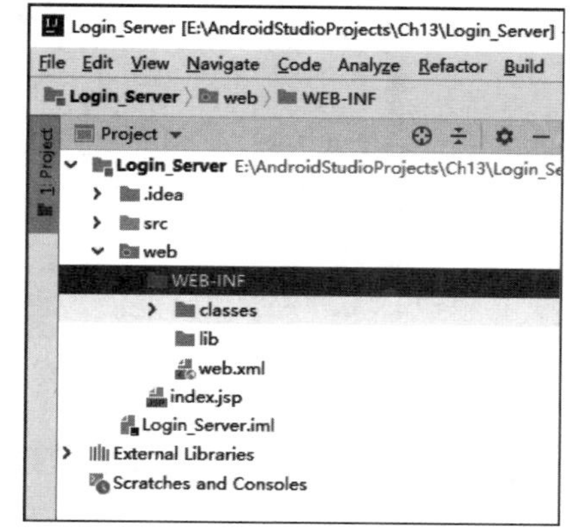

图 13-12　新建项目的目录结构

(2) 在右侧选项卡中选择 Dependencies 项，单击右边的"＋"号(Add (Alt＋Insert))选择新增 JARS 路径项(JARs or directories...)；弹出选项对话框，指定该新增 JARS 路径为本项目的 lib 目录；然后指定该目录所包括的类别为 Jar Directory；最后单击 OK 按钮。这一系列操作分别如图 13-14～图 13-16 所示。

(3) 单击左侧 Artifacts 选项卡，指定运行输出目录为 Tomcat 安装路径下 webapps 目录的子目录，目录名与新建项目同名，如图 13-17 所示，然后单击 OK 按钮。

3) 设置项目的 Tomcat 运行时配置属性

依次选择菜单 Run→Edit Configurations...，进入 Run/Debug Configurations 对话框，一般选择默认配置，只有少数情况下需要修改。例如需要修改 URL 地址，或修改端口号，如图 13-18 所示。如果要指定 Web 应用的资源名等部署信息，选择 Deployment 标签，如图 13-19 所示。

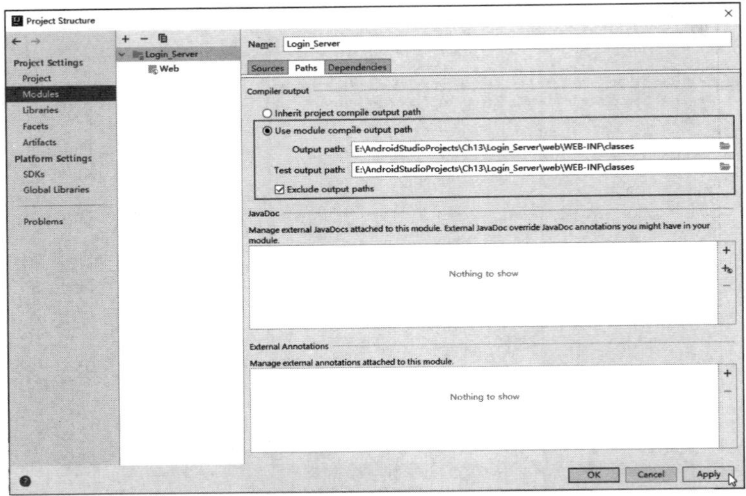

图 13-13　新建项目的目录结构

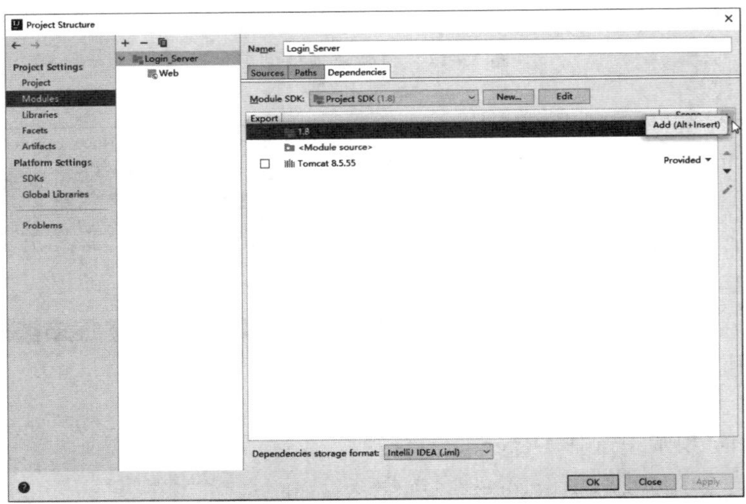

图 13-14　新增项目的 JARS 路径

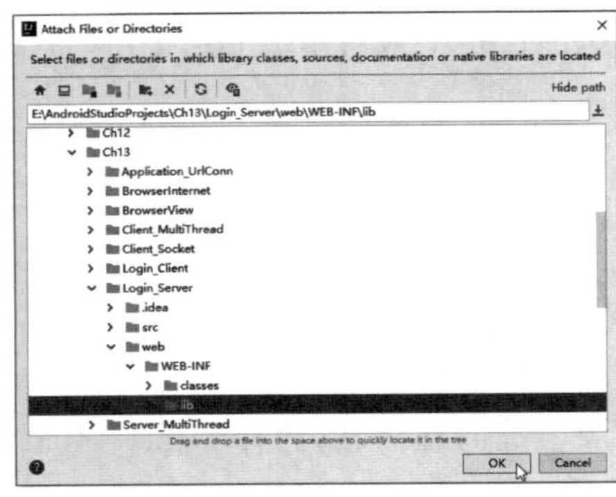

图 13-15　指定 JARS 对应的目录

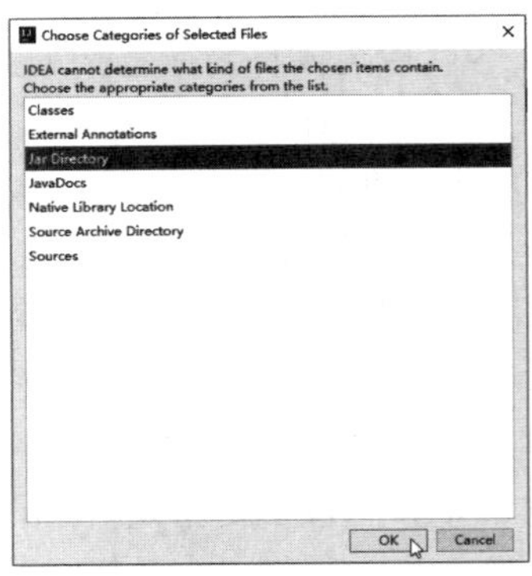

图 13-16　指定目录所包括的类别

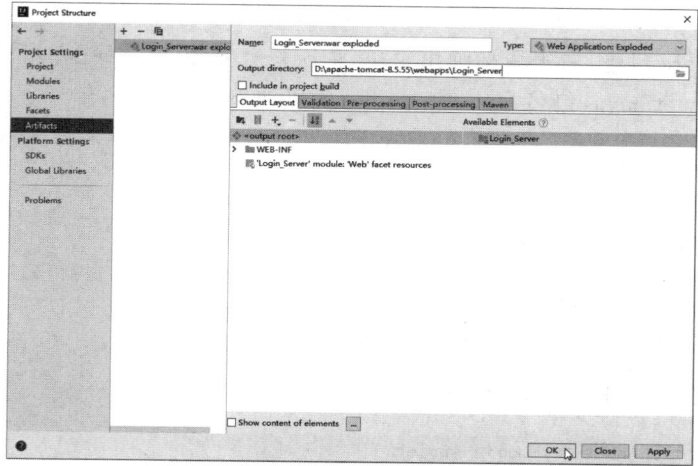

图 13-17　指定项目的运行输出目录

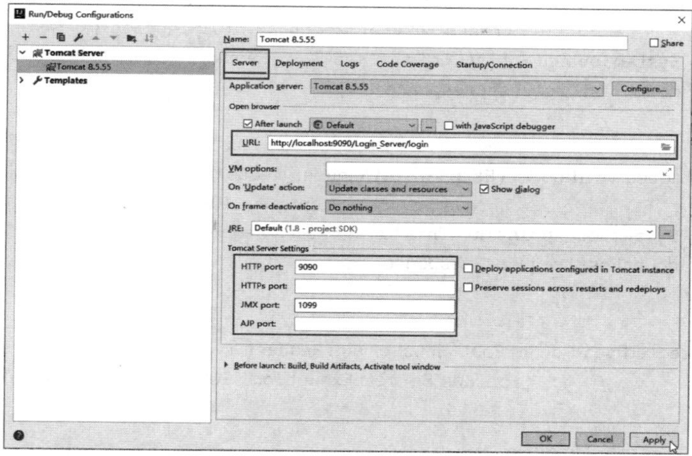

图 13-18　设置 Tomcat 服务器的 URL 地址及端口号

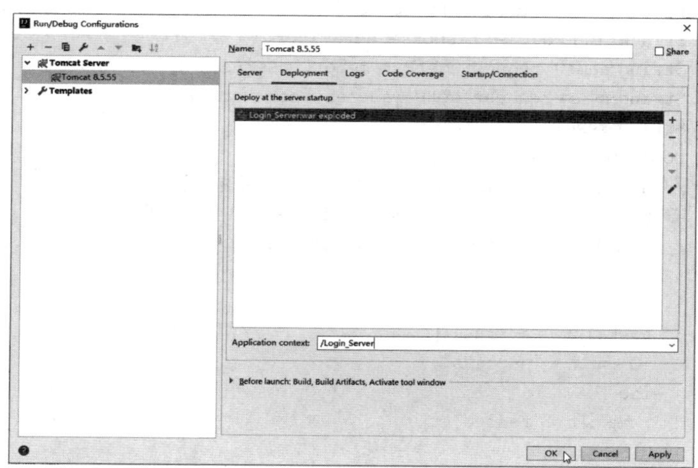

图 13-19　设置 Tomcat 服务器的部署信息

4）创建编写 Servlet 代码

单击项目的 src 目录，右击鼠标弹出菜单，依次选择菜单 New→Servlet，创建 Servlet 文件名为 LoginServlet. java，包名为 ee. example，代码如下。

```
package ee.example;

import javax.servlet.ServletException;
import javax.servlet.annotation.WebServlet;
import javax.servlet.http.HttpServlet;
import javax.servlet.http.HttpServletRequest;
import javax.servlet.http.HttpServletResponse;
import java.io.IOException;
import java.io.PrintWriter;

@WebServlet(name = "LoginServlet")
public class LoginServlet extends HttpServlet {

    protected void doPost(HttpServletRequest request, HttpServletResponse response)
                            throws ServletException, IOException {
        System.out.println("This is servlet, Request: POST");
        doGetPost(request,response);
    }

    protected void doGet(HttpServletRequest request, HttpServletResponse response)
                            throws ServletException, IOException {
        System.out.println("This is servlet, Request: GET");
        doGetPost(request,response);
    }

    protected void doGetPost(HttpServletRequest request, HttpServletResponse response)
                            throws ServletException, IOException{

        //设置编码集
        response.setContentType("text/html;charset = utf - 8");
        request.setCharacterEncoding("utf - 8");
        response.setCharacterEncoding("utf - 8");
        PrintWriter out  =  response.getWriter();
        String name = request.getParameter("username");
        String pwd = request.getParameter("password");
        String result = "";
        boolean isExist = false;
        //通过客户端返回的用户名和密码进行登录验证
        if ("admin".equals(name)&&"123456".equals(pwd)) {
            System.out.println("Login Succes!");
            isExist  =  true;
        }
        if (isExist  ==  true) {                //用户登录验证成功
            result = "success";
        } else {                                //用户名或密码错误
            result = "failed";
        }
        out.write(result);
        out.flush();
```

```
            out.close();
            System.out.println(result);
        }

}
```

5）编辑配置文件 web.xml

在项目的 web/WEB-INF 目录下有一个 web.xml 文件，是 Servlet 项目的配置文件，用于指定< servlet >和< servlet-mapping >属性。其中包括 Servlet 的全称、包名，以及映射的名称和 Servlet 的资源名。该资源名< url-pattern >就是出现在 URL 中最后一个“/”之后的名称。web.xml 代码如下。

```
<?xml version = "1.0" encoding = "UTF - 8"?>
< web - app xmlns = "http://xmlns.jcp.org/xml/ns/javaee"
         xmlns:xsi = "http://www.w3.org/2001/XMLSchema - instance"
         xsi:schemaLocation = "http://xmlns.jcp.org/xml/ns/javaee http://xmlns.jcp.org/xml/ns
                                   /javaee/web - app_4_0.xsd"
         version = "4.0">
    < servlet >
         < servlet - name > LoginServLet </servlet - name >
         < servlet - class > ee.example.LoginServlet </servlet - class >
    </servlet >
    < servlet - mapping >
         < servlet - name > LoginServLet </servlet - name >
         < url - pattern >/login </url - pattern >
    </servlet - mapping >
</web - app >
```

运行结果：运行本案例分两部分进行。首先必须启动服务器端程序，然后才是运行客户端程序。

1）启动服务器端程序的两种方式

方式一：首先在 Tomcat 的保存文件夹的 bin 子文件夹下（本书为 D:\apache-tomcat-8.5.55\bin），打开 startup.bat 批处理文件启动 Tomcat；然后在浏览器的地址栏输入全路径资源名：http://localhost:9090/Login_Server/login。

方式二：在 IntelliJ IDEA 中，对于已经配置好 Tomcat 属性的项目，单击 Run 按钮，启动服务器端程序，如图 13-20 所示。

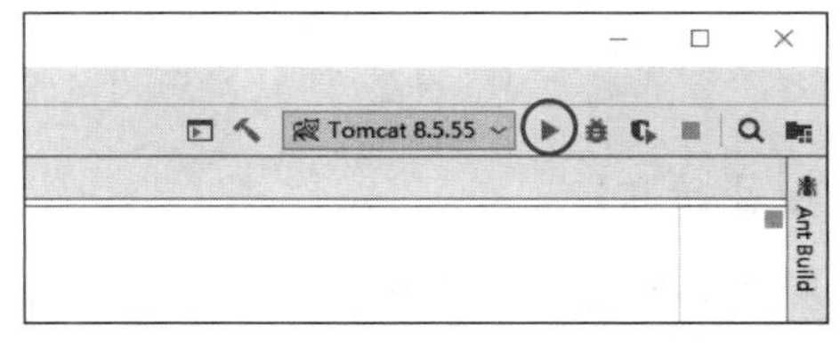

图 13-20 启动服务器的 Servlet 程序

2）运行客户端程序

在 Android Studio 支持的模拟器上运行 Login_Client 项目。在项目初始界面中输入用户名为 admin，输入密码为 123456，如图 13-21(a)所示。单击“登录(GET)”按钮，是发送 GET 请求及参数到服务器；单击“登录(POST)”按钮，是发送 POST 请求及参数流到服务器。单击两个按钮完成的功能是一样的。如果输入的用户名和密码正确，服务器端返回 success，可在客户端看到“登录成功！”的提示消息，如图 13-21(b)所示。

同时，在 IntelliJ IDEA 的输出栏窗格，也可以看到服务器接收请求和响应的记录，如

图 13-21(c)所示。

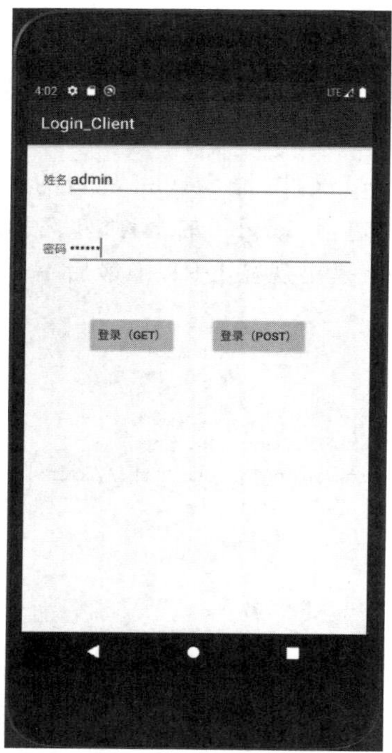

(a) 初始登录界面

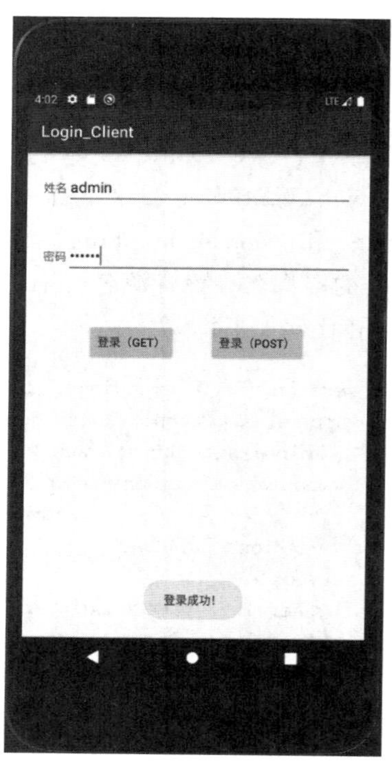

(b) 反馈登录成功信息

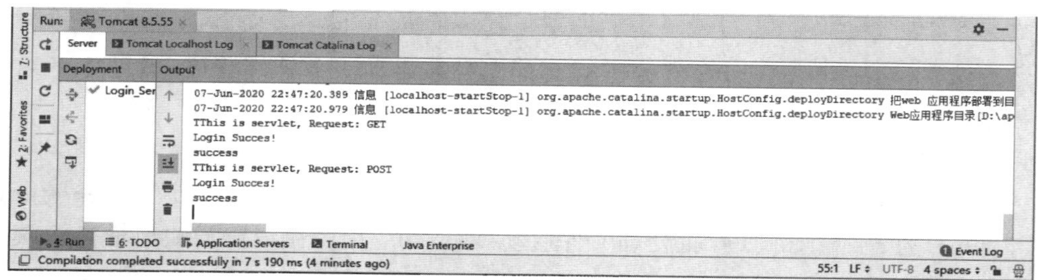

(c) IntelliJ IDEA的输出窗口

图 13-21　用户登录应用客户端-服务器端的运行效果

3. 上传文件

从 App 向服务器上传文件也是网络通信常见的应用，特别是对于社交类 App 和政务类 App，上传文件功能不可缺少。上传文件涉及服务器必须处理接收文件内容并保存文件到指定地址，所以服务器端程序的逻辑处理工作量也不少。有许多第三方机构对于一些常用的功能提供了方法支持库。

例如，Apache 软件基金会为 HTTP 服务器、Tomcat 提供了大量的开源的、安全的、高效的可扩展的方法库。在 Servlet 中使用 Apache 提供的方法，可以完成将上传文件存储到指定地址。在服务器端，将接收到的文件存储到指定的路径中，需要添加 commons-fileupload-1.4.jar 和 commons-io-2.7.jar 两个 JAR 包。

加载JAR包的步骤如下。

1）下载JAR包

Apache软件基金会官网网址为：http://commons.apache.org/。在首页的Apache Commons Proper的组件列表中单击FileUpload项，进入Apache Commons FileUpload页，选择相应版本的JAR包commons-fileupload-1.4-bin.zip，下载。

同样，在首页的Apache Commons Proper的组件列表中单击IO项，进入Apache Commons IO页，选择相应版本的JAR包commons-io-2.7-bin.zip，下载。

2）复制JAR包

解压下载的压缩包，将commons-fileupload-1.4-bin下的commons-fileupload-1.4.jar文件复制到Tomcat本机文件夹的lib文件夹下，即复制到D:\apache-tomcat-8.5.55\lib下。同样，将commons-io-2.7-bin下的commons-io-2.7.jar文件复制到D:\apache-tomcat-8.5.55\lib下。

3）加载JAR包

将commons-fileupload-1.4.jar和commons-io-2.7.jar两个JAR包，加载到服务器Servlet项目中。在IntelliJ IDEA内，加载JAR包。依次选择菜单File→Project Structure...，进入Project Structure对话框，在左侧选择Libraries，在右侧单击"+"按钮，然后在D:\apache-tomcat-8.5.55\lib文件夹下依次选择刚刚复制的JAR包文件commons-fileupload-1.4.jar和commons-io-2.7.jar，如图13-22所示。然后单击OK按钮，即可完成JAR包的加载。此时在项目的目录结构中可以看到刚刚加载的新JAR包项。

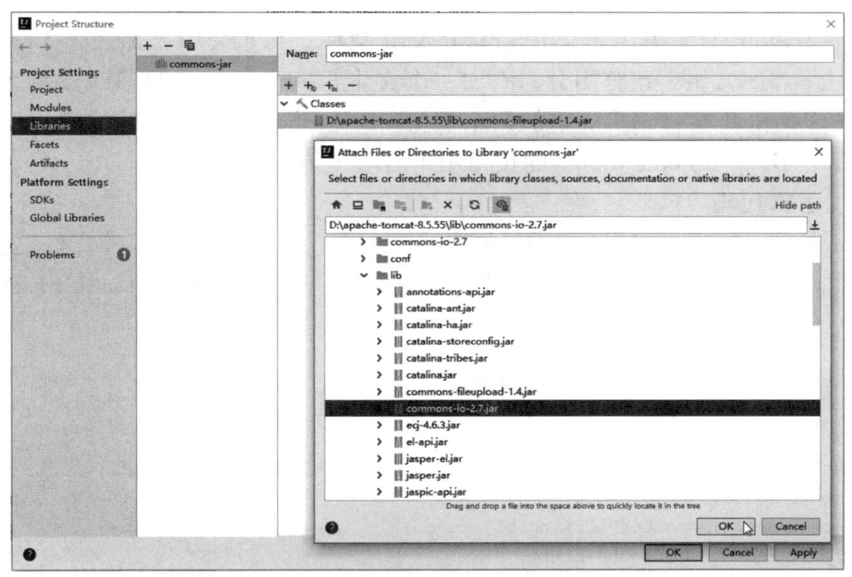

图13-22　加载JAR包到项目中

下面以上传图片文件为例，来介绍如何实现上传应用项目的开发。

【案例13.5】 从移动设备的图库资源中选择一个图片文件，上传给服务器；再由服务器实现保存图片到服务器本机磁盘中。

说明：本案例需要实现两个项目，一是使用Android开发客户端应用项目，二是服务器端的 Servlet 项目。

Android 客户端应用项目主要实现用户打开图库、选择图片，并可以重新指定保存图片的文件名，以及上传到服务器等功能；服务器端 Servlet 项目主要实现接收文件，并应用第三方提供的方法存储文件到 D:\uploads 文件夹中并返回信息给客户端，如果该文件夹不存在则创建之。

本案例使用本地机作为服务器主机，其 IP 地址为：192.168.101.10，端口号为：8080。

客户端开发步骤及解析：过程如下。

1）创建项目

在 Android Studio 中创建一个名为 Upload_Client 的项目。其包名为 ee. example. upload_client。

2）声明访问网络、访问 SD 卡权限

在 AndroidManifest. xml 中的< application >元素前添加访问网络权限声明，代码如下。

```
<uses-permission android:name="android.permission.INTERNET" />
<uses-permission android:name="android.permission.READ_EXTERNAL_STORAGE"/>
```

在< application >元素内，不仅要添加允许明文流量网络请求的属性，还要添加开启旧的存储模式请求的属性，代码如下。

```
android:usesCleartextTraffic="true"
android:requestLegacyExternalStorage="true"
```

3）设计布局

编写 res/layout 目录下的 activity_main. xml 文件。在布局中添加一个 ImageView 控件，ID 为 imageView，用于预览图片；添加一个文本框和一个输入框，ID 分别为 Textv 和 editText，用于为上传文件命名；添加两个按钮，ID 为 choose_image 的按钮用于启动选择图库里的图片，ID 为 upload_image 的按钮用于启动上传。

4）开发逻辑代码

在 java/ee. example. upload_client 包下编写 MainActivity. java 程序文件，主要负责用户交互操作；对于网络通信和上传文件操作，专门编写一个工具类 UploadUtil 来实现。

MainActivity. java 代码如下。

```
package ee.example.upload_client;

import android.Manifest;
import android.app.AlertDialog;
import android.app.Dialog;
import android.content.DialogInterface;
import android.content.Intent;
import android.content.pm.PackageManager;
import android.database.Cursor;
import android.net.Uri;
import android.os.Bundle;
import android.os.Handler;
import android.os.Message;
import android.provider.MediaStore;
import android.view.View;
```

```
import android.view.View.OnClickListener;
import android.widget.Button;
import android.widget.EditText;
import android.widget.ImageView;
import android.widget.Toast;

import androidx.appcompat.app.AppCompatActivity;
import androidx.core.app.ActivityCompat;
import androidx.core.content.ContextCompat;

import java.io.File;

public class MainActivity extends AppCompatActivity implements OnClickListener {

    private EditText editTextName;
    private ImageView imageView;
    private Button btnchoose,btnupload;

    private int RESULT_LOAD_IMG = 1;
    private String imgPath;

    private static String requestURL = "http://192.168.101.10:8080/Upload_Server/upload";
    private Handler handler;
    @Override
    protected void onCreate(Bundle savedInstanceState) {
        super.onCreate(savedInstanceState);
        setContentView(R.layout.activity_main);
        checkPermissionAndPhotoAlbum();

        imageView = (ImageView) this.findViewById(R.id.imageView);
        editTextName = (EditText) findViewById(R.id.editText);
        btnchoose = (Button) findViewById(R.id.choose_image);
        btnupload = (Button) findViewById(R.id.upload_image);
        btnchoose.setOnClickListener(this);
        btnupload.setOnClickListener(this);

        handler = new Handler(getMainLooper()) {
            @Override
            public void handleMessage(Message msg) {
                String result = (String)msg.obj;
                if (msg.what == 0x11) {
                    if ("success".equals(result)) {
                        Toast.makeText(getApplicationContext(), "上传成功!",
                            Toast.LENGTH_LONG).show();
                    } else if("failed".equals(result)) {
                        Toast.makeText(getApplicationContext(), "上传失败!",
                            Toast.LENGTH_LONG).show();
                    }else{
                        Toast.makeText(getApplicationContext(), "上传异常!",
                            Toast.LENGTH_LONG).show();
                    }
                }
            }
        };
```

```
    }

    private static final int PERMISSION_ALBUM_REQUEST_CODE = 1;    //申请相册读写权限
                                                                   //请求码
    //检查权限,授权后调用相册选择图片
    private void checkPermissionAndPhotoAlbum(){
        int hasPhotoAlbumPermission = ContextCompat.checkSelfPermission(getApplication(),
                                      Manifest.permission.READ_EXTERNAL_STORAGE);
        if (hasPhotoAlbumPermission == PackageManager.PERMISSION_GRANTED) {
            //有权限
        } else {
            //若没有权限,则申请权限
            ActivityCompat.requestPermissions(this,
                    new String[ ]{Manifest.permission.READ_EXTERNAL_STORAGE},
                    PERMISSION_ALBUM_REQUEST_CODE);
        }
    }

    @Override
    public void onClick(View v) {
        switch (v.getId()) {
            case R.id.choose_image:
                chooseImage();
                break;
            case R.id.upload_image:
                if (imgPath != null && !imgPath.isEmpty()) {
                    File file = new File(imgPath);
                    if (file.exists()) {
                        uploadImage(file);
                        return;
                    }
                }
                Toast.makeText(getApplicationContext(), "请选择图片!",
                        Toast.LENGTH_LONG).show();
                break;
            default:
                break;
        }
    }


    public void chooseImage() {
        //从相册中选择图片
        Intent galleryIntent = new Intent(Intent.ACTION_PICK,
                MediaStore.Images.Media.EXTERNAL_CONTENT_URI);
        startActivityForResult(galleryIntent, RESULT_LOAD_IMG);
    }

    //当图片被选中时的返回结果
    @Override
    protected void onActivityResult(int requestCode, int resultCode, Intent data) {
        super.onActivityResult(requestCode, resultCode, data);
        try {
            if (requestCode == RESULT_LOAD_IMG && resultCode == RESULT_OK
```

```
                              && data != null) {
                Uri imguri = data.getData();
                String[] filePathColumn = {MediaStore.Images.Media.DATA};
                //获取游标
                Cursor cursor = getContentResolver().query(imguri, filePathColumn,
                                        null, null, null);
                if (cursor != null) {
                    cursor.moveToFirst();
                    int columnIndex = cursor.getColumnIndex(filePathColumn[0]);
                    imgPath = cursor.getString(columnIndex);
                    cursor.close();
                    imageView.setImageURI(imguri);
                    editTextName.setText(new File(imgPath).getName());
                } else{
                    alert();
                }
            } else {
                Toast.makeText(this, "你没有选择图片",
                        Toast.LENGTH_LONG).show();
            }
        } catch (Exception e) {
            Toast.makeText(this, "发生异常", Toast.LENGTH_LONG).show();
        }
    }

    private void alert() {
        Dialog dialog = new AlertDialog.Builder(this).setTitle("提示")
                .setMessage("您选择的不是有效的图片")
                .setPositiveButton("确定", new DialogInterface.OnClickListener() {
                    public void onClick(DialogInterface dialog, int which) {
                        imgPath = null;
                    }
                }).create();
        dialog.show();
    }

    private void uploadImage(File f){
        final File imgfile = f;
        new Thread() {
            @Override
            public void run() {
                String result = "";
                try {
                    String newname = editTextName.getText().toString();
                                                    //传到服务器的文件命名
                    result = UploadUtil.upload(imgfile, newname, requestURL);
                } catch (Exception e) {
                    e.printStackTrace();
                    result = "error";
                }
                //返回信息
                Message msg = new Message();
                msg.what = 0x11;
                msg.obj = result;
```

```
173                 handler.sendMessage(msg);
174             };
175         }.start();
176     }
177     @Override
178     protected void onDestroy() {
179         super.onDestroy();
180   }
181
182 }
```

（1）第 52～69 行，定义 Handler 传入消息，并进行相应处理操作。

（2）第 72～84 行，判断是否有 SD 卡的读权限，如果没有则申请授权即动态请求权限。

（3）第 109～114 行，实现自定义方法 chooseImage()。其中 Intent.ACTION_PICK 就是获取本机相册图片的 Intent。

（4）第 118～143 行，重写 onActivityResult()方法。当用户在相册中选择了一个图片文件，便会自动调用该方法。其中第 125～133 行，通过游标的方式获取图片的全路径文件名，第 132 行，实现显示该文件名。

（5）第 145～154 行，动态生成一个消息对话框并显示。在对话框中分别定义了标题、提示信息和按钮事件。

（6）第 156～176 行，实现自定义方法 uploadImage()。在该方法中将调用网络通信模块。

（7）访问网络的操作要在子线程中完成，所以第 158～175 行，创建一个子线程。重写 run()方法，实现子线程的主要操作。第 163 行，获取 UI 中文本输入框内的字符串，作为将上传给服务器保存的文件名，第 164 行，调用 UploadUtil 类的上传方法，其中 imgfile 是原始文件路径，newname 是将在服务器端存储的文件名，requestURL 是服务器的 URL；request 是从服务器端返回的信息。第 170～173 行，创建 Message 对象获取 request 返回信息，传递给 handler 对象。

UploadUtil.java 是网络通信操作的工具类实现，代码如下。

```
package ee.example.upload_client;

import java.io.DataOutputStream;
import java.io.File;
import java.io.FileInputStream;
import java.io.IOException;
import java.io.InputStream;
import java.io.OutputStream;
import java.net.HttpURLConnection;
import java.net.MalformedURLException;
import java.net.URL;
import java.util.UUID;

public class UploadUtil {
    private static final String TAG = "uploadUtil";
    private static final int TIME_OUT = 10 * 1000;     //超时时间为 10 秒
    private static final String CHARSET = "utf - 8";   //设置编码

    public static String upload(File file, String filename, String uploadUrl) {
```

```
        String rspstr = "";
        String BOUNDARY = UUID.randomUUID().toString();       //边界标识,随机生成
        String PREFIX = "--" , LINE_END = "\r\n";
        String CONTENT_TYPE = "multipart/form-data";          //内容类型

        try {

            URL url = new URL(uploadUrl);
            HttpURLConnection conn = (HttpURLConnection) url.openConnection();
            conn.setReadTimeout(TIME_OUT);
            conn.setConnectTimeout(TIME_OUT);
            conn.setDoInput(true);                            //允许输入流
            conn.setDoOutput(true);                           //允许输出流
            conn.setUseCaches(false);                         //不允许使用缓存
            conn.setRequestMethod("POST");                    //请求方式
            conn.setRequestProperty("Charset", CHARSET);      //设置编码
            conn.setRequestProperty("connection", "keep-alive");
            conn.setRequestProperty("Content-Type", CONTENT_TYPE + ";
                                    boundary=" + BOUNDARY);

            OutputStream outputSteam = conn.getOutputStream();
            DataOutputStream dos = new DataOutputStream(outputSteam);

            //写入上传文件的头部
            StringBuffer sb = new StringBuffer();
            sb.append(PREFIX);
            sb.append(BOUNDARY);
            sb.append(LINE_END);
            /**
                写入上传文件的头部
                这里重点注意:
                name 里面的值为服务器端需要 key,只有这个 key 才可以得到对应的文件
                filename 是文件的名字,包含扩展名
                如果服务器端有文件类型的校验,必须明确指定 ContentType
            **/
            sb.append("Content-Disposition: form-data; name=\"img\"; filename=\"" + filename
                        + "\"" + LINE_END);
            sb.append("Content-Type: " + CONTENT_TYPE + "; charset=" + CHARSET
                        + LINE_END);
            sb.append(LINE_END);
            dos.write(sb.toString().getBytes());     //写入头部(包含普通参数,文件的标识等)

            //写入文件
            FileInputStream is = new FileInputStream(file);
            byte[] bytes = new byte[1024];
            int len = -1;
            while ((len = is.read(bytes)) != -1) {
                dos.write(bytes, 0, len);
            }
            is.close();

            //写入尾部
            dos.write(LINE_END.getBytes());
            byte[] end_data = (PREFIX + BOUNDARY + PREFIX + LINE_END).getBytes();
```

```
71                dos.write(end_data);
72
73                dos.flush();
74                dos.close();
75
76                //读取返回数据
77                if (conn.getResponseCode() == HttpURLConnection.HTTP_OK) {
78                    StringBuffer buffer = new StringBuffer();
79                    InputStream returnis = conn.getInputStream();
80                    int ch;
81                    while ((ch = returnis.read()) != -1) {
82                        buffer.append((char) ch);
83                    }
84                    rspstr = buffer.toString();
85                    returnis.close();
86                } else {
87                    rspstr = "failed";
88                }
89            } catch (MalformedURLException e) {
90                e.printStackTrace();
91            } catch (IOException e) {
92                e.printStackTrace();
93            }
94            return rspstr;
95        }
96    }
```

（1）第27～28行，创建HttpURLConnection对象conn。

（2）第29～37行，设置conn对象相关参数和请求属性。本例是要进行图片文件的上传，将发送比较大的输出流，所以选择使用POST请求。

（3）第43～57行，上传文件头部，上传文件必须包含头部信息，告诉服务器该文件的类型和文件标识信息；第60～66行，上传文件体；第69～74行，上传文件尾部，以便告诉服务器，该文件上传完毕。

（4）第77～88行，读取由服务器传回的消息。

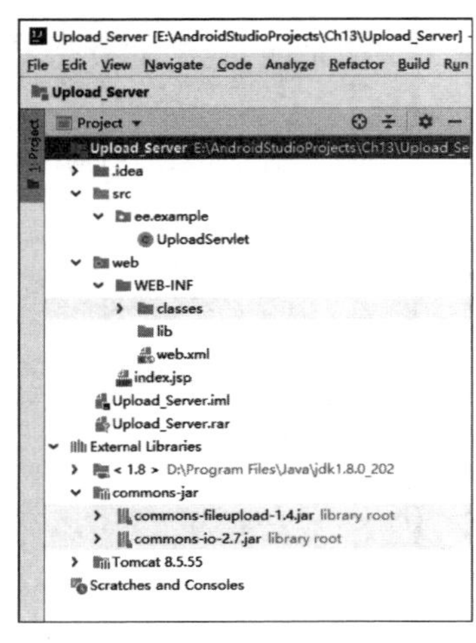

图13-23　Upload_Server的目录结构

服务器端开发步骤及解析：过程如下。

使用IntelliJ IDEA作为集成开发工具，进行服务器Servlet项目的开发。由于该Servlet项目需要接收文件并保存文件，本案例使用Apache的commons-fileupload-1.4.jar和commons-io-2.7.jar这两个JAR包提供方法来实现。在创建Servlet项目时需要添加这两个JAR包。

1）创建Servlet项目并添加JAR包

在IntelliJ IDEA中创建一个名为Upload_Server的项目。并添加commons-fileupload-1.4.jar和commons-io-2.7.jar这两个JAR包。从项目的目录结构中可以看到新添加的JAR包，如图13-23所示。

2）创建编写 Servlet 代码

单击项目的 src 目录，右击鼠标弹出菜单，依次选择菜单 New→Servlet，创建 Servlet 文件名为 UploadServlet.java，包名为 ee.example，代码如下。

```
package ee.example;

import javax.servlet.ServletException;
import javax.servlet.annotation.WebServlet;
import javax.servlet.http.HttpServlet;
import javax.servlet.http.HttpServletRequest;
import javax.servlet.http.HttpServletResponse;
import java.io.*;
import java.util.List;

import org.apache.commons.fileupload.FileItem;
import org.apache.commons.fileupload.disk.DiskFileItemFactory;
import org.apache.commons.fileupload.servlet.ServletFileUpload;

@WebServlet(name = "UploadServlet")
public class UploadServlet extends HttpServlet {


    protected void doGet(HttpServletRequest request, HttpServletResponse response)
                        throws ServletException, IOException {
    }

    protected void doPost(HttpServletRequest request, HttpServletResponse response)
                        throws ServletException, IOException {

        System.out.println("img");

        request.setCharacterEncoding("utf-8");        //设置编码
        //获得磁盘文件条目工厂
        DiskFileItemFactory factory = new DiskFileItemFactory();
        //设定文件需要上传到的路径
        File file = new File("D://uploads");
        if (!file.exists()) {
            file.mkdirs();
        }
        factory.setRepository(file);
        //设置缓存的大小为 1M
        factory.setSizeThreshold(1024 * 1024);

        //文件上传处理
        ServletFileUpload upload = new ServletFileUpload(factory);
        try {
            //可以接收多个上传文件
            List<FileItem> list = (List<FileItem>) upload.parseRequest(request);
            for (FileItem item : list) {
                //获取属性名字
                String name = item.getFieldName();
                //如果获取的表单信息是普通的文本信息时
                if (item.isFormField()) {
```

```
                    //获取用户具体输入的字符串,因为表单提交过来的是字符串类型
                    String value = item.getString();
                    request.setAttribute(name, value);
                } else {
                    //获取路径名
                    String value = item.getName();
                    //索引到最后一个反斜杠
                    int start = value.lastIndexOf("\\");
                    //截取上传文件的文件名,start +1 表示从最后的反斜杠之后开始截取
                    String filename = value.substring(start + 1);
                    request.setAttribute(name, filename);
                    File destFile = new File(file.getPath(), filename);
                    if (destFile.exists()){
                        destFile.delete();
                    }
                    //写到磁盘上
                    item.write(destFile);                    //第三方提供的 zgx
                    System.out.println("Upload success: " + filename);
                    response.getWriter().print("success");   //将上传结果返回给客户端
                }
            }
        } catch (Exception e) {
            System.out.println("Upload failed");
            response.getWriter().print("failed");                //将上传结果返回给客户端
            e.printStackTrace();
        }
    }
}
```

3）编辑配置文件 web. xml

代码如下。

```
<?xml version="1.0" encoding="utf-8"?>
<web-app xmlns="http://xmlns.jcp.org/xml/ns/javaee"
         xmlns:xsi="http://www.w3.org/2001/XMLSchema-instance"
         xsi:schemaLocation="http://xmlns.jcp.org/xml/ns/javaee http://xmlns.jcp.org
                             /xml/ns/javaee/web-app_4_0.xsd"
         version="4.0">

    <servlet>
         <description>This is the description of my J2EE component</description>
         <display-name>This is the display name of my J2EE component</display-name>
         <servlet-name>UploadServlet</servlet-name>
         <servlet-class>ee.example.UploadServlet</servlet-class>
    </servlet>

    <servlet-mapping>
         <servlet-name>UploadServlet</servlet-name>
         <url-pattern>/upload</url-pattern>
    </servlet-mapping>

</web-app>
```

运行结果：运行本案例分两部分进行。首先必须启动服务器端程序，然后才是运行客

户端程序。

1）启动服务器端程序

在 IntelliJ IDEA 中，单击 Run 按钮，启动服务器端程序。

2）运行客户端程序

在 Android Studio 支持的模拟器上，运行 Upload_Client 项目。项目初始界面如图 13-24(a)所示。单击"选择图片"按钮，即刻进入本机的相册列表，如图 13-24(b)所示。用户可以进入子相册中选择图片，如图 13-24(c)所示。当选择成功后，回到前一个页面，可以看到选择的图片以及该图片的文件名，此时用户可以修改文件名，或保留原文件名。然后单击"上传图片"按钮后，如果上传成功，服务器返回成功的信息，如图 13-24(d)所示。

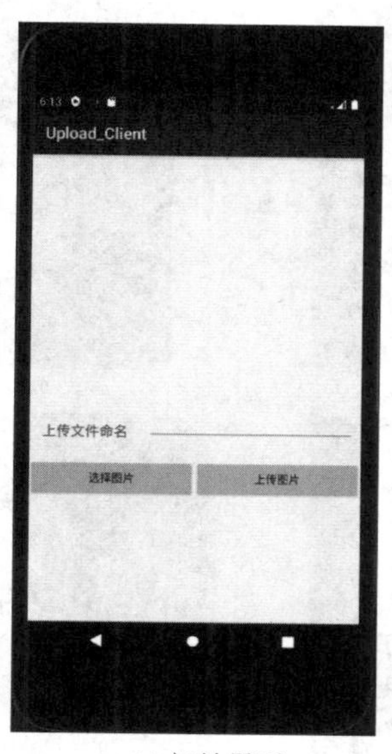

(a) 初始界面

(b) 相册列表信息

(c) 子相册图片列表

(d) 上传成功反馈

图 13-24　上传图片文件应用客户端-服务器端的运行效果

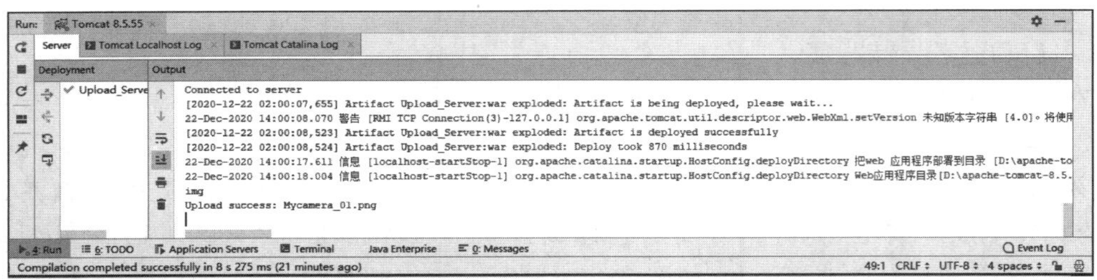

(e) IntelliJ IDEA的输出窗口

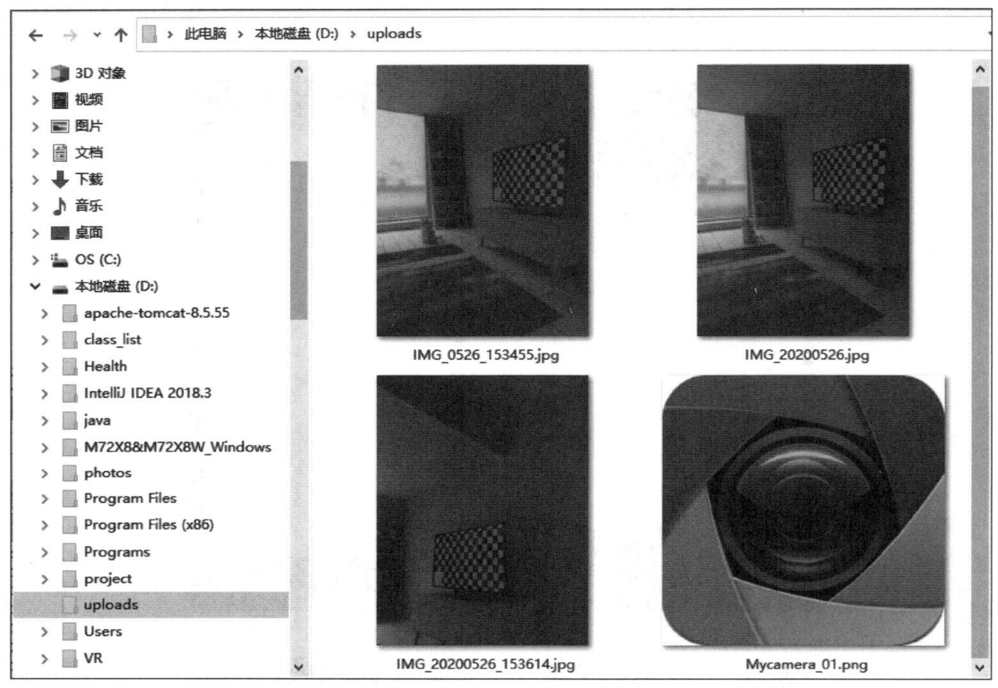

(f) 服务器指定文件夹中的上传文件

图 13-24 （续）

同时，在 IntelliJ IDEA 的输出栏窗格，也可以看到服务器接收请求和响应的记录，如图 13-24(e)所示。在服务器指定的文件夹中查看上传的文件，如图 13-24(f)所示。

13.3.4 OkHttp 网络请求框架

OkHttp 是 Android 的 HTTP 客户端网络请求轻量级框架，非常高效，支持 SPDY、连接池、GZIP 和 HTTP 缓存。OkHttp 必须在 Android 5.0(API 21)及以上版本、Java 8 及以上版本上使用。目前 OkHttp 最新版本是 3.12.x，OkHttp 源码的开源地址为 https://github.com/square/okhttp。

1. OkHttp 的优点

目前，该封装库支持发送有参数和无参数的 GET、POST 请求，发送 POST 的 JSON 数据；支持基于 HTTP 的文件上传、下载和进度回调，加载图片；支持请求回调，直接返回对象、对象集合；支持 Session 的保持；支持自签名网站 HTTPS 的访问；支持取消某个请求；等等。与传统网络请求框架相比，主要优点如下。

（1）支持 SPDY，可以合并多个连接到同一个主机地址的请求，提高请求效率。

（2）OkHttp 使用 Okio 来大大简化数据的访问与存储，Okio 是一个增强 java.io 和 java.nio 的库。通过共享 Socket，减少对服务器的请求次数；通过连接池，减少了请求延迟；使用缓存响应数据，避免重复的网络请求，减少服务器负荷。减少了对数据流量的消耗。

（3）默认情况下，OkHttp 会自动处理常见的网络问题，像二次连接、SSL 的握手问题。

（4）自动处理 GZip 压缩。

2. OkHttp 内部构成

1）OkHttp 内部封装功能

OkHttp 内部封装了许多网络通信方面的功能，主要功能如下。

（1）Dispatcher 作为调度器，内部有线程池，负责调度同步请求和异步请求队列。

（2）Proxy 代理设置，相当于完全信任的中间代理。

（3）List < Protocol > protocols，支持的具体 HTTP 协议版本，有 HTTP/1.0，HTTP/1.1 等。

（4）Interceptors、networkInterceptors，这两个是 OkHttp 中重要的拦截器。

（5）Cache 的存储配置，默认没有。如果需要用，自己填写存储文件位置以及存储空间大小。

（6）SocketFactory，使用默认的 Socket 工厂产生 Socket，TCP 传输层有关，三次握手。

（7）SSLSocketFactory，带上 SSL 的 Socket 的工厂，就是整个 HTTPS 连接的过程。

（8）ConnectionPool，连接池。

（9）DNS 域名解析器，根据域名获得 IP 地址。

（10）retryOnConnectionFailure，在请求失败后是否自动重试。

（11）connectTimeout，连接超时时间。

2）创建 OkHttpClient

创建 OkHttpClient 的代码片段如下。

```
OkHttpClient okHttpClient = new OkHttpClient.Builder()
        .readTimeout(20, TimeUnit.SECONDS)
        .build();
```

官方建议使用单例创建 OkHttpClient，即一个进程中只创建一次即可，以后的每次交易都使用该实例发送交易。这是因为 OkHttpClient 拥有自己的连接池和线程池，这些连接池和线程池可以重复使用，这样做利于减少延迟和节省内存，如果每次发送交易都创建一个 OkHttpClient，将会浪费很多内存资源。

3）创建 Call 对象

在 OkHttp，每次网络请求就是一个 Request，在 Request 里填写应用需要的 url、header 等其他参数，再通过 Request 构造出 Call，Call 内部去请求参数，得到回复，并将结果告诉调用者。创建 Call 的代码片段如下。

```
val request = Request.Builder()
        .url("")
        .build();
okHttpClient.newCall(request);
```

创建 Call 对象时传进去了一个 Request 对象，Request 对象表示用户的请求参数，并传入 OkHttpClient 对象。从 Request 类的实现代码中能看到 Request 对象包含的内容，Request 类部分代码如下。

```
public final class Request {
    private final HttpUrl url;                    //请求 url
    private final String method;                  //请求的 get/post/put/delete 方式
    private final Headers headers;                //请求头
    private final RequestBody body;               //请求体
    private final Object tag;

    private volatile URI javaNetUri;
    private volatile CacheControl cacheControl;

    private Request(Builder builder) {
        this.url = builder.url;
        this.method = builder.method;
        this.headers = builder.headers.build();
        this.body = builder.body;
        this.tag = builder.tag != null ? builder.tag : this;
    }
}
```

一个 Call 对象表示一次请求，每一次请求都会生成一个新的 Call，Call 其实是一个接口，它的具体实现类是 RealCall。RealCall 构造方法如下。

```
protected RealCall(OkHttpClient client, Request originalRequest){
    this.client = client;
    this.originalRequest = originalRequest;     }
```

4）执行请求

OkHttp 中提供了两种请求方式：一种是同步请求，第二种是异步请求。同步请求调用 call.execute()方法，异步请求调用 call.enqueue(Callback callback)方法。

执行请求最终通过 Dispatcher 类来调用，Dispatcher 是 OkHttp 的任务调度核心类，负责管理同步和异步的请求，管理每一个请求任务的请求状态，并且其内部维护了一个线程池用于执行相应的请求。

同步和异步的请求可以这么理解：把 Dispatcher 当成生产者，把线程池当成消费者，如果生产的线程小于可消费的范围，则立即加入消费队列；而当生产者生产的线程大于消费者所能承受的最大范围，就把未能及时执行的任务保存在 readyAsyncCalls 队列中，当时机成熟，也就是线程池有空余线程可以执行时，会调用 promoteCall()方法把等待队列中的任务取出放到线程池中执行，并且把这个任务转移到 runningAsyncCalls 队列中去。

5）拦截器链

拦截器是一种强大的机制。它可以监视、重写和重试调用。获取 OkHttpClient 中设置的各个 Intercepter 拦截器，通过拦截器链对请求数据和返回数据进行处理，内部采用责任链模式，将每一个拦截器对应负责的处理任务进行严格分配，最后将交易结果返回并回调到暴露给调用者的接口上。

3. Gradle 配置

由于 OkHttp 是第三方提供的处理网络请求支持包，如果 Android 应用要使用 OkHttp，必须在应用项目的模块级 build.gradle 文件的 dependencies{ }中添加 OkHttp 依赖，代码片段如下。

```
dependencies {
    ...
implementation 'com.squareup.okhttp3:okhttp:4.9.0'
}
```

4. OkHttp 发送请求

OkHttp 向服务器发送请求主要包括无参、有参的 GET 请求；无参、有参的 POST 请求；发送 POST 的 Json 数据。下面先来看发送 GET 请求的代码实现。

1）发送 GET 请求

OkHttp 向服务器发送请求有同步和异步之分。

(1) 使用 GET 发送同步请求，代码片段如下。

```
public String syncGet(String url) {
    String result = null;
    OkHttpClient client = new OkHttpClient();
    Request request = new Request.Builder()
        .url(url)
        .build();
    try {
        Response response = client.newCall(request).execute();
        if (response.isSuccessful()) {
            result = response.body().string();
        } else {
            throw new IOException("Unexpected code " + response);
        }
    } catch (IOException e) {
        e.printStackTrace();
    }
    return result;
}
```

(2) 使用 GET 发送异步请求，代码片段如下。

```
public void nonSyncGet(String url, Callback responseCallback) {
    String result = null;
    OkHttpClient client = new OkHttpClient();
    Request request = new Request.Builder()
        .url(url)
```

```
            .build();
        Call call = client.newCall(request);
        call.enqueue(responseCallback);
    }
```

(3) 使用 GET 获取图片

获取网络图片有 2 种方式，一是获取 byte 数组，二是获取输入流。注意，onResponse()方法应在子线程中执行。

```
OkHttpClient client = new OkHttpClient();
Request build = new Request.Builder().url(url).build();
client.newCall(build).enqueue(new Callback() {
    @Override
    public void onResponse(Response response) throws IOException {
//      byte[] bytes = response.body().bytes();            //声明数组变量 bytes
        InputStream is = response.body().byteStream();     //声明输入流对象 is
        Options options = new BitmapFactory.Options();
        options.inSampleSize = 8;
//      Bitmap bitmap = BitmapFactory.decodeByteArray(bytes, 0, bytes.length,options);
                                                           //将图片传入数组
        Bitmap bitmap = BitmapFactory.decodeStream(is, null, options);
                                                           //将图片传入输入流
        Message msg = handler.obtainMessage();
        msg.obj = bitmap;
        handler.sendMessage(msg);
    }

    @Override
    public void onFailure(Request arg0, IOException arg1) {
        System.out.println("wisely fail:" + arg1.getCause().getMessage());
    }
});
```

2) 发送 POST 请求

比起 GET 请求，OkHttp 的 POST 请求一般需要传入一些参数，例如 header、body 等。这些参数封装在 body 中，使用 post(body)实现传入。参数包括一些 header，传入参数或者 header 包括发送 POST 数据。

(1) 传入 POST 参数，代码片段如下。

```
RequestBody formBody = new FormEncodingBuilder()
    .add("platform", "android")
    .add("name", "bug")
    .add("subject", "XXXXXXXXXXXXXXX")
    .build();
Request request = new Request.Builder()
    .url("https://api.github.com/repos/square/okhttp/issues")
    .header("User-Agent", "OkHttp Headers.java")
    .addHeader("Accept", "application/json; q=0.5")
    .addHeader("Accept", "application/vnd.github.v3+json")
    .post(formBody)
    .build();
```

(2) 发送同步请求,代码片段如下。

```
public String syncPost(String url) {
    String result = null;
    String json = "test";
    OkHttpClient client = new OkHttpClient();
    MediaType JSON = MediaType.parse("application/json; charset = utf - 8");
    RequestBody body = RequestBody.create(JSON, json);
    Request request = new Request.Builder()
        .url(url)
        .post(body)
        .build();
    Response response = null;
    try {
        response = client.newCall(request).execute();
        if (response.isSuccessful()) {
            result = response.body().string();
        } else {
            throw new IOException("Unexpected code " + response);
        }
    } catch (IOException e) {
        e.printStackTrace();
    }
    return result;
}
```

(3) 发送异步请求,代码片段如下。

```
public void nonSyncPost(String url, Callback responseCallback) {
    String json = "test";
    OkHttpClient client = new OkHttpClient();
    MediaType JSON = MediaType.parse("application/json; charset = utf - 8");
    RequestBody body = RequestBody.create(JSON, json);
    Request request = new Request.Builder()
        .url(url)
        .post(body)
        .build();
    try {
        client.newCall(request).enqueue(responseCallback);
    } catch (Exception e) {
        e.printStackTrace();
    }
}
```

(4) 表单提交请求。

表单提交是最常用的发送数据方式,代码片段如下。

```
OkHttpClient client = new OkHttpClient();
RequestBody body = new FormEncodingBuilder()
    .add("userName", "13363114390")
    .add("password", "200820e3227815ed1756a6b531e7e0d2").build();

Request build = new Request.Builder().url(url).post(body).build();
client.newCall(build).enqueue(new Callback() {
```

```
        @Override
        public void onResponse(Response response) throws IOException {
            String lenght = response.header("Content-Length");
            System.out.println("wisely-- lenght:" + lenght);

            LoginResponseBean loginResponse =
                           new Gson().fromJson(response.body().charStream(), LoginResponse.class);
            System.out.println("wisely---" + loginResponse.getMessage());
        }

        @Override
        public void onFailure(Request arg0, IOException arg1) {
            System.out.println("wisely-----fail");
        }
    });
```

上面是一个简单的登录表单的提交，其中将返回的Json数据封装到了一个bean中。除了能够获取Json数据外，还能获取到各个消息头。例如创建一个LoginResponseBean.java，代码如下。

```
public static class LoginResponseBean{
    private String tokeId;
    private boolean result;
    private String message;

    public String getTokeId() {
        return tokeId;
    }
    public void setTokeId(String tokeId) {
        this.tokeId = tokeId;
    }
    public boolean isResult() {
        return result;
    }
    public void setResult(boolean result) {
        this.result = result;
    }
    public String getMessage() {
        return message;
     }
    public void setMessage(String message) {
        this.message = message;
    }
}
```

(5) 上传图片请求。

OkHttp上传图片的功能强大，支持多图片上传。多图片上传的代码片段如下。

```
private MediaType PNG = MediaType.parse("application/octet-stream");

OkHttpClient client = new OkHttpClient();
RequestBody body = new MultipartBuilder()
        .type(MultipartBuilder.FORM)
```

```
            .addPart(Headers.of("Content-Disposition","form-data;
                        name=\"files\";filename=\"img1.jpg\""),RequestBody.create(PNG, file1))
            .addPart(Headers.of("Content-Disposition","form-data;
                        name=\"files\";filename=\"img2.jpg\""),RequestBody.create(PNG, file2))
            .build();
    Request request = new Request.Builder()
       <span style="white-space:pre"></span>.url(url)
       .post(body).build();
    client.newCall(request).enqueue(new Callback() {
        @Override
        public void onResponse(Response response) throws IOException {
            if(response.isSuccessful()){
                UploadPNGResponse uploadPNGResponse = new Gson().fromJson(response.body().charStream(),
                                                                 UploadPNGResponse.class);
                  String msg = uploadPNGResponse.getMsg();
                  List<String> list = uploadPNGResponse.getList();
                  for (String string : list) {
                      System.out.println("wisely---path:" + string);
                  }
              }
          }

          @Override
          public void onFailure(Request arg0, IOException arg1) {
              System.out.println("wisely---fail--" + arg1.getCause().getMessage());
          }
     });

     class UploadPNGResponse{
          String msg;
          boolean result;
          List<String> list;
          public String getMsg() {
              return msg;
          }
          public void setMsg(String msg) {
              this.msg = msg;
          }
          public boolean isResult() {
              return result;
          }
          public void setResult(boolean result) {
              this.result = result;
          }
          public List<String> getList() {
              return list;
          }
          public void setList(List<String> list) {
              this.list = list;
          }
     }
```

本章以上章节介绍的都是基于浏览器-服务器间的 HTTP 协议的移动网络应用。其实，移动设备还有其他类型的网络应用，比如点对点的即时通信应用，例如国内流行的 QQ、微信，国外的 Twitter、Facebook 等都属于点对点的即时通信应用。在网络访问通信时，为什么能准确无误地进行传输？实际上与网络通信协议分不开。计算机网络的通信协议是 TCP/IP 协议。

13.4 基于 TCP 协议的 Socket 通信

TCP/IP 通信协议是一种可靠的网络协议，其通信的基础是 Socket，中文称套接字。Socket 属于网络传输层的一种技术，在网络连接的客户端和服务器端都要用到。为了更好地理解和应用 Socket 技术，先应该了解 TCP/IP 协议。

13.4.1 TCP/IP 协议概述

TCP/IP 全称为传输控制协议/因特网互联协议（Transmission Control Protocol/Internet Protocol），简称网络通信协议。它是 Internet 最基本的协议，定义了电子设备如何连入因特网，以及数据如何在它们之间传输的标准。

协议采用了 4 层结构，从下至上分别是网络接口层、网络层、传输层、应用层。每一层都呼叫它的下一层所提供的网络来完成自己的需求。最底层是网络接口层，包括 OSI 七层参考模型的物理层和数据链路层，比如以太网和 WiFi 等通信信道，充当软件和硬件之间的桥梁作用；网络层使用的是 IP 协议，它使用 IP 地址作为标识，将数据通过网络独立处理和分发，从而发送到目的地；传输层主要包括两个协议，TCP 和 UDP，这两种协议都建立在网络层所提供的服务基础上；应用层集中实现了 OSI 参考模型中的会话层、表示层和应用层，它包括了一些服务，负责与终端的用户做一些认证、数据处理及压缩工作，将应用项目的数据处理之后交给传输层。

TCP/IP 通信协议是一种可靠的网络协议，其通信的基本模型如图 13-25 所示。

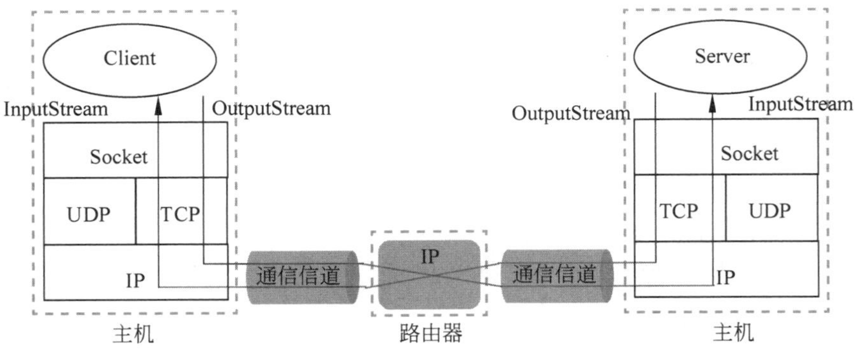

图 13-25 TCP/IP 协议网络通信模型

13.4.2 Socket 通信

Socket 被称为套接字，是应用层与 TCP/IP 协议簇通信的中间抽象层，在程序内部提供了与外界通信的端口。在程序设计中，Socket 把复杂的 TCP/IP 协议簇的内容隐藏在套接字端口后面，用户无须关心协议的实现，只需建立 Socket 连接，即可使用通信双方的数据传输通道。Socket 是端口级通信，数据丢失率低，使用简单且易于移植。

Socket 通信可以采用两种模式：TCP 可靠通信和 UDP 不可靠通信。UDP 称为用户数据包协议，是一种无连接的、效率高的通信方式，可以实现一对一、一对多、多对多的交互通信，但是容易丢失数据包，一般用于小数据包（不超过 64K 字节）数据通信。TCP 称传输控制协议，每一条 TCP 有且只有两个端点，是一对一的关系，提供面向字节流的既可传输又可接收的全双工通信。所以 TCP 是一种可靠的面向连接传输协议，同时它也是一种 Client-Server 模式的协议。当在网络上传递比较大的数据时，比如传输图片、歌曲、电影时通常使用 TCP 传输模式。

在 Android 平台下进行 Socket 开发和在 Java 平台下的开发比较类似。Java 在包 java.net 中提供了两个类 Socket 和 ServerSocket，分别用来表示双向连接的客户端和服务端。使用 Socket、ServerSocket 编程方式可以说是比较底层的，其他的高级协议（如 HTTP）都是建立在此基础之上的，而且 Socket 编程是跨平台的编程，可以在异构语言之间进行通信。所以掌握 Socket 网络编程是基础。

1. Socket 类

使用 Socket 来与一个服务器通信，必须先在客户端创建一个 Socket，并在 Socket 对象中指定服务器的 IP 地址和端口，这也是使用 Socket 通信的第一步。创建 Socket 的构造方法如下。

```
Socket(inetAddress remoteAddress ,int remotePort)
```

其中，参数 remoteAddress 是远程服务器的 IP 地址，这里，IP 地址可以由一个字符串来定义，这个字符串可以是数字型的地址（如 192.168.1.1），也可以是主机名（如 ee.example）。参数 remotePort 是远程服务器的端口号，其有效范围是 0～65535。端口号最好选用大于 1024 的数字，因为 0～1023 是系统预留的号码。

利用 Socket 的构造函数，可以在创建一个 TCP 套接字后，先连接到指定的远程地址和端口号上。由于使用 Socket 编程是一种网络通信应用，所以要在 AndroidManifest.xml 文件中的< manifest >元素内添加网络访问权限，代码如下。

```
< uses - permission android:name = "android.permission.INTERNET"/>
```

在编程中，经常用到的 Socket 方法见表 13-8。

表 13-8 Socket 类常用的方法及说明

方 法	说 明
connect(SocketAddress endpoint)	endpoint 为要连接的 IP 和端口。该方法用于客户端连接服务器端
connect(SocketAddress endpoint, int timeout)	在 timeout 延迟时间内，连接指定 IP 和端口
getInputStream()	获得网络连接输入，同时返回一个 InputStream 对象实例
getOutputStream()	使得连接的另一端得到输入，同时返回一个 OutputStream 对象实例

续表

方　法	说　明
getInetAddress()	获取网络地址对象，该对象是一个 InetAddress 实例
isConnected()	判断 Socket 对象是否连接上
isClosed()	判断 Socket 对象是否关闭
close()	关闭 Socket 对象

2. ServerSocket 类

ServerSocket 类是实现一个服务器端的 Socket，利用这个类可以监听来自网络的请求。创建 ServerSocket 的构造方法如下。

```
ServerSocket(int localPort [,int queueLimit])
```

其中，参数 localPort 为端口号。在创建 ServerSocket 对象时必须指定一个端口号，以便客户端能够向该端口号发送连接请求。如果把端口设为 0，那么将由系统为服务器分配一个端口（称为匿名端口）。参数 queueLimit 为可选项，用来显式地设置连接请求队列的长度，它将覆盖操作系统限定的队列的最大长度。

常用的 ServerSocket 方法见表 13-9。

表 13-9　ServerSocket 类常用的方法及说明

方　法	说　明
accept()	开始接收客户端的连接。有客户端连上时就返回一个 Socket 对象。如果需要持续侦听连接，则在循环语句中调用该方法
getInetAddress()	返回一个 IP 地址，它使得 ServerSocket 获得服务器绑定的 IP 地址
getLocalPort()	返回一个 int 值，它使得 ServerSocket 获得服务器绑定的端口号
isClosed()	判断 Socket 服务器是否关闭
close()	关闭 Socket 服务器

accept()方法是一种阻塞性方法，所谓阻塞性方法就是该方法被调用后将等待客户的请求，直到有一个客户启动并请求连接到相同的端口，然后 accept()返回一个对应于客户的 Socket。如果没有连接请求，accept()方法将阻塞并等待。

close()方法用于服务器释放占用的端口，并且断开与所有客户的连接。当一个服务器程序运行结束时，即使没有执行 ServerSocket 的 close()方法，系统也会释放这个服务器占用的端口。因此，服务器程序并不一定要在结束之前执行 close()方法。在某些情况下，如果希望及时释放服务器的端口，以便让其他程序能占用该端口，则可以显式地调用 ServerSocket 的 close()方法。

3. Socket 通信过程

Socket 通信过程主要分服务器端和客户端。

服务器端，首先使用 ServerSocket 监听指定的端口，等待客户连接请求，客户连接后，会话产生；在完成会话后，关闭连接。

客户端,使用 Socket 对网络上某一个服务器的某一个端口发出连接请求,一旦连接成功,打开会话；会话完成后,关闭 Socket。

如果是多个客户同时连接服务器,服务器中主程序监听一个端口,等待客户接入；同时构造一个线程类,准备接管会话。当一个 Socket 会话产生后,将这个会话交给线程处理,然后主程序继续监听。

在 Android Socket 的通信中,中间的管道连接是通过 InputStream 和 OutputStream 流实现的,一旦管道建立起来就可以进行通信,如果对同一个 Socket 创建重复管道会异常。关闭管道的同时意味着关闭 Socket。

注意,在开发 Socket 程序时一定要关注通信的顺序。服务器端首先得到输入流,然后将输入流信息输出到其各个客户端；客户端首先建立连接,然后写入输出流,最后再获得输入流。如果开发的顺序不对则会产生 EOFException 的异常。为此,给出使用 Socket 的开发步骤。

1）服务器端编程步骤

(1）创建服务器端套接字并绑定到一个端口上。

(2）套接字设置监听模式等待连接请求。

(3）接受连接请求后进行通信。

(4）返回,等待下一个连接请求。

2）客户端编程步骤

(1）创建客户端套接字(指定服务器端 IP 地址与端口号)。

(2）连接(Android 创建 Socket 时会自动连接)。

(3）与服务器端进行通信。

(4）关闭套接字。

下面通过具体的案例来介绍 Socket 通信的应用。

13.4.3　Socket 通信应用

接下来通过两个案例来学习 Socket 通信的应用。

【案例 13.6】 使用 Socket 进行服务器与客户端之间的通信：当客户连接服务器成功后服务器向控制台传送一条字符串信息。

说明：本案例需要分别开发客户端和服务器端程序。客户端程序是 Android Application,服务器端程序是 Java Application,都在 Android Studio 中开发实现。

客户端设计两个按钮,单击第一个按钮执行连接服务器,单击第二个按钮执行发送一条固定文字串给服务器；服务器端设计一个循环,在循环中不断监听客户端的连接请求,如果接收到请求就在子线程中执行输出客户端传来的字符串。

本案例使用本地机作为服务器主机,其 IP 地址为：192.168.101.10,端口号为：9999。

客户端开发步骤及解析：过程如下。

1）创建项目

在 Android Studio 中创建一个名为 Client_Socket 的项目。其包名为 ee.example.client_socket。

2）声明访问网络权限

在 AndroidManifest.xml 中的< application >元素前添加访问网络权限声明，代码如下。

```
<uses-permission android:name = "android.permission.INTERNET" />
```

3）设计布局

编写 res/layout 目录下的 activity_main.xml 文件。在布局中添加两个按钮，ID 为 butConnect 的按钮用于连接服务器，ID 为 butSend 的按钮用于发送字符串给服务器。

4）开发逻辑代码

在 java/ee.example.client_socket 包下编写 MainActivity.java 程序文件，代码如下。

```
package ee.example.client_socket;

import androidx.appcompat.app.AppCompatActivity;
import android.os.Bundle;
import android.util.Log;
import android.view.View;
import android.widget.Button;

import java.io.DataOutputStream;
import java.io.IOException;
import java.net.InetSocketAddress;
import java.net.Socket;
import java.net.UnknownHostException;

public class MainActivity extends AppCompatActivity implements View.OnClickListener {

    private Socket socket;

    @Override
    protected void onCreate(Bundle savedInstanceState) {
        super.onCreate(savedInstanceState);
        setContentView(R.layout.activity_main);
        Button butConnect = findViewById(R.id.butConnect);
        Button butSend = findViewById(R.id.butSend);
        butConnect.setOnClickListener(this);
        butSend.setOnClickListener(this);
    }

    @Override
    public void onClick(View v) {
        switch (v.getId()) {
            case R.id.butConnect:
                Thread conn = new Thread(new NetConn());
                conn.start();
                break;

            case R.id.butSend:
                    new Thread() {
                    @Override
                    public void run() {
```

```
41                        try {
42                            DataOutputStream writer =
                                      new DataOutputStream(socket.getOutputStream());
43                            writer.writeUTF("嘿嘿,你好啊,服务器."); //上传服务器的信息
44                            System.out.println("发送消息");
45                        } catch (IOException e) {
46                            e.printStackTrace();
47                        }
48                    }
49                }.start();
50                break;
51
52            default:
53                break;
54        }
55    }
56
57    protected class NetConn implements Runnable {
58        @Override
59        public void run() {
60            try{
61                socket = new Socket();
62                socket.connect(new InetSocketAddress("192.168.101.10", 9999),5000);
63                Log.i("Android", "与服务器建立连接:" + socket);
64            } catch (UnknownHostException e) {
65                e.printStackTrace();
66            } catch (IOException e) {
67                e.printStackTrace();
68            }
69        }
70    }
71 }
```

(1) 执行 Socket 网络连接和通信的代码需要放在子线程中进行。在此代码程序中使用两种方式来创建子线程。

(2) 第 32～35 行,重写第一个按钮("连接服务器"按钮)的 onClick()方法。在该方法中通过显式方式创建子线程,该子线程的执行体 run()方法在后面的 Runnable 子类 NetConn 中定义。

(3) 第 57～70 行,定义 Runnable 子类 NetConn。在该子类中,首先创建一个 Socket 对象,然后调用 connect()方法连接服务器,这里允许连接服务器的延迟时间为 5 秒。

(4) 第 37～50 行,重写第二个按钮("发送信息"按钮)的 onClick()方法。在该方法中通过匿名方式创建子线程,该子线程的执行体 run()方法直接在 new Thread(){}内重写。

(5) 第 40～48 行,定义匿名子线程执行体 run()方法。在该方法内,创建一个 DataOutputStream 对象 writer,然后将"嘿嘿,你好啊,服务器."字符串以 UTF-8 编码格式写入 writer 对象。writeUTF(String str)是 java.io.DataOutputStream 的一个常用方法,其功能是以可移植的方式使用 UTF-8 编码将字符串 str 写入输出流。

服务器端开发步骤及解析:过程如下。

服务器端的是 Java Application。在 Android Studio 中创建 Java Application 项目是在

创建一个空模板项目的基础上，具体步骤如下。

1）创建服务器项目

（1）在 Android Studio 中依次选择菜单 File→New→New Project…，然后选用 Empty Activity 空模板，在 Create New Project 对话框中输入项目名称，其他的可选择默认。如图 13-26 所示，然后单击 Finish 按钮。

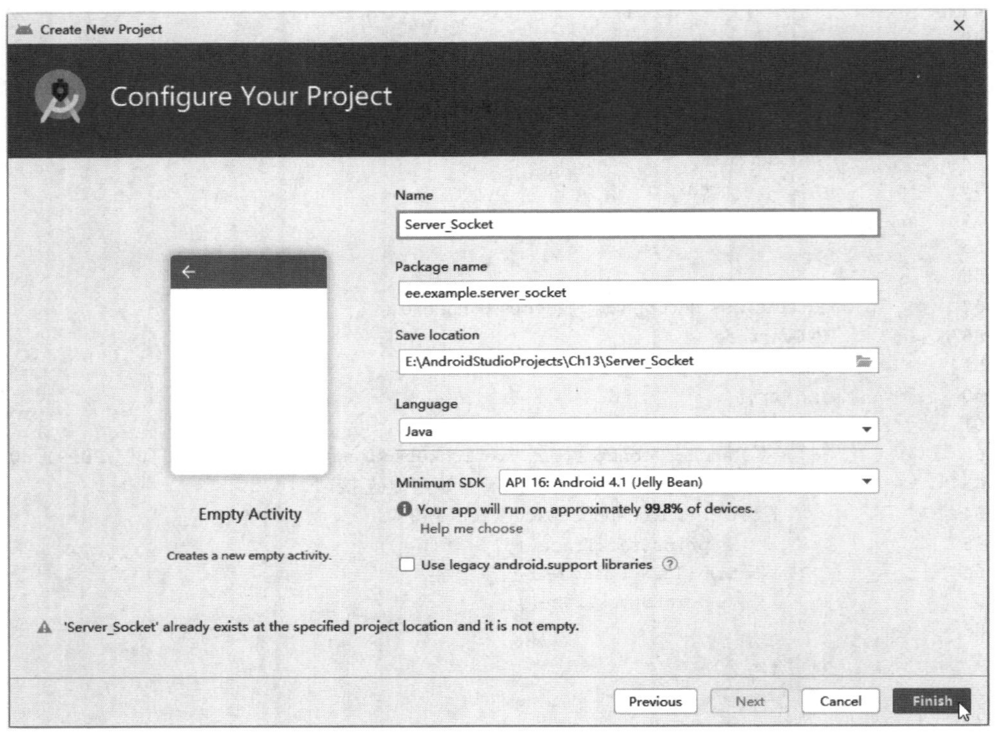

图 13-26　Android 创建新项目对话框

（2）创建 Java 模块。在 Android Studio 项目目录中选择 app，右击弹出菜单，依次选择菜单 New→Module，然后选择 Java or Kotlin Library 模块类型，如图 13-27 所示。然后单击 Next 按钮。为 Library name 和 Class name 命名，如图 13-28 所示。然后单击 Finish 按钮。完成模块创建后得到的目录结构如图 13-29 所示。

2）开发服务器逻辑代码

在 java/ee.example.server 包下编写 MyServer.java 程序文件，代码如下。

```
package ee.example.server;

import java.io.DataInputStream;
import java.io.IOException;
import java.net.InetAddress;
import java.net.ServerSocket;
import java.net.Socket;

public class MyServer {

```

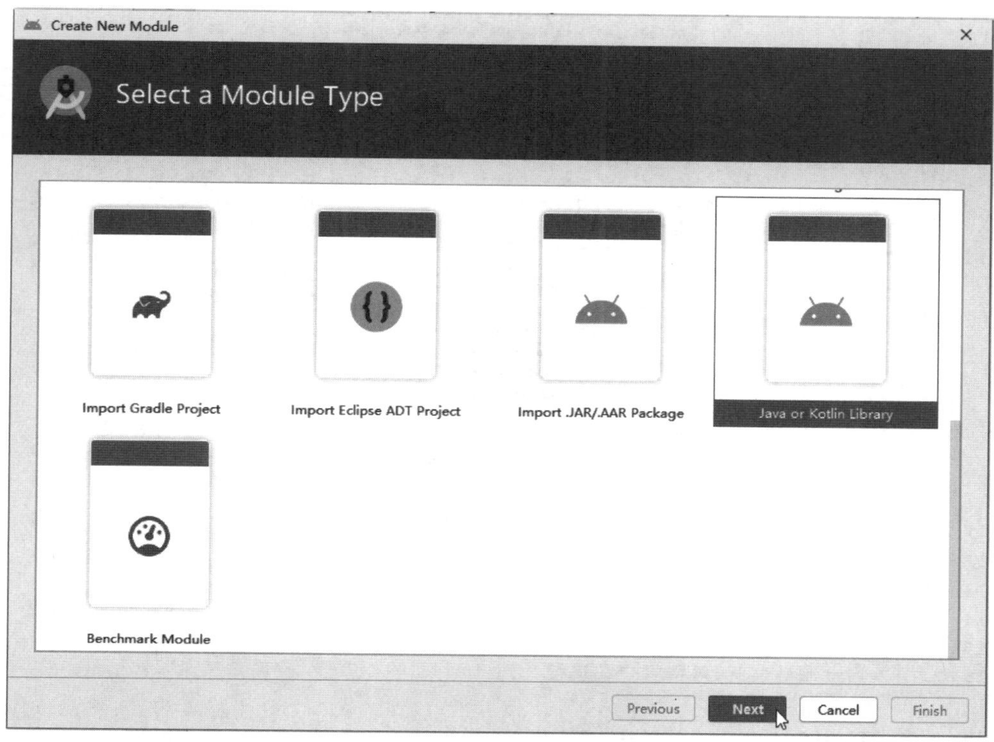

图 13-27　选择模块类型

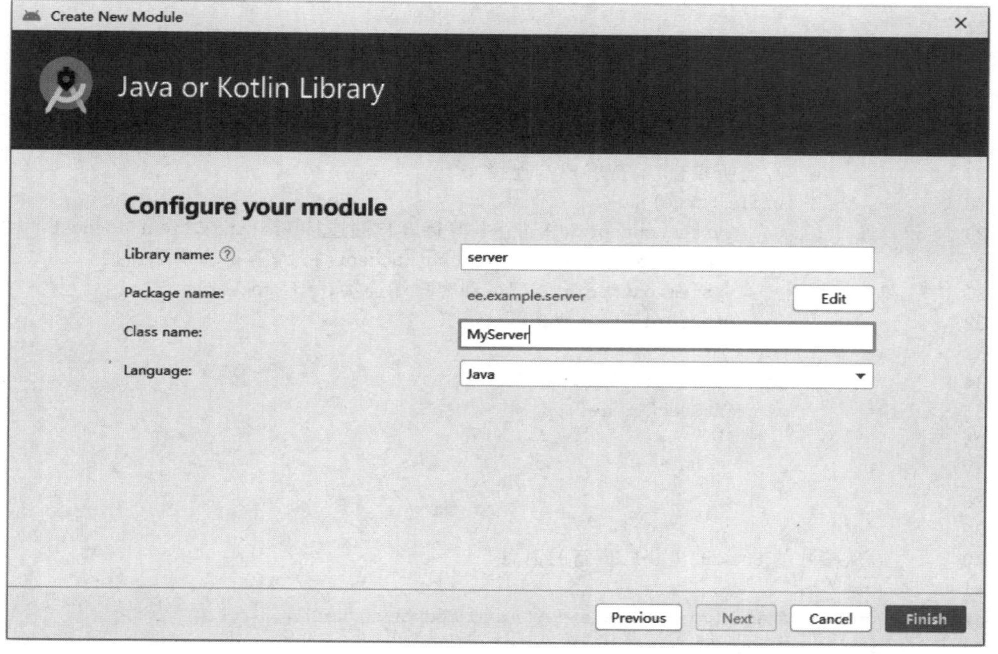

图 13-28　为模块和类文件命名

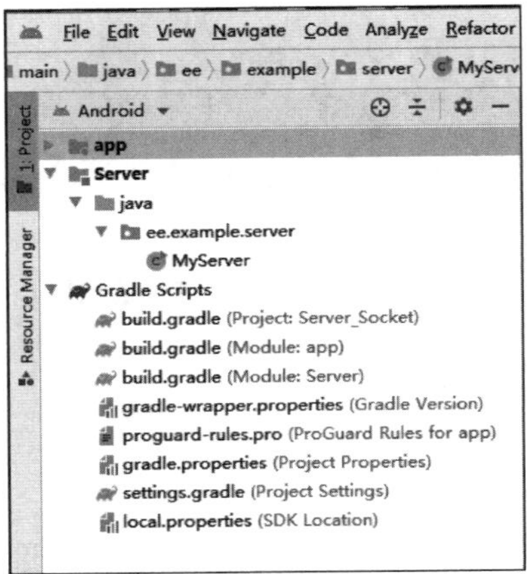

图 13-29　Socket 服务器端项目的目录结构

```
    public static void main(String[ ] args) {
        startService();
    }

    /**
     启动服务监听,等待客户端连接
     **/
    private static void startService() {
        try {
            InetAddress addr = InetAddress.getLocalHost();
            System.out.println("local host:" + addr);

            //创建 ServerSocket
            ServerSocket serverSocket = new ServerSocket(9999);
            System.out.println("--开启服务器,监听端口 Listener Port: 9999--");

            //监听端口,等待客户端连接
            while (true) {
                System.out.println("--等待客户端连接 Waiting Client connect--");
                Socket socket = serverSocket.accept(); //等待客户端连接
                System.out.println("得到客户端连接:" + socket);
                startReader(socket);
            }
        } catch (IOException e) {
            e.printStackTrace();
        }
    }

    /**
     从参数的 Socket 里获取最新的消息
     **/
    private static void startReader(final Socket socket) {
        new Thread() {
            @Override
            public void run() {
                DataInputStream reader;
```

```
47                     try {
48                         //获取读取流
49                         reader = new DataInputStream(socket.getInputStream());
50                         while (true) {
51                             System.out.println(" * 等待客户端输入 * ");
52                             //读取数据
53                             String msg = reader.readUTF();
54                             System.out.println("获取到客户端的信息:" + msg);
55                         }
56                     } catch (IOException e) {
57                         e.printStackTrace();
58                     }
59                 }
60             }.start();
61         }
62     }
```

(1) 第 20 行,使用 InetAddress.getLocalHost()方法获取本机的 IP 地址。

(2) 第 24 行,创建端口号为 9999 的 ServerSocket 对象,用于监听该端口号的客户端请求。

(3) 第 28～33 行,在 while(true)循环中不断地监听端口。其中第 30 行的 accept()方法是阻塞式方法,当没有接收到客户端的连接请求时会一直等待,直到有一个客户发起连接请求为止,接收客户端的 Socket 对象。第 32 行,为 Socket 对象调用 startReader()方法,该方法中创建子线程。由此可以说明,对每一个 Socket,都会有一个单独的子线程来处理其请求。

(4) 第 42～61 行,定义 startReader()自定义方法。在该方法中创建一个子线程,在子线程中读取从客户端 Socket 对象中传入的输入流。第 53 行,使用 readUTF()方法将输入流对象中的数据转换成 UTF-8 编码格式的字符串。

运行结果:运行本案例分两部分进行。首先必须启动服务器端程序,然后运行客户端程序。

1) 启动服务器端程序有两种方式

方式一:在 Android Studio 中选择 MyServer.java,然后右击鼠标,弹出菜单。在菜单中选择 Run 'MyServer.main()'项,如图 13-30 所示。

方式二:在 Android Studio 的工具栏中,首先在运行项下拉列表框中选择 MyServer 项,然后单击"运行"按钮,如图 13-31 所示。

当运行服务器项目后,在 Android Studio 下部的 Run 窗格可以看到运行结果。有时可能会出现中文乱码的情况,这是因为 Android Studio 环境下的输出字符编码格式可能不是 UTF-8。解决中文乱码问题,只需要在 vmoption 中添加 encoding=UTF-8 即可,具体操作是在搜索栏输入 vmoption,选择第一个 Edit Custom VM Options...项,如图 13-32 所示。如果此前 Android Studio 中没有配置过该文件,则需要新建,出现如图 13-33 所示的对话框,选择 Create。在该文件的最后一行,添加-Dfile.encoding=UTF-8 即可,如图 13-34 所示。

在完成这一系列配置之后,重新运行 MyServer。此时乱码问题解决了。在 Run 窗格可以看到服务器端程序运行的初始结果,如图 13-35 所示。

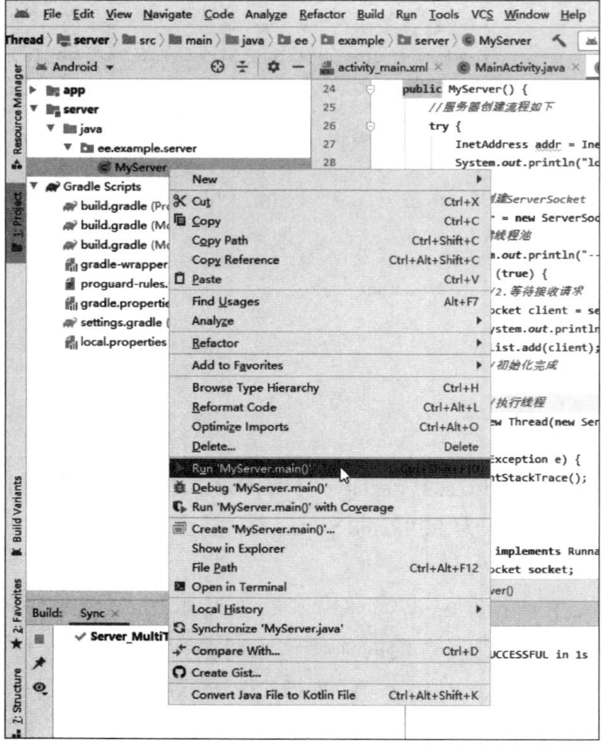

图 13-30　从弹出菜单中选择“运行”

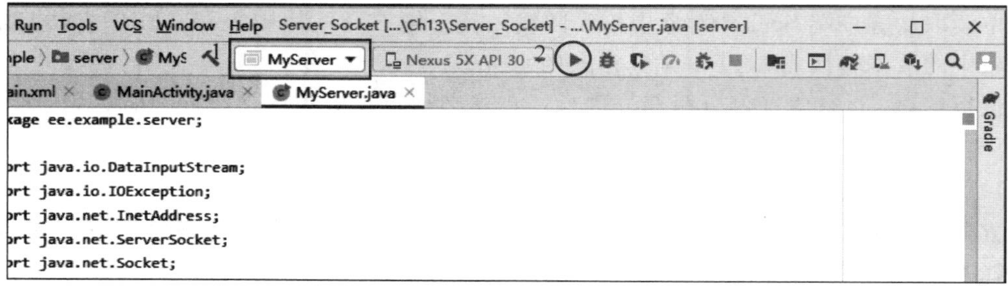

图 13-31　从工具栏中选择“运行”

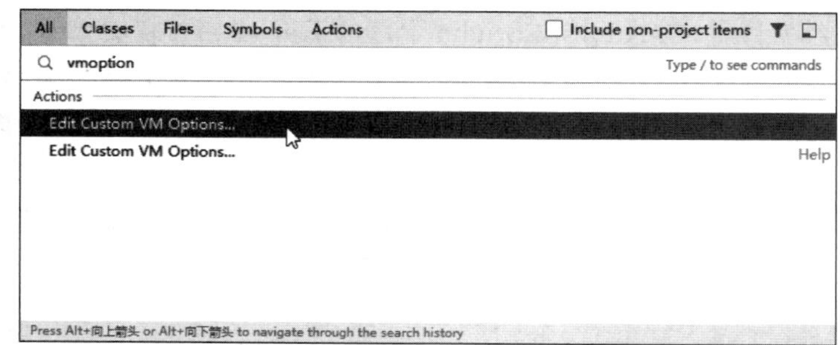

图 13-32　选择编辑 vmoption 配置文件

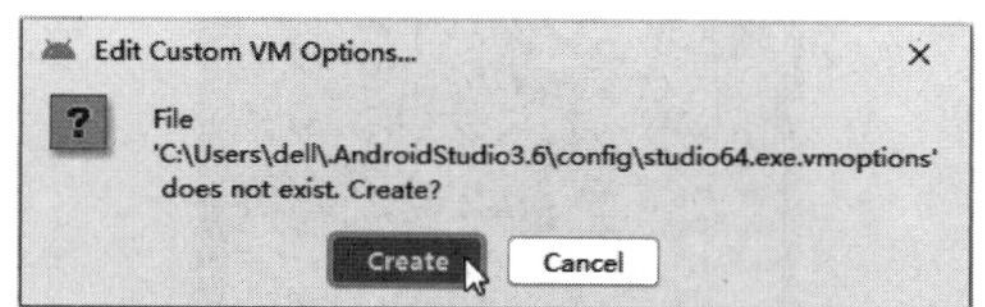

图 13-33　创建 vmoption 配置文件对话框

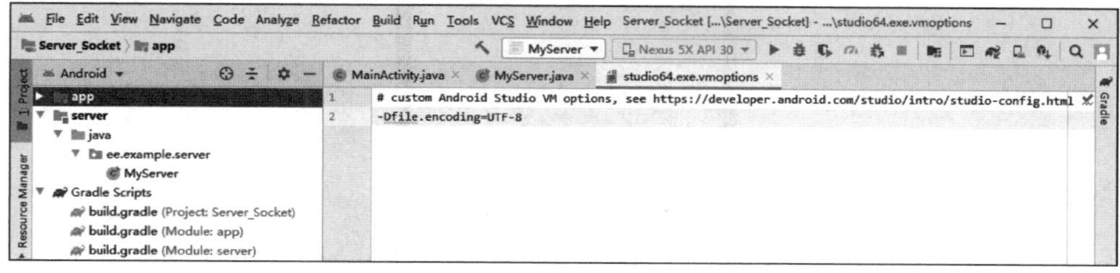

图 13-34　添加编码 UTF-8

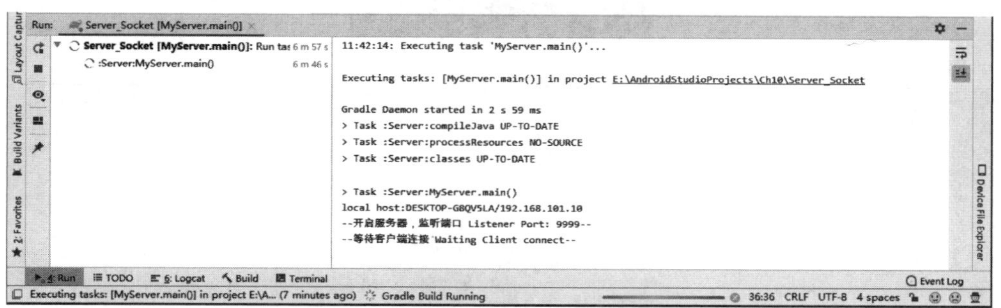

图 13-35　服务器启动后的初始运行结果

2）运行客户端程序

在 Android Studio 支持的模拟器上，运行 Client_Socket 项目。项目初始界面只有两个按钮，如图 13-36(a)所示。单击客户端的“连接服务器”按钮，在服务器端的 Run 窗格可以看到连接请求的响应结果，如图 13-36(b)所示。单击客户端的“发送消息”按钮，在服务器端的 Run 窗格可以看到发送的“嘿嘿，你好啊，服务器.”文字串内容，如图 13-36(c)所示。

从案例 13.6 可以看到，对于 Socket 网络通信，端口号非常重要，服务器就是靠端口号与客户端建立连接的。换句话说，如果有多个客户端，对应到同一个网址的同一个端口号，那么服务器可以与多个客户端进行通信对话。

服务器端-客户端的应用都是一对多的应用，这就意味着一个服务器要接受多个客户的请求。为了保证多个用户同时访问服务器而不被阻塞，服务器为每个 Socket 单独启动一个子线程，每一个线程只与一个客户端进行通信。

下面模拟一个简单的聊天室应用，来说明服务器端多线程的编程技术。

【案例 13.7】 使用 Socket 技术开发一个简易聊天室，可以显示客户端的 IP 地址，可以统计在线的人数，显示客户端聊天的内容。当客户输入 bye 时退出聊天室。

说明：本案例同样需要分别开发客户端和服务器端程序。

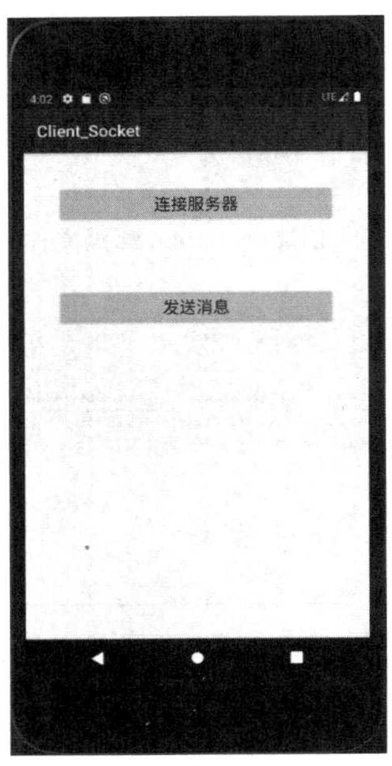

(a) 客户端界面

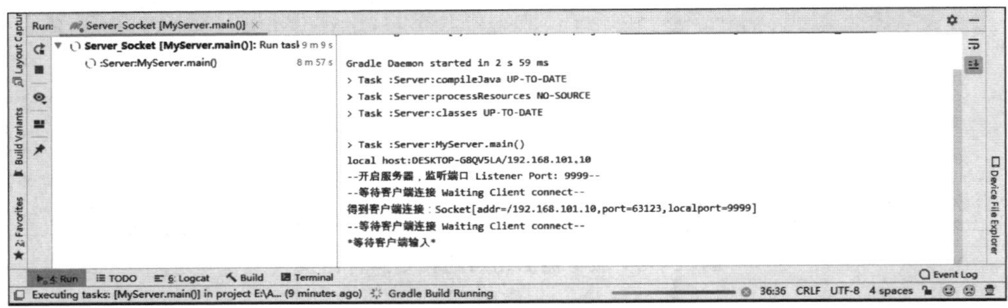

(b) 客户端发出连接请求后的服务器响应

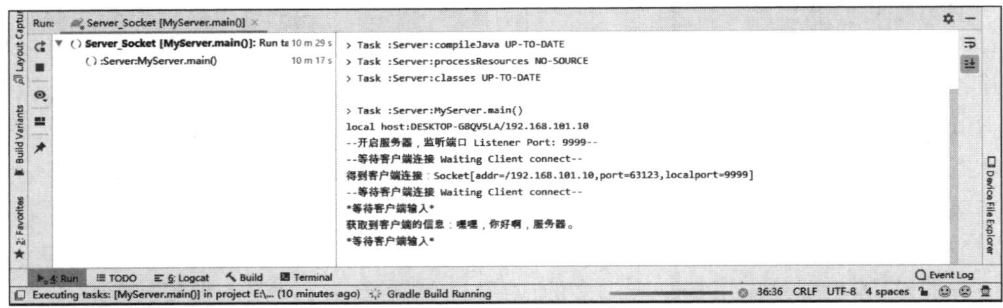

(c) 客户端发出消息后的服务器响应

图 13-36　Socket 客户端-服务器端的运行结果

在客户端，设计一个较大的区域用于显示聊天内容，在屏幕底部留出一行用于输入聊天内容以及发送按钮。在服务器端设计一个循环，在循环中不断监听客户端的连接请求，使用 List 来保存所有的 Socket，对每个 Socket 都创建一个线程来处理接收到的输入流信息。

本案例使用本地机作为服务器主机，其 IP 地址为 192.168.101.10，端口号为 8899。

客户端开发步骤及解析：过程如下。

1）创建项目

在 Android Studio 中创建一个名为 Client_MultiThread 的项目。其包名为 ee.example.client_multithread。

2）声明访问网络权限

在 AndroidManifest.xml 中的<application>元素前添加访问网络权限声明。

3）设计布局

编写 res/layout 目录下的 activity_main.xml 文件。在布局中先添加一个按钮，ID 为 butConn，单击它连接服务器。然后设计一个 RelativeLayout，ID 为 chatview，其内添加一个 TextView、一个 EditText 和一个 Button。在初始运行时设置 RelativeLayout 可视属性为 gone，即隐藏，在向服务器发送连接请求后，再显示。TextView 的 ID 为 txtmsg，用于显示聊天内容；EditText（ID 为 sendmsg）与 Button（ID 为 butSend）位于屏幕底部，用于输入聊天内容和发送聊天内容。

4）开发逻辑代码

在 java/ee.example.client_multithread 包下编写 MainActivity.java 程序文件主要负责用户交互操作逻辑，而子线程的执行体专门编写一个 Runnable 子类 ClientThread 来实现。

MainActivity.java 代码如下。

```java
package ee.example.client_multithread;

import androidx.appcompat.app.AppCompatActivity;
import android.os.Bundle;
import android.os.Handler;
import android.os.Message;
import android.util.Log;
import android.view.View;
import android.widget.Button;
import android.widget.EditText;
import android.widget.RelativeLayout;
import android.widget.TextView;

import java.io.BufferedReader;
import java.io.PrintWriter;
import java.net.Socket;

public class MainActivity extends AppCompatActivity implements View.OnClickListener {

    private TextView txtshow;
    private EditText editsend;
    private Button btnconn, btnsend;
    private RelativeLayout chatv;
    private static final String HOST = "192.168.101.10";
    private static final int PORT = 8899;
    private Socket socket = null;
    private BufferedReader in = null;
    private PrintWriter out = null;
```

```
    private String content = "";
    private StringBuilder sb = null;
    private Handler handler;
    private ClientThread clientThread;

    @Override
    protected void onCreate(Bundle savedInstanceState) {
        super.onCreate(savedInstanceState);
        setContentView(R.layout.activity_main);
        sb = new StringBuilder();
        txtshow = findViewById(R.id.txtmsg);
        editsend = findViewById(R.id.sendmsg);
        btnconn = findViewById(R.id.butConn);
        btnsend = findViewById(R.id.butSend);
        chatv = findViewById(R.id.chatview);

        sb.append(txtshow.getText().toString());

        btnconn.setOnClickListener(this);
        btnsend.setOnClickListener(this);

        //用于发送接收到的服务器端的消息,显示在界面上
        handler = new Handler(getMainLooper()) {
            @Override
            public void handleMessage(Message msg) {
                //如果消息来自子线程
                if (msg.what == 0x123) {
                    sb.append("\n" + msg.obj);
                    Log.i("wotainanl", "@@" + sb);
                    txtshow.setText(sb.toString());   //将 sb 里的内容添加到文本框中
                }
            }
        };
    }

    @Override
    public void onClick(View v) {
        switch (v.getId()) {
            case R.id.butConn:
                btnconn.setVisibility(View.GONE);
                chatv.setVisibility(View.VISIBLE);
                clientThread = new ClientThread(handler); //
                new Thread(clientThread).start();
                break;

            case R.id.butSend:                          //为发送按钮设置单击事件
                try {
                    Message msg = new Message();
                    msg.what = 0x456;
                    msg.obj = editsend.getText().toString();
                    clientThread.toserverHandler.sendMessage(msg);
                    editsend.setText("");               //清空输入文本框
                } catch (Exception e) {
                    e.printStackTrace();
```

```
83                    };
84                    break;
85            }
86        }
87
88    }
```

(1) 第 32 行，声明一个 ClientThread 的对象，这个 ClientThread 就是自定义的另一个类，该类继承了 Runnable，在其中定义了子线程的若干个功能实现方法。后面会给出代码及说明。

(2) 第 45 行，将 TextView 控件 txtshow 原有的文本串添加到 StringBuilder 类型的变量 sb 中。StringBuilder 是一种可变的字符串类型，适用于不断改变其中内容的应用场景。

(3) 第 51～61 行，创建一个 Handler 实例，用于从子线程中获取服务器端传入的消息，并将消息转换成字符串，以另起行添加到 txtshow 控件中。

(4) 第 65～86 行，重写两个按钮的 onClick()方法。注意，这两个按钮是有先后执行顺序的。因为初始界面只有 btnconn 按钮可见；当单击了 btnconn 按钮后，又只有 btnsend 按钮可见。

(5) 第 67～72 行，定义 btnconn 按钮的单击事件。btnconn 按钮被单击后即隐藏(见第 68 行)，由于该按钮是首先出现在界面中的，被单击顺序在先，所以需要创建子线程，并启动。

(6) 第 74～85 行，定义 btnsend 按钮的单击事件。第 78 行将输入文本框内的文字串赋值给 msg 消息的 obj 属性，第 79 行将 msg 发送到子线程中的 Handler 对象 toserverHandler 中。

ClientThread.java 是实现一个 Runnable 子类的代码，用于定义子线程执行体的操作，代码如下。

```
package ee.example.client_multithread;

import android.os.Handler;
import android.os.Looper;
import android.os.Message;
import android.util.Log;

import java.io.BufferedReader;
import java.io.BufferedWriter;
import java.io.IOException;
import java.io.InputStreamReader;
import java.io.OutputStreamWriter;
import java.io.PrintWriter;
import java.net.Socket;
import java.net.SocketTimeoutException;

public class ClientThread implements Runnable {

    private static final String HOST = "192.168.101.10";
    private static final int PORT = 8899;
    private Socket socket = null;
    private Handler toclientHandler;        //向 UI 线程发送消息的 Handler 对象
    public Handler toserverHandler;         //接收 UI 线程消息的 Handler 对象
    private BufferedReader in = null;
    private PrintWriter out = null;

    public ClientThread(Handler myhandler) {
```

```
            this.toclientHandler = myhandler;
        }

        public void run() {
            try {
                socket = new Socket(HOST, PORT);    //建立连接到远程服务器的 Socket
                //初始化输入输出流
                in = new BufferedReader(new InputStreamReader(socket.getInputStream(), "utf-8"));
                out = new PrintWriter(new BufferedWriter(new OutputStreamWriter(
                        socket.getOutputStream(), "utf-8")), true);
                Log.i("wotainanl", "in" + in + "@@" + out);
                //创建子线程
                new Thread() {
                    @Override
                    public void run() {
                        String fromserver = null;
                        try {
                            while ((fromserver = in.readLine()) != null) {
                                Message servermsg = new Message();
                                servermsg.what = 0x123;
                                servermsg.obj = fromserver;
                                toclientHandler.sendMessage(servermsg);
                            }
                        } catch (IOException e) {
                            e.printStackTrace();
                        }
                    }
                }.start();

                Looper.prepare();       //在子线程中初始化一个 Looper 对象,即为当前线程创
                                        //建消息队列
                toserverHandler = new Handler() {
                    @Override
                    public void handleMessage(Message msg) {
                        if (msg.what == 0x456) {
                            try {
                                out.println(msg.obj);     //将输出流包装为打印流
                            } catch (Exception e) {
                                e.printStackTrace();
                            }
                        }
                    }
                };
                Looper.loop();      //启动 Looper,运行刚初始化的 Looper 对象,循环取消息队
                                    //列的消息
            } catch (SocketTimeoutException el) {
                System.out.println("网络连接超时!");
            } catch (Exception e) {
                e.printStackTrace();
            }
        }
    }
```

(1) 本类有两个 Handler 对象，一个是 toclientHandler，是向客户端的 UI 发送消息的

Handler；另一个是 toserverHandler，是向服务器端发送消息的 Handler。

(2) 第 27～29 行，实现类的构造方法。

(3) 第 31～76 行，重写 run()方法。该方法是本子类的主体部分，在该方法内，创建 socket 对象，初始化输入流 in、输出流 out。然后分别向客户端和服务器端传递消息。

(4) 第 40～55 行，在该 run()方法内再创建一个子线程，用于从服务器端接收输入流，如果输入流非空，便通过 toclientHandler 发送回客户端。

(5) 第 57～70 行，使用 Looper 来封装一个 Android 线程中的消息循环。调用 Looper. prepare()给线程创建一个消息循环，调用 Looper. loop()使消息队列循环起作用。也就是说，消息队列的循环体是位于 Looper. prepare()和 Looper. loop()之间的代码。在消息循环中创建一个 toserverHandler 的 Handler 对象，将消息传送到输出流中，向服务器端输出。

服务器端开发步骤及解析：过程如下。

在 Android Studio 中创建 Java 项目 Server_MultiThread，在 java/ee. example. server 包下创建服务器端逻辑代码文件 MyServer. java，代码如下。

```java
package ee.example.server;

import java.io.BufferedReader;
import java.io.BufferedWriter;
import java.io.IOException;
import java.io.InputStreamReader;
import java.io.OutputStreamWriter;
import java.io.PrintWriter;
import java.net.InetAddress;
import java.net.ServerSocket;
import java.net.Socket;
import java.util.ArrayList;
import java.util.List;

public class MyServer {
    public static final int PORT = 8899;   //端口号
    private List<Socket> mList = new ArrayList<Socket>();
    private ServerSocket server = null;

    public static void main(String[] args) {
        new MyServer();
    }

    public MyServer() {
        //服务器创建流程如下
        try {
            InetAddress addr = InetAddress.getLocalHost();
            System.out.println("local host:" + addr);

            //1.创建 ServerSocket
            server = new ServerSocket(PORT);
            //创建线程池
            System.out.println("--服务器开启中--");
            while (true) {
                //2.等待接收请求，这里接收客户端的请求
```

```
                    Socket client = server.accept();
                    System.out.println("得到客户端连接:" + client);
                    mList.add(client);
                    //初始化完成

                    //执行线程
                    new Thread(new Service(client)).start();
                }
            } catch (Exception e) {
                e.printStackTrace();
            }
        }

        class Service implements Runnable {
            private Socket socket;
            private BufferedReader in = null;
            private String content = "";

            public Service(Socket clientsocket) {
                this.socket = clientsocket;
                try {
                    //3.接收请求后创建链接 socket
                    //4.通过 InputStream 和 outputStream 进行通信
                    in = new BufferedReader(new InputStreamReader(socket.getInputStream(),"utf-8"));

                    content = "用户:" + this.socket.getInetAddress() + "~加入了聊天室"
                            + "当前在线人数:" + mList.size();
                    this.sendmsg();
                } catch (IOException e) {
                    e.printStackTrace();
                }
            }

            @Override
            public void run() {
                try {
                    while ((content = in.readLine()) != null) {
                        for (Socket s : mList) {
                            System.out.println("从客户端接收到的消息为:" + content);
                            if (content.equals("bye")) {
                                System.out.println("~~~~~~~~~~~~~~");
                                mList.remove(socket);
                                in.close();
                                content = "用户:" + socket.getInetAddress()
                                        + "退出:" + "当前在线人数:" + mList.size();
                                //5.关闭资源
                                socket.close();
                                this.sendmsg();
                                break;
                            } else {
                                content = socket.getInetAddress() + " 说: " + content;
                                this.sendmsg();
                            }
                        }
```

```
90                     }
91                 }catch (Exception e) {
92                     e.printStackTrace();
93                 }
94             }
95 
96             //为连接上服务器端的每个客户端发送信息
97             public void sendmsg() {
98                 System.out.println(content);
99                 int num = mList.size();
100                for (int index = 0; index < num; index++) {
101                    Socket mSocket = mList.get(index);
102                    PrintWriter pout = null;
103                    try {
104                        //PrintWriter 和 BufferWriter 使用方法相似
105                        pout = new PrintWriter(new BufferedWriter(
106                                new OutputStreamWriter(mSocket.getOutputStream(), "utf-8")), true);
107                      pout.println(content); //将输出流包装为打印流
108                  } catch (IOException e) {
109                      e.printStackTrace();
110                  }
111              }
112          }
113 
114      }
115  }
```

(1) 第 24～47 行,是该类的构造方法,在构造方法中初始化服务器。其中第 34～43 行,是一段 while(true)循环,在此循环中不停地执行监听请求,处理请求操作。第 38 行,当接收到客户端请求后,将该客户端的 Socket 对象添加到 List 对象 mList 中。服务器程序就是通过该 mList 对象里的 Socket 对象序列来创建线程处理多用户请求的线程池。

(2) 第 49～114 行,定义一个继承 Runnable 的内部子类 Service。这是服务器端程序的主体内容。

(3) 第 54～67 行,是该子类的构造方法,在构造方法中初始化从客户端传入的输入流对象,统计 mList 中的对象个数,mList 中的对象个数就是当前在线的人数。然后调用自定义方法 sendmsg()。

(4) 第 70～94 行,重写 Service 类的执行体 run()方法。在该方法中处理从客户端传入的数据流信息。在此将遍历 mList 中的所有 Socket 对象的输入信息,如果输入信息为 bye,则从 mList 列表中移动该 Socket 对象,重新统计当前在线人数,关闭资源;否则调用方法 sendmsg()。

(5) 第 97～112 行,定义 sendmsg()方法。该方法具体实现向连接上服务器端的每个客户端发送消息(见第 100～111 行的 for 循环),消息是通过 PrintWriter 对象 pout 调用 println()输出的。

在这里,服务器就好比一个消息中转站,向连接到服务器上的所有客户端转发某个客户端发送的信息,这就是聊天室的基本功能。

运行结果:运行本案例分两部分进行。首先启动服务器端程序,然后再运行客户端程序。

在 Android Studio 支持的模拟器上,运行 Server_MultiThread 项目,如图 13-37(a)所

示；再运行 Client_ MultiThread 项目，初始界面如图 13-37(b)所示。单击“连接服务器”按钮，在服务器端的 Run 窗格可以看到连接上服务器的信息，如图 13-37(c)所示。在客户端同步可以看到连接信息，如图 13-37(d)所示。在客户端输入一段文字，如图 13-37(e)所示，然后单击“发送消息”按钮，即可在服务器端和客户端都看到发出的信息内容，分别如图 13-37(f)和图 13-37(g)所示。当客户端输入 bye 时，该客户端即刻退出服务器连接；服务器会重新统计在线人数，并向聊天室发出该客户已退出的信息，分别如图 13-37(h)和图 13-37(i)所示。

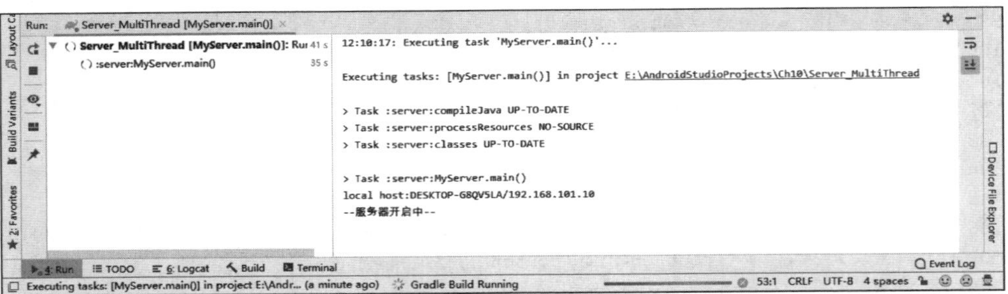

(a) 服务器启动后的初始运行结果

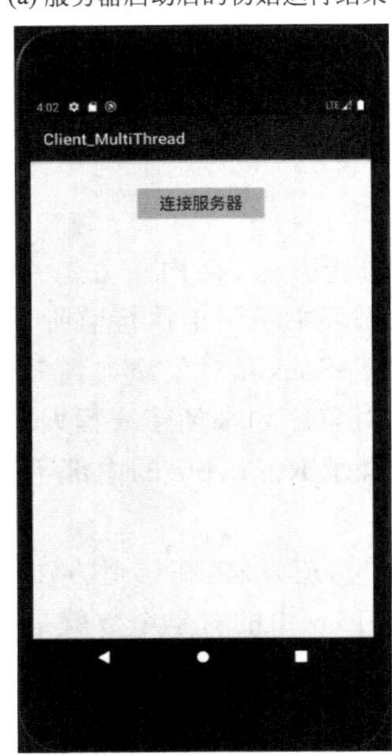

(b) 客户端初始运行界面

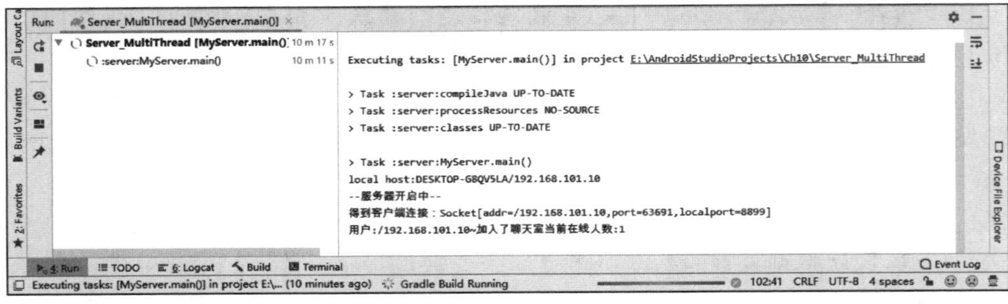

(c) 服务器端接收到客户端的连接请求

图 13-37　简单聊天室服务器端-客户端的运行结果

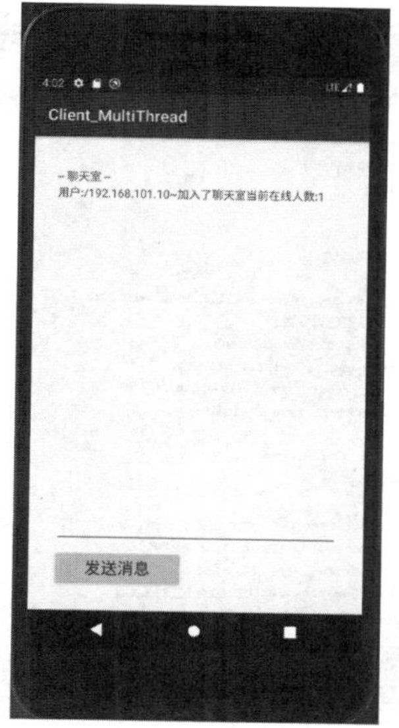

(d) 客户端收到来自服务器端的信息

(e) 客户端输入文字信息

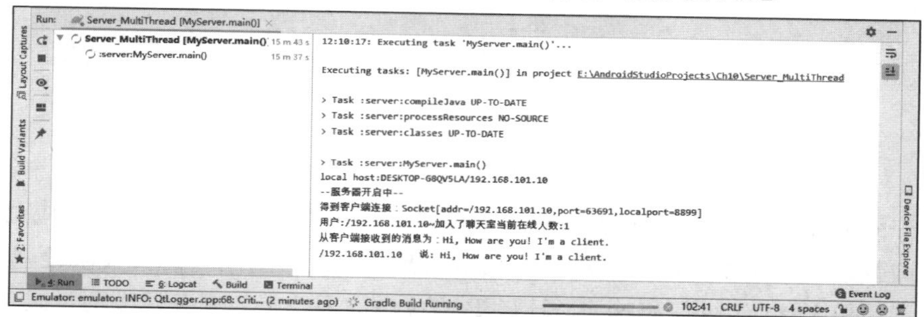

(f) 服务器端接收到客户端发送的信息

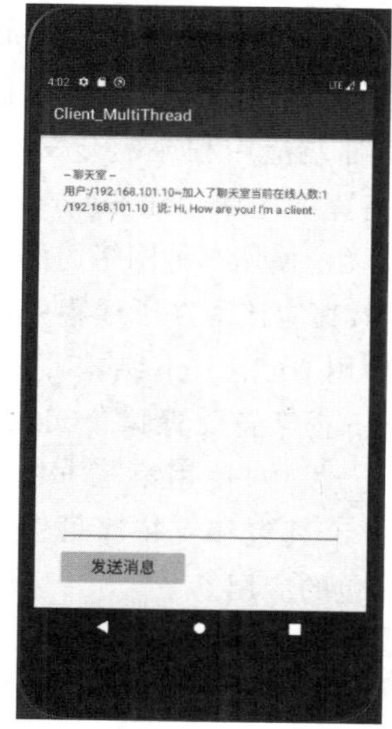

(g) 客户端收到服务器端广播的信息

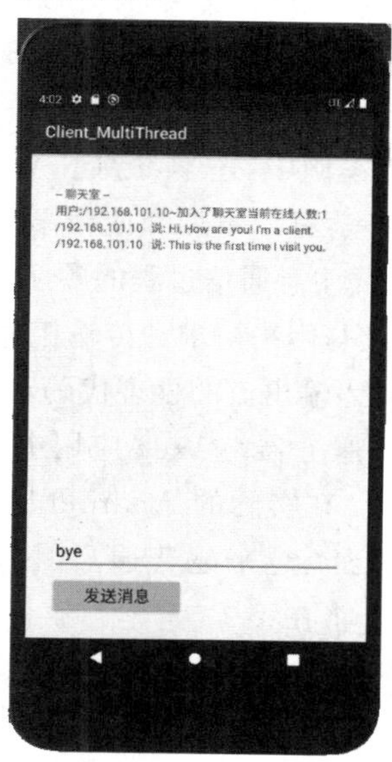

(h) 客户端输入退出信息bye

图 13-37 （续）

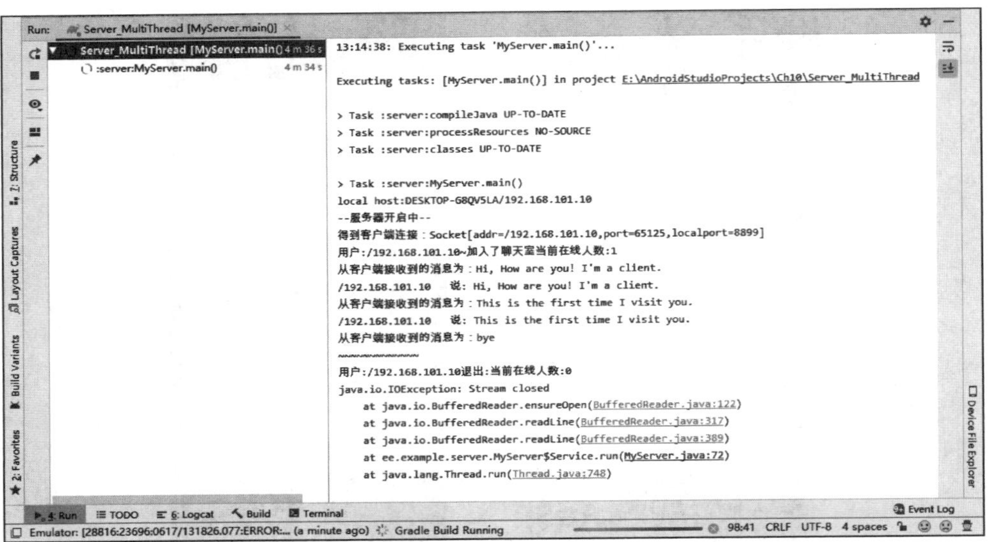

(i) 服务器端对客户退出做出的响应处理

图 13-37 （续）

小结

本章主要介绍了在 Android 平台下网页浏览、HTTP 协议下的网络访问应用、网络通信协议下的数据传输应用等网络应用有关技术。具体涉及对应用的网页权限设置；使用手机浏览器和使用 WebView 浏览网页；HTTP 协议中的客户端与服务器端的请求-响应工作模式，Tomcat 服务器简介与安装，使用 HttpURLConnection 接口和使用 OkHttp 框架来发送网络请求的编程技术；使用 Socket 和 ServerSocket 实现 TCP 协议下的数据通信编程技术等。并通过案例来学习这些技术的用法，掌握 GET 和 POST 上传数据的区别。在访问 Web 服务器端应用案例中，还学习到了 Web 服务器端应用程序的开发环境配置和 Web 服务器端应用的部署，学习了添加和使用第三方支持库的应用。在 Socket 通信案例中，知道了使用 Socket 开发时要注意通信过程的顺序，等等。无论使用哪种网络通信方式，在编程细节上都要注意：凡是涉及与网络请求和传输相关的代码，都应该写在子线程中，防止阻塞产生；凡是连接服务器、对输入输出流的处理代码，都应该使用 try...catch 结构等编程技术。

在开发客户端-服务器端 Web 应用中，OkHttp 框架备受青睐。OkHttp 是由 Square 组织开发的第三方库。在实际的 Android 应用项目开发中，使用第三方机构或组织提供的开源工具包和支持库已经越来越普遍。因为第三方工具包和支持库可以方便地扩展应用功能，提升开发效能。将在第 14 章学习第三方工具包的应用。

练习

1. 设计“我的音乐盒”的用户进入应用，可使用案例 13.4 中的服务器端 Login_Server

应用项目，Tomcat 服务器端口号为 9090。要求：初次进入时需要登录，实现将用户名(admin)和密码(123456)上传到服务器端，当登录成功后自动记住用户输入的登录信息，进入主页。再次登录时直接进入欢迎页面，5 秒后进入主页。

2. 设计“我的音乐盒”的歌手信息上传应用，可使用案例 13.5 中的服务器端 Upload_Server 应用项目，Tomcat 服务器端口号为 8080。输入的信息包括歌手头像、姓名、出生日期、简介等内容，要求如下。

(1) 歌手头像从本机的图库中选取，上传到服务器上指定文件夹中存储。

(2) 适当修改 Upload_Server 项目，将接收到的歌手相关文本信息，以与头像文件名同名的 XML 文件，保存到同一指定文件夹中。(选做)

第 14 章
第三方SDK应用

随着智能手机的广泛应用、功能的日益丰富，越来越多的第三方服务商向 Android 推出了大量的 SDK，这些 SDK 是第三方服务公司为了便于开发人员使用其提供的服务而开发的工具包，封装了一些复杂的逻辑实现以及请求，响应解析的 API。第三方的 SDK 包含的服务十分广泛，例如地图、广告、支付、统计、社交、推送等服务类别，利用第三方 SDK 开发相应的应用非常高效快捷。

本章将对应用较普遍的几类 SDK 的开发应用进行介绍，包括：基于地图的定位服务、POI 服务、导航服务，基于语音的识别服务、合成服务，基于即时通信的微信等。

14.1 地图 SDK

地图是人们在日常生活中不可或缺的应用。手机地图服务最开始由 Google 公司推出的 Google Map 提供，在中国，人们更多地使用由百度推出的百度地图、阿里巴巴推出的高德地图提供 SDK 来开发应用。无论是使用百度地图 SDK，还是使用高德地图 SDK，在开发前必须做好三个步骤：获取密钥；下载开发包；配置开发环境。

14.1.1 获取密钥

获取相应资源的密钥是使用地图 SDK 的第一步。每申请一个密钥，只能对应一个应用项目。在申请密钥时，需要提供 Android 应用项目的开发版和发布版的 SHA1 信息，还需要提供项目的包名。

1. 获取 Android 项目相关信息

SHA1 是一种安全散列算法，由一串十六进制数表示，主要用来验证数据的完整性。一个 Android 项目有两个 SHA1 码，一个是开发版的 SHA1，它与 Android 的 SDK 版本相关；另一个是发布版的 SHA1，它与项目的 APK 签名文件相关。

1）获取开发版 SHA1

开发版的 SHA1 与 Android 的 SDK 版本相关，如果项目都使用相同的 SDK，那么开发版 SHA1 也相同。查看 SHA1 码的方式有两种：一种是在命令行状态使用 keytool-list-v -keystore 命令；另一种是在 Android Studio 中的 Gradle 任务中查看。

方式一：在命令行状态环境下查看。在 Android Studio 中的 Terminal 窗格，就像在 Windows 环境下运行 cmd 命令一样，使用命令进行操作。打开 Terminal 窗格有两种方式，一是利用菜单，依次选择菜单 View→Tool Windows→Terminal；二是利用 Android Studio 底部的工具栏，单击 Terminal，即可打开 Terminal 窗格。

查看开发版 SHA1，需要指定 debug. keystore 文件的完整路径。该文件存储在 Android 的安装文件夹. android 下。本书所用计算机的 debug. keystore 存储路径是：C:\Users\dell\. android，debug. keystore 的原始密码默认是 android。在 Terminal 中通过 keytool 命令查找。查看命令如下。

```
keytool -list -v -keystore C:\Users\dell\.android\debug.keystore
```

输入该命令后按 Enter 键，接下来要输入密码，原始密码为 android。当输入的密码验证通过后即可列出相关信息，其中包括开发版的 SHA1 编码，执行结果如图 14-1 所示。

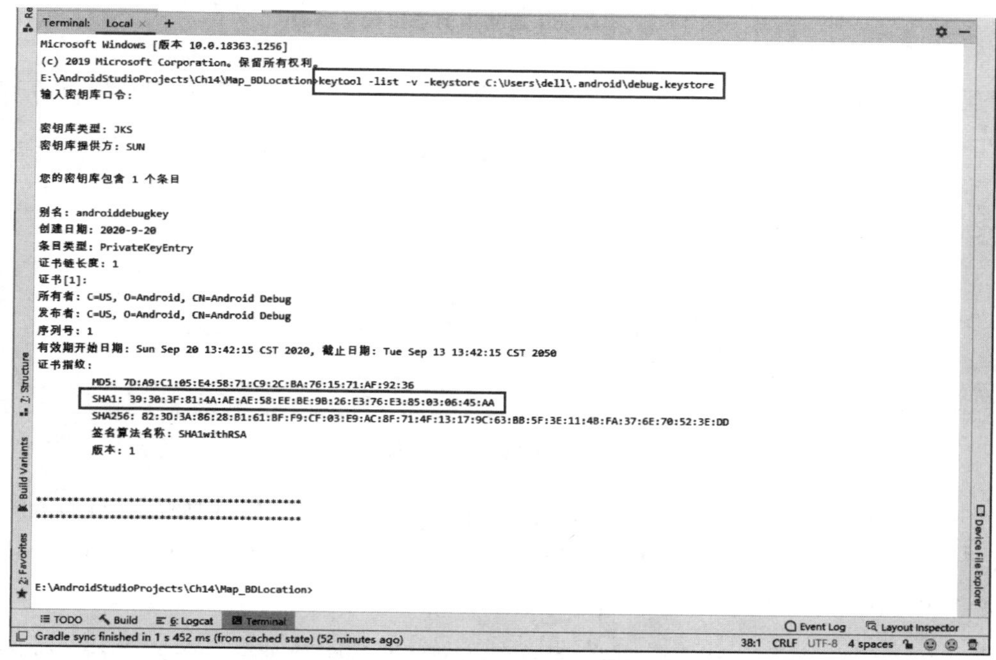

图 14-1 Terminal 窗格中的开发版 SHA1 信息

方式二：在 Android Studio 的右侧工具栏中，单击 Gradle，打开 Gradle 窗格，展开项目的 Tasks 结点下的 android，如图 14-2 所示。双击 signingReport 项，即可在 Android Studio 下部的 Run 窗格看到签名报告的运行结果，其中包括了开发版的 SHA1 编码，如图 14-3 所示。

2）获取发布版 SHA1

获取发布版的 SHA1，需要 Android 项目的 APK 签名文件。创建项目的签名文件就是一个签名打包过程，在 Android Studio 中依次选择菜单 Build→Generate Signed Bundle/APK...进行生成。具体操作在第 1 章的 1.4.4 节已经介绍了，在此不赘述。注意，在生成签名文件时要记住设置的密码，因为在查看签名信息时需要输入该密码。

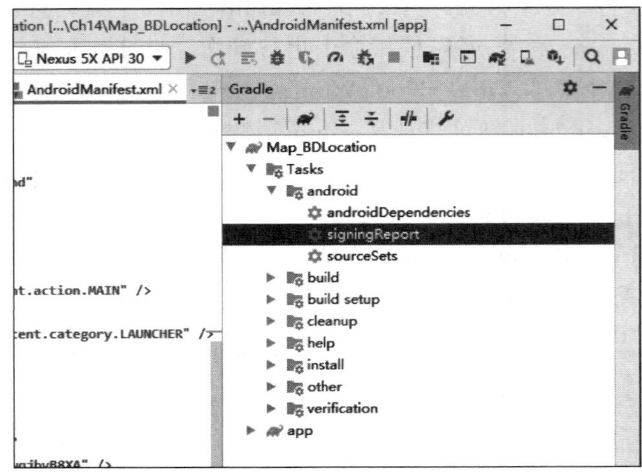

图 14-2　Gradle 窗格中的项目任务目录

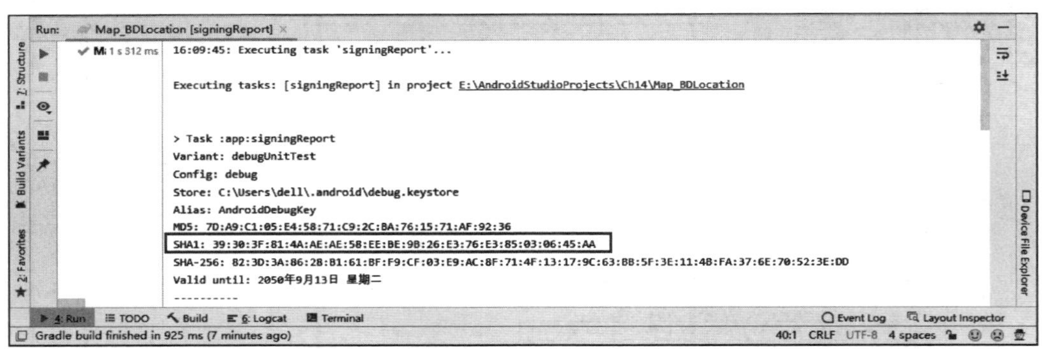

图 14-3　Run 窗格中的签名报告信息

查看发布版 SHA1，需要指定项目的签名文件的全路径文件名。例如本项目的 APK 签名文件全路径文件名为 E:\AndroidStudioProjects\Ch14\Map_BDLocation\Map_BDLocation.jks，查找命令如下。

```
keytool -list -v -keystore E:\AndroidStudioProjects\Ch14\Map_BDLocation\Map_BDLocation.jks
```

输入该命令之后按 Enter 键，接下来要输入签名文件的访问密码，该密码是在创建签名文件时设置的。当输入的密码验证通过后即可列出相关信息，如图 14-4 所示。

3）获取 Android 项目的包名

打开 Android 项目的 AndroidManifest.xml 清单文件，package 属性所对应的内容即为应用项目的包名。

在 Android Studio 中，项目的包名还可以通过模块的 build.gradle 文件里的 applicationId 来配置包名，如果在 build.gradle 文件中配置的包名和 AndroidMainfest.xml 中的 package 属性对应的内容不同时，项目包名以模块 build.gradle 文件的 applicaionId 中定义为准，如图 14-5 所示。

```
Terminal:  Local  +
Microsoft Windows [版本 10.0.18363.1256]
(c) 2019 Microsoft Corporation。保留所有权利。

E:\AndroidStudioProjects\Ch14\Map_BDLocation>keytool -list -v -keystore E:\AndroidStudioProjects\Ch14\Map_BDLocation\Map_BDLocation.jks
输入密钥库口令:

密钥库类型: JKS
密钥库提供方: SUN

您的密钥库包含 1 个条目

别名: test
创建日期: 2020-12-20
条目类型: PrivateKeyEntry
证书链长度: 1
证书[1]:
所有者: CN=BdMap, OU=Person, O=StudyG, L=GD, ST=GZ, C=cn
发布者: CN=BdMap, OU=Person, O=StudyG, L=GD, ST=GZ, C=cn
序列号: 7a7ebd08
有效期开始日期: Sun Dec 20 16:53:06 CST 2020, 截止日期: Thu Dec 14 16:53:06 CST 2045
证书指纹:
         MD5: 5E:C3:B6:B5:F1:21:94:8D:EC:9A:D9:5D:FE:86:DE:80
         SHA1: 25:F0:87:D4:AD:58:FC:8F:37:CF:6A:38:FF:3A:0D:09:3C:B7:E6:9F
         SHA256: 3D:67:9A:F8:B8:6B:D1:6E:94:4D:39:1A:56:71:0A:D6:45:6C:54:29:A3:A0:9C:A5:A2:D8:C0:33:C0:48:71:65
         签名算法名称: SHA256withRSA
         版本: 3
Generate Signed APK: APK(s) generated successfully for module 'app' with 1 build variant: // Build variant 'r... (5 minutes ago)  41:12  CRLF  UTF-8  4 spaces
```

图 14-4 Terminal 窗格中的发布版 SHA1 信息

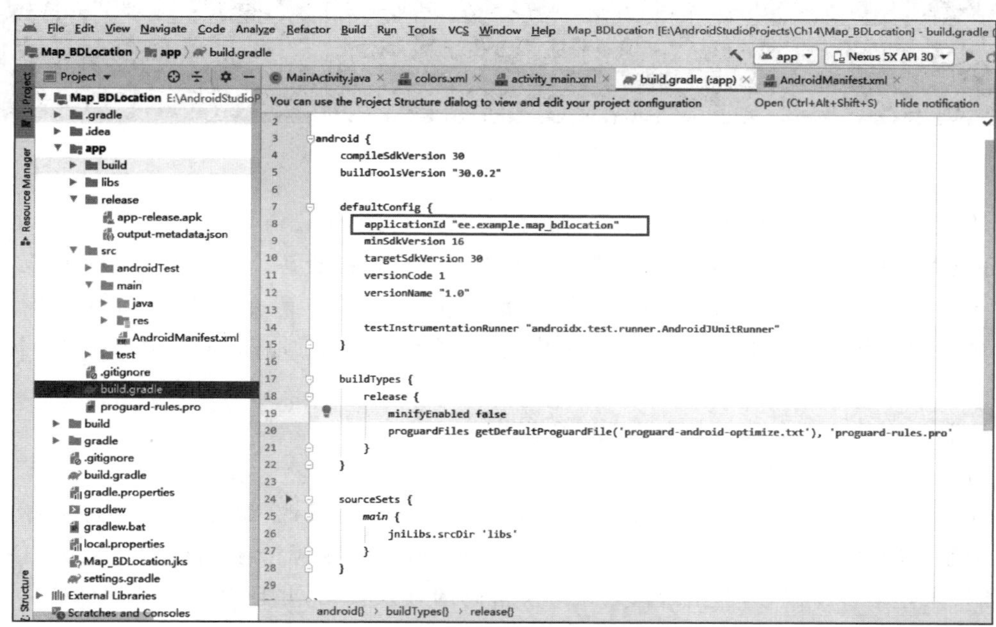

图 14-5 模块级 build. gradle 中的包名信息

2. 获取地图密钥

如果使用百度地图 SDK 开发应用,需要在百度地图开放平台上获取 AK 密钥。如果使用高德地图 SDK 开发应用,需要在高德开放平台上获取 Key 密钥。

1) 获取百度 AK 密钥

百度地图为开发者提供了一个开放平台,平台网址为 http://lbsyun. baidu. com/。进入开放平台后请仔细浏览平台首页关于开发者的相关信息，这样对以后的开发有帮助。首次使用该平台，需要注册百度账号，并申请成为百度地图开发者。

针对不同的开发环境，百度地图都不同程度地提供了相关的SDK，如图14-6所示。其中，为Android开发提供的SDK，是一套基于Android 4.0及以上版本设备的应用接口。开发者使用该套SDK可以轻松访问百度地图服务和数据，构建功能丰富、交互性强的地图类应用项目。

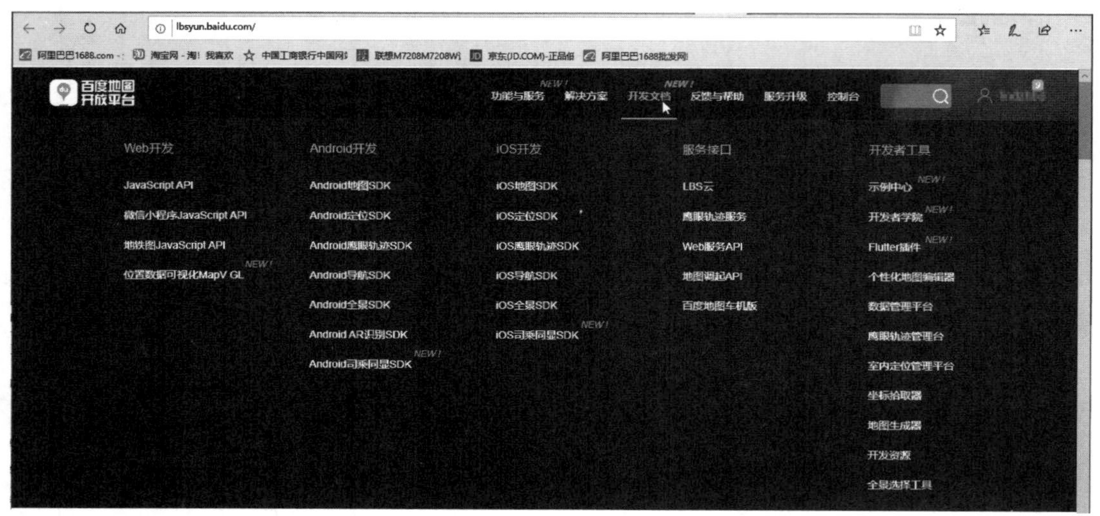

图 14-6　百度地图开放平台提供的 SDK

从Android 4.0版本开始，引入百度地图开放平台的SDK需要AK验证，通过AK密钥，开发者可以更方便、更安全地配置自身使用的百度地图资源。每一个AK密钥仅且唯一对于一个应用验证有效，多个应用(包括多个包名)需要申请多个AK；如果在同一个项目中同时使用多个应用，例如Android定位SDK和Android地图SDK，可以使用同一个AK。

获取AK的流程大致可分为如下四个步骤。

(1) 登录API控制台

在百度的API控制台可以创建新的AK，也可以查看、修改、删除之前所创建的AK。API控制台网址为：http://lbsyun.baidu.com/apiconsole/key。输入网址，进入API控制台，如果未登录百度账号，会出现输入账号及密码的登录页面，并且该账号必须拥有百度地图开放平台的开发者身份，只有这个身份才能正常进入API控制台。

(2) 创建应用

进入API控制台后，在控制台看板中展开“应用管理”项，选择“我的应用”，如图14-7所示。然后单击“创建应用”按钮，开始创建新的应用。

(3) 配置SHA1和包名

单击创建应用，将会进入如图14-8所示的页面，在这个页面中，开发者需要填写应用名称、选择应用类型和配置SHA1及包名。

在输入应用名称时，开发者可以自行定义，建议与Android的应用名称一致，便于管理。

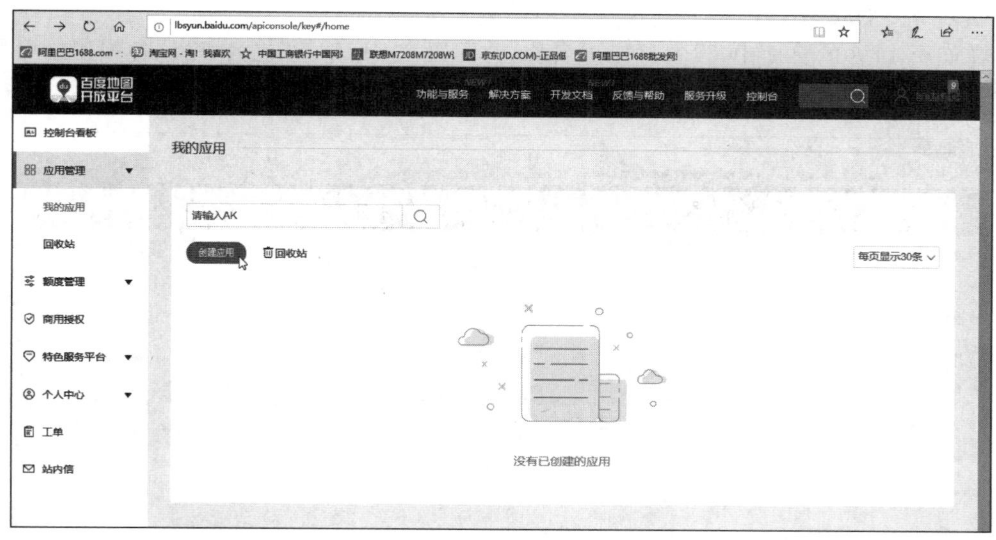

图 14-7　百度 API 控制台

图 14-8　配置百度应用的名称、SHA1 及包名信息

对于应用类型，请选择 Android SDK，如果选择了其他类型，将导致所生成的 AK 不可用。在启用服务项，按默认的全部启用即可。

(4) 提交生成 AK

在如图 14-8 的页面中，设置好应用名称，并配置完 SHA1 和包名，确认各项信息填写无误后，单击“提交”按钮，即可生成 AK 的密钥，如图 14-9 所示。

图 14-9　百度地图开放平台创建应用的 AK 密钥

注意：由此获取的 AK 密钥，只为 Android 当前的应用项目所拥有，其他的 Android 项目使用无效。

2) 获取高德 Key 密钥

高德开放平台为开发者提供了一个开发高德地图应用的资源平台，相关 Android 的开发指南网址为：https://lbs.amap.com/api/android-sdk/summary/。首次使用该平台，需要注册高德开发者账号。

获取 Key，在高德开放平台的控制台中进行，获取步骤同百度获取 AK 相似，同样也是四个步骤：①登录控制台；②创建应用；③配置 SHA1 和包名；④提交生成 Key。

进入控制台，可从高德开放平台 Android 开发指南网页最上端导航栏的右侧，单击“控制台”进入，如图 14-10 所示。

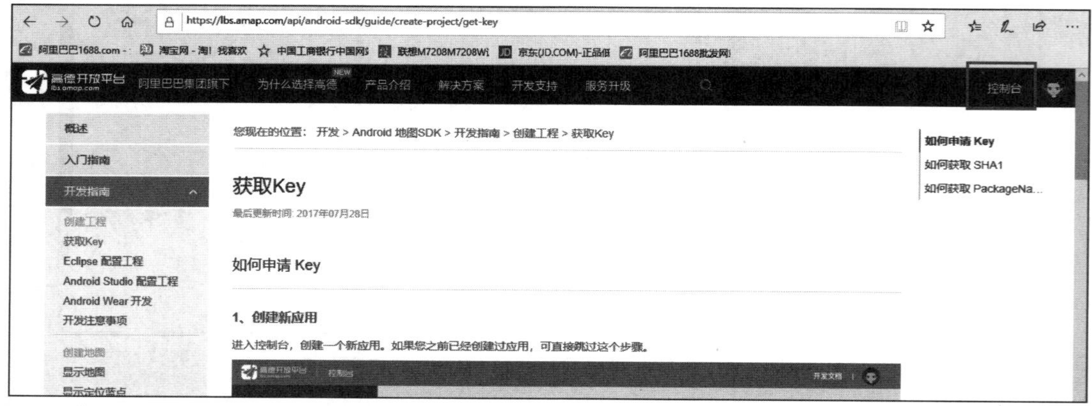

图 14-10　高德开放平台的 Android 开发指南页面

然后创建应用，并命名 Key 名称，配置 SHA1 和包名等信息。配置后的界面如图 14-11 所示。

图 14-11　配置高德应用的名称、SHA1 及包名信息

单击“提交”按钮，即可在“我的应用”网页看到新创建的应用和 Key 值，如图 14-12 所示。同样，每一个 Android 应用项目只对应一个 Key。

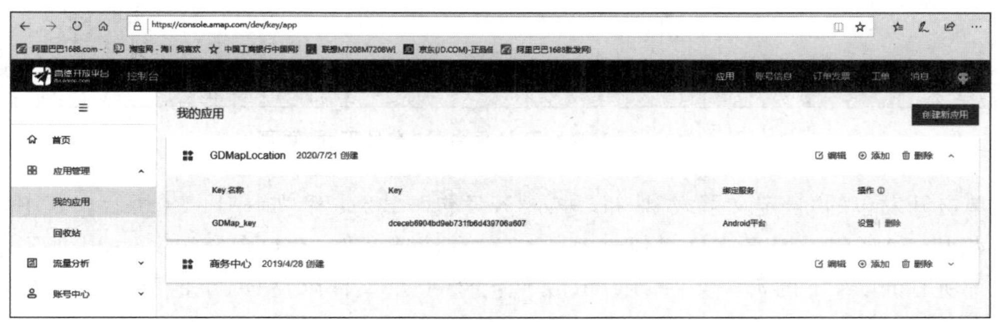

图 14-12　高德开放平台创建应用的 Key 密钥

14.1.2　下载开发包

地图应用的第二步，是下载相应的开发包。下面分别介绍百度地图和高德地图的开发包下载位置和相关选项。

1. 百度地图开发包

使用百度地图 SDK，需要从百度地图开放平台上下载应用需要的开发包，网址为 http://lbsyun.baidu.com/index.php?title=androidsdk/sdkandev-download。在下载页选择“产品下载”，在右边有三个下载按钮，其中“开发包下载”是我们需要的，同时还有一些供开发者学习的 Demo 源代码下载、地图组件下载，都是对于开发有帮助的资源，如图 14-13 所示。

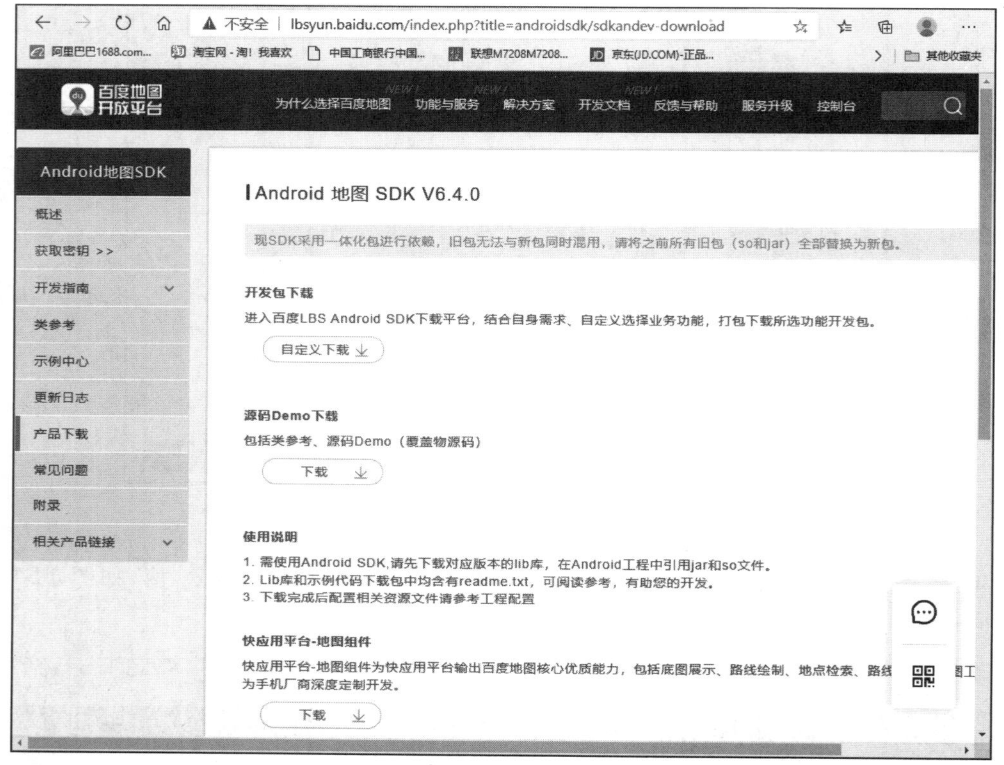

图 14-13　百度地图开放平台的“产品下载”页面

单击开发包的“自定义下载”按钮，进入“开发包下载”页面，如图 14-14 所示。其中包括 JAR 包和 AAR 包。JAR 文件只包含 class 文件和清单文件，不包含资源文件，比如不包括图片等所有的 res 下的资源文件。AAR 文件包含 class 文件以及 res 下的所有资源文件。

在如图 14-14 的页面中，选择了需要的应用功能，将它们集成在一个 JAR 开发包中，然后单击“开发包”按钮，开始下载。再单击“示例代码”按钮，下载一些示范案例，帮助读者快速学习如何使用这些 SDK 进行开发。

2. 高德地图开发包

高德开放平台 Android 开发包的下载网址为 https://lbs.amap.com/api/android-sdk/download，在该网页上提供了三个版本的开发包，即 Android 地图 SDK、Android 地图 SDK 旧版和 AAR 包下载。其中 Android 地图 SDK 提供 Android 2.3.0 版本及以上的 SDK，

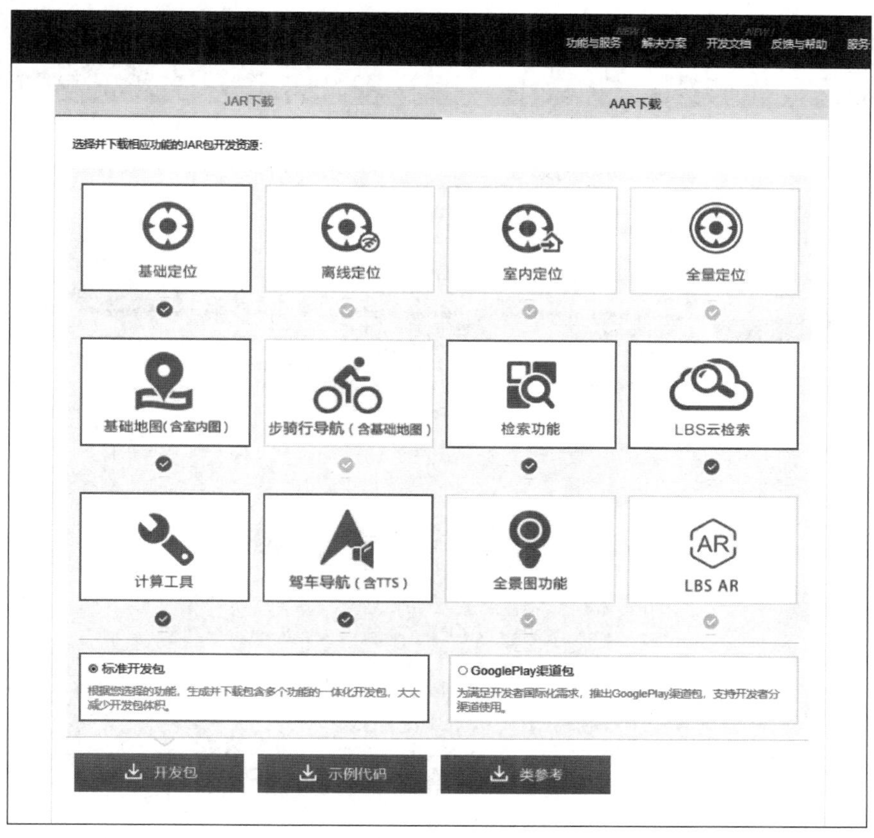

图 14-14　百度地图开放平台的"开发包下载"页面

Android SDK 旧版提供低于 Android 2.3.0 版本的 SDK,如图 14-15 所示。

从网页上可以看到,支持 Android 2.3.0 以上的版本有比较丰富的开发包,可根据自己的应用来选择性地下载。本书选择"一键下载",其中包括 2D、3D 地图包、搜索包示例源代码和相关文档。基本上涵盖了地图定位、POI 搜索、地图导航等常用的地图应用。

14.1.3　配置开发环境

前两步只是获取开发地图应用的开发资源,关键是要配置到开发环境中,才能实现地图的应用。本书以配置 Android Studio 为例,大致可分为如下三个步骤:复制相关的开发包到项目中的 libs 目录,配置模块 build.gradle 文件,配置 AndroidManifest.xml 清单文件。

1. 复制相关开发包

在 Android Studio 的 Project 视图窗格,选择 Project 展示方式,即可以看到 libs 目录,libs 目录在项目的 app 目录下,与 src 目录在同一级。如果目录结构中没有 libs,可以自行创建。

如果 Project 窗格选择的是 Android 展示方式,得到的相应目录是 jniLibs。

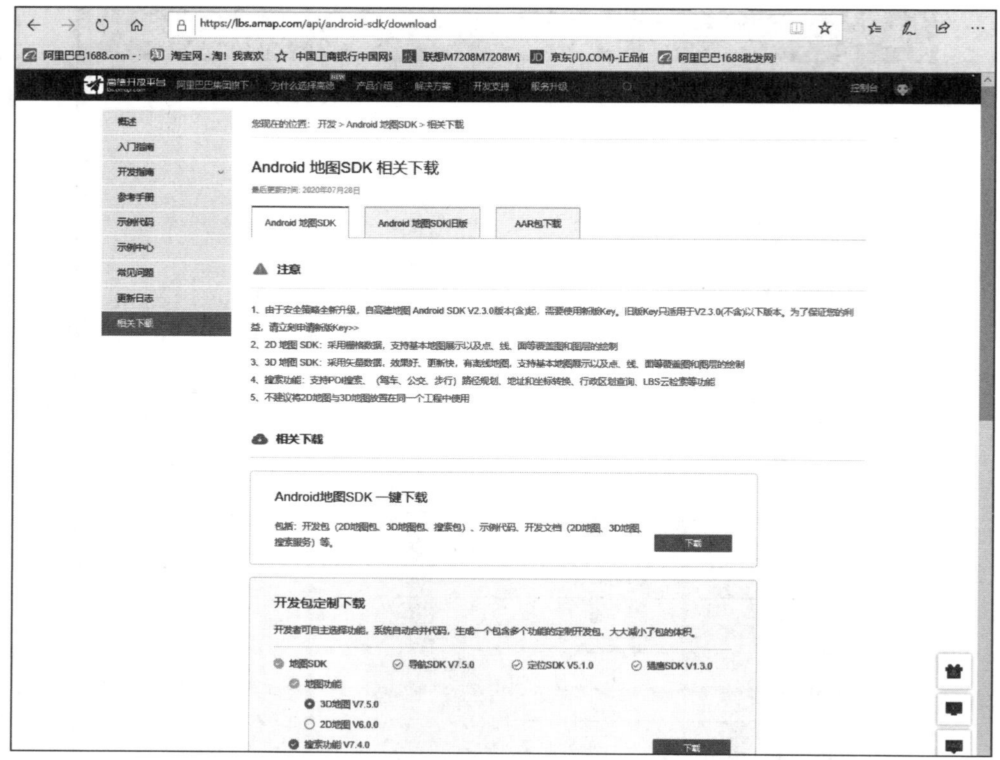

图 14-15　高德开放平台的"开发包下载"页面

使用地图 SDK 开发，需要将相关的开发包复制到应用项目的 libs 目录中。其中包括 SO 包和 JAR 包。

1）复制 SO 包

不同的应用设备需要使用不同的 SO 包。如果所开发的地图应用要求能够运行在各种 Android 设备上，必须支持所有设备的 CPU 类型。常见的 Android 设备的 CPU 类型分以下几种。

armeabiv-v7a：第 7 代及以上的 ARM 处理器，是目前主流版本。2011 年以后生产的大部分 Android 设备都使用它。

arm64-v8a：第 8 代、64 位 ARM 处理器，以三星 Galaxy S6 为代表的设备使用该处理器。

armeabi：第 5 代、第 6 代的 ARM 处理器，早期的手机用得比较多。

x86：平板电脑、模拟器用得比较多。

x86_64：64 位的平板电脑。

不同架构的 CPU 会加载不同的 .so 文件，支持地图 SDK 的相应 .so 文件必须加载到项目中。这些文件包含在以 CPU 类型命名的目录中。如果需要所开发的地图应用在各种 Android 设备上都可以运行，那么需要将这些目录复制到项目的 libs 目录中。相关的 SO 文件会包含在 AAR 压缩包里，相关的 SO 文件也可以从下载例程的 libs 目录中获得。

2）复制 JAR 包

不同的应用也需要使用不同的地图 JAR 包。百度 JAR 包在下载之前提供了开发功能选择，高度集成在一个 JAR 包文件中，即 BaiduLBS_Android.jar 文件。

如果要开发百度地图应用，并且能运行在任意 Android 设备上，需要复制相关的 SO 包和 JAR 包，如图 14-16 所示。

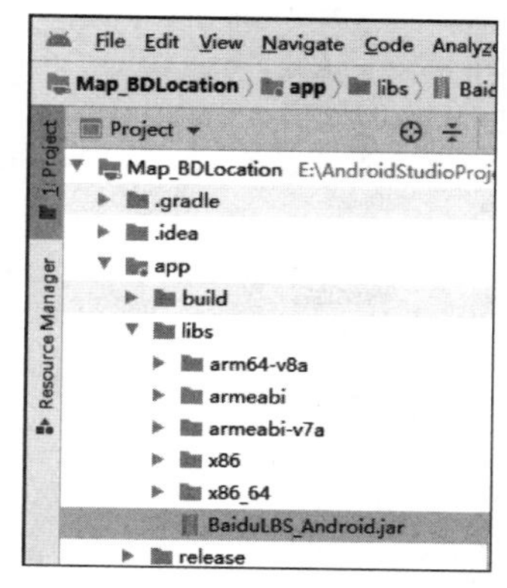

图 14-16　复制到项目 libs 目录中的开发包

与百度地图 JAR 包不同，高德地图的 JAR 包区分了功能，即每一个 JAR 包只包括一类应用的功能。示例如下。

Amap_2DMap_V6.0.0_20191106.jar 主要针对 2D 地图的栅格数据，支持基本地图展示以及点、线、面等覆盖图和图层的绘制。

Amap_location_V5.1.0_202000708.jar 主要针对位置信息服务，可以获取定位结果、地址文字描述以及地理围栏等功能。

Amap_Search_V7.3.0_20200331.jar 支持 POI 搜索、(驾车、公交、步行)路径规划、地址和坐标转换、行政区划查询、LBS 云检索等功能。

使用高德地图开发应用时，根据需求复制相应的 JAR 包。一般不建议将 2D 地图与 3D 地图放置在同一个项目中使用。

2. 配置 gradle

在 Android Studio 中，配置 gradle 非常重要，有时候代码编写没问题，但 gradle 配置不同步，会造成编译失败。

将复制到项目的 libs 目录下的.jar 文件添加到本地项目库中，操作方法是选择对应的文件，然后右击，在弹出菜单中选择 Add As Library…。在执行该命令之后，依次选择菜单 File→Sync Project with Gradle Files，使得项目与 gradle 配置同步。此时在 Android Studio 的 Project 视窗可以看到 BaiduLBS_Android.jar 文件包可以展开，如图 14-17 所示。

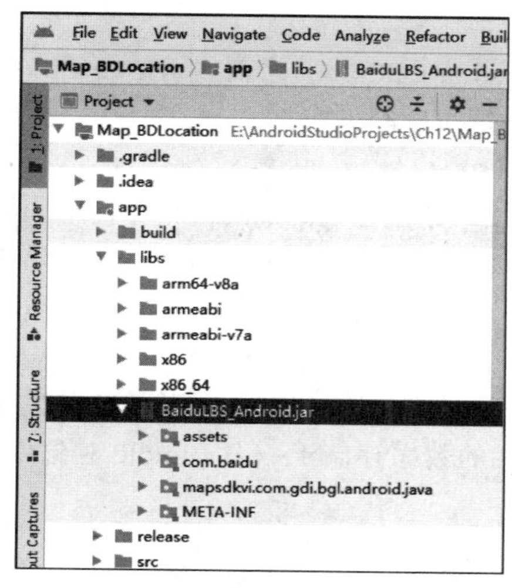

图 14-17　展开 BaiduLBS_Android.jar 文件包的内容

相关的配置信息可以在模块级 build.gradle 文件中看到。以配置百度地图应用为例，配置完成后的模块 build.gradle 文件内容如图 14-18 所示。

复制包含.so 文件的 arm64-v8a、armeabiv-v7a 等目录到 libs 目录下，在 build.gradle 文件的 android{}内需要手动添加指定目录说明配置，代码如下。

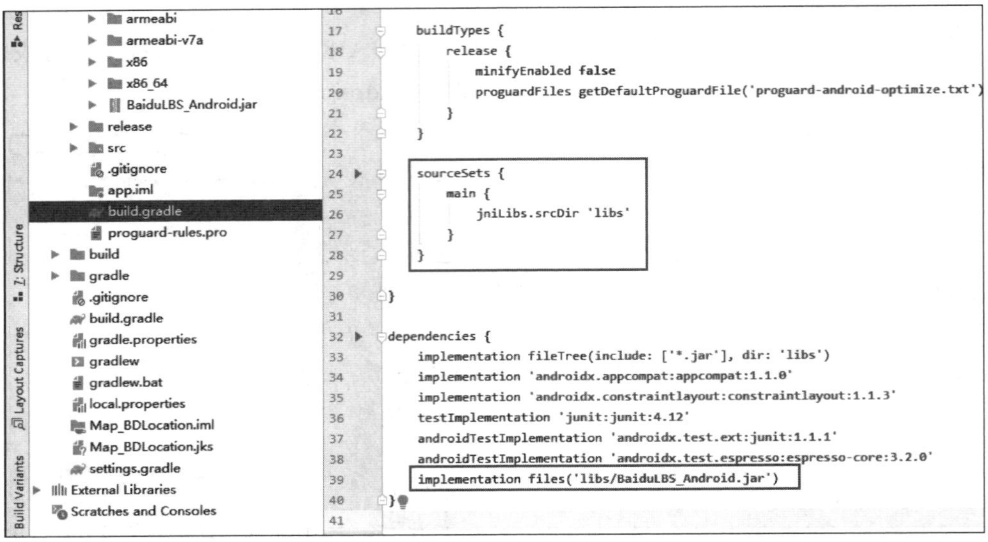

图 14-18　模块级 build. gradle 文件内容

```
sourceSets {
    main {
        jniLibs. srcDir 'libs'
    }
}
```

复制百度 JAR 包文件 BaiduLBS_Android. jar 到 libs 目录下，然后选择 Add As Library...添加到项目库中，在 build. gradle 文件的 dependencies 中会自动增加如下依赖配置。

```
implementation files('libs/BaiduLBS_Android. jar')
```

3. 配置 AndroidManifest. xml

完成上述操作之后，还需要配置 AndroidManifest. xml 文件，将之前申请的密钥和相关的地图应用所需要的权限添加到相应位置，才能开始编写代码工作。

1）添加密钥

为了保证地图 SDK 的功能正常使用，需要将前面申请的密钥添加到清单文件中。如果使用百度地图 SDK，则添加 AK 密钥；如果使用高德地图 SDK，则添加 Key 密钥。配置方法是在清单文件的< application >元素内增加< meta-data >子元素。

< meta-data >子元素直译为“元数据”，可为< activity >、< activity-alias >、< application >、< provider >、< receiver >、< service >等组件提供附加数据项。任何组件元素可以包含任意数量的< meta-data >子元素。系统将 meta-data 配置的数据存储于一个 Bundle 对象中，可以通过 PackageItemInfo. metaData 字段获取。

添加百度 AK，在< application >元素内增加如下内容。

```
<meta - data
    android:name = "com. baidu. lbsapi. API_KEY"
    android:value = "v9VFG7u2EbcYIHMvnBZ0LgpwqjbvB8XA" />
```

添加高德 Key,在<application>元素内增加如下内容。

```
<meta-data
    android:name="com.amap.api.v2.apikey"
    android:value="dceceb6904bd9eb731fb6d439706a607" />
```

其中属性 value 的值,就是所获取的密钥。

2）声明权限

开发地图应用,会用到多种权限。例如,进行地图相关业务数据请求,包括地图数据、路线规划、POI 检索等,则需要申请访问网络权限；根据网络状态,确定是否进行数据请求,还是转换网络请求,则需要申请读取网络状态权限；如果在定位或导航中需要用到 WiFi 网络,则需要申请获取 WiFi 状态,更改 WiFi 连接状态权限；如果要使用 GPS 定位,则需要申请获取精确 GPS 位置和获取粗略 GPS 位置权限；如果开发者使用了 so 动态加载功能并且把 so 文件存放在外置存储区域,则需要申请读取外置存储权限；如果开发者使用了离线地图,并且数据写在外置存储区域,则需要申请写外置存储权限；如果允许程序访问额外的定位提供者指令获取模拟定位信息,则需要申请访问额外定位命令权限；在手机定位应用中,往往会用到读写手机本身的状态和身份信息,通常也会需要申请读写手机状态权限；等等。

一般情况下,使用地图 SDK 开发应用,在清单文件会声明如下权限。具体的编程时,开发者可根据应用项目的需求进行增加或删减。

```
<uses-permission android:name="android.permission.INTERNET" />
<uses-permission android:name="android.permission.ACCESS_NETWORK_STATE" />
<uses-permission android:name="android.permission.CHANGE_WIFI_STATE"/>
<uses-permission android:name="android.permission.ACCESS_WIFI_STATE"/>
<uses-permission android:name="android.permission.ACCESS_FINE_LOCATION"/>
<uses-permission android:name="android.permission.ACCESS_COARSE_LOCATION"/>
<uses-permission android:name="android.permission.READ_EXTERNAL_STORAGE" />
<uses-permission android:name="android.permission.WRITE_EXTERNAL_STORAGE" />
<uses-permission android:name="android.permission.ACCESS_LOCATION_EXTRA_COMMANDS"/>
<uses-permission android:name="android.permission.READ_PHONE_STATE"/>
```

注意：在实际开发中,需要根据应用的功能需求范围,在 AndroidManifest.xml 中确定要声明的相关权限。对于部分被列入危险权限的,还需要在代码中动态请求授权。

14.1.4　地图应用

Android 地图应用较多的功能包括地图定位、搜索兴趣点、路径规划、电子围栏等。使用百度地图、高德地图或其他的地图服务商提供的开发包都可以实现,开发过程也大同小异。本书将给出两个案例来说明开发过程。

1. 显示地图并定位

手机本身是具有定位功能的,在第 12 章就介绍过手机定位的技术和实现。但是仅有手机的功能,只能给出一串长长的数据来表示当前位置,缺乏直观的效果,用户体验不好。现在可以使用第三方的地图 SDK 开发应用,于是,在 2D 或 3D 的地图上标注定位成为需求。

下面利用百度地图 SDK,来开发地图定位案例。

【案例 14.1】 使用百度地图 SDK，在地图上标注当前的位置。要求本应用能适用大多数的 Android 手机和平板计算机。

说明：在构建布局时，使用百度地图来作为背景。调用地图定位，需要设置定位的若干属性，以及设置定位监听等操作，使用一个定位服务类来专门负责管理此类操作。在主逻辑文件中使用监听机制来处理定位结果的展示。本例将采用定位标记点和文字两种方式显示定位。

开发步骤及解析：过程如下。

1）创建项目

在 Android Studio 中创建一个名为 Map_BDLocation 的项目。其包名为 ee. example. map_bdlocation。

2）生成 APK 签名

创建 APK 签名文件名为 Map_BDLocation. jks，存储在 E:\AndroidStudioProjects\Ch14\Map_BDLocation 路径下，并且记住启用该签名文件的密码。

3）获取 AK 密钥

在百度地图开放平台控制台中新建项目，配置本 Android 项目的发布版 SHA1、Android SDK 的开发版 SHA1 和本项目的包名 ee. example. map_bdlocation。生成本项目的 AK 密钥为 v9VFG7u2EbcYIHMvnBZ0LgpwqjbvB8XA。

4）下载开发包

因为本案例要求能适用于各种 Android 设备，因此需要下载 AAR 包，其中包括有适配各种 Android 设备的. so 文件包，以及地图相应功能的. jar 文件包。

5）配置 gradle

解压下载包，将包含. so 文件的 arm64-v8a、armeabiv-v7a、armeabi、x86、x86_64 目录以及开发功能包 BaiduLBS_Android. jar 文件都复制到本项目的 libs 目录下；然后将 BaiduLBS_Android. jar 文件添加到项目库中（右击目录中文件名，在弹出菜单中选择 Add As Library...）；同步项目与 gradle 配置（依次选择菜单 File→Sync Project with Gradle Files），配置模块 build. gradle。执行完上述操作后，项目的模块 build. gradle 文件代码如下。

```
apply plugin: 'com.android.application'

android {
    compileSdkVersion 29
    buildToolsVersion "29.0.0"

    defaultConfig {
        applicationId "ee.example.map_bdlocation"
        minSdkVersion 16
        targetSdkVersion 29
        versionCode 1
        versionName "1.0"

        testInstrumentationRunner "androidx.test.runner.AndroidJUnitRunner"
    }

    buildTypes {
```

```
18            release {
19                minifyEnabled false
20                proguardFiles getDefaultProguardFile('proguard-android-optimize.txt'),
                          'proguard-rules.pro'
21            }
22        }
23
24        sourceSets {
25            main {
26                jniLibs.srcDir 'libs'
27            }
28        }
29
30    }
31
32    dependencies {
33        implementation fileTree(include: ['*.jar'], dir: 'libs')
34        implementation 'androidx.appcompat:appcompat:1.1.0'
35        implementation 'androidx.constraintlayout:constraintlayout:1.1.3'
36        testImplementation 'junit:junit:4.12'
37        androidTestImplementation 'androidx.test.ext:junit:1.1.1'
38        androidTestImplementation 'androidx.test.espresso:espresso-core:3.2.0'
39        implementation files('libs/BaiduLBS_Android.jar')
40    }
```

(1) 第 24～28 行，指定.so 文件目录。新版的 Gradle 拥有自动打包编译 so 文件的功能，并且为 so 文件指定的默认目录是 app/src/main/jniLibs，只需要将 so 文件复制到该文件夹下，编译运行即可。但是为了更好地管理第三方库文件，建议将 jar 文件和 so 文件放在一起，统一复制在 app/libs 目录下，为此，需要在 build.gradle 的 android{}中添加这段命令。

(2) 第 33 行，声明一个本地依赖，其含义是将当前项目中 libs 目录下的所有扩展名为.jar 的文件都添加到项目的构建路径当中去。请注意，该语句只是本地依赖声明，而不是本地依赖的添加。

(3) 第 39 行，添加本地依赖。在完成对 libs 目录下的 BaiduLBS_Android.jar 文件的 Add As Library 选项后，该文件才被添加到本地依赖中。

6) 声明权限与配置 AK 密钥

在 AndroidManifest.xml 中添加本应用需要的权限，代码片段如下。

```
<!-- 访问网络,进行地图相关业务数据请求 -->
<uses-permission android:name="android.permission.INTERNET" />
<!-- 获取网络状态,根据网络状态切换进行数据请求网络转换 -->
<uses-permission android:name="android.permission.ACCESS_NETWORK_STATE" />
<!-- 获取精确 GPS 位置 -->
<uses-permission android:name="android.permission.ACCESS_FINE_LOCATION"/>
<!-- 获取粗略 GPS 位置 -->
<uses-permission android:name="android.permission.ACCESS_COARSE_LOCATION"/>
<!-- 更改 WiFi 连接状态 -->
<uses-permission android:name="android.permission.CHANGE_WIFI_STATE"/>
<!-- 获取 WiFi 状态 -->
<uses-permission android:name="android.permission.ACCESS_WIFI_STATE"/>
```

```
<!-- 允许程序读写手机状态和身份 -->
<uses-permission android:name="android.permission.READ_PHONE_STATE"/>
```

在 AndroidManifest.xml 的<application>元素的子元素 meta 内，写入百度地图应用（模块）的 Key，代码片段如下。

```
<application
    ...>
    ...
<meta-data
    android:name="com.baidu.lbsapi.API_KEY"
    android:value="v9VFG7u2EbcYIHMvnBZ0LgpwqjbvB8XA" />
</application>
```

7）准备图片资源

将事先准备的用于标记点的图标，复制到 res/drawable 目录中。

8）准备颜色资源

编写 res/values 目录下的 colors.xml 文件，添加半透明黑色 halfblack 和 white 等颜色。

9）设计布局

编写 res/layout 目录下的 activity_main.xml 文件。应用界面选用<FrameLayout>布局框架，实现多层叠加布局。在底层使用百度地图作为整个应用的背景图，在屏幕的顶部使用线性布局，该线性布局使用半透明色，在其中显示当前定位点的经度、纬度以及详细地址说明。在布局中添加百度地图控件，其代码片段如下。

```
<com.baidu.mapapi.map.MapView
    android:id="@+id/bmapView"
    android:layout_width="match_parent"
    android:layout_height="match_parent"
    android:clickable="true" />
```

10）开发逻辑代码

在 java/ee.example.map_bdlocation 包下编写两个代码程序，一个是主逻辑控制代码程序 MainActivity.java 文件，另一个是提供地图定位属性设置、地图显示生命周期等管理服务类代码程序 LocatingService.java 文件。

在 MainActivity.java 程序文件中，其中有两个主要方法，其一是实现自定义方法 initLocation()，代码片段如下。

```
private void initLocation() {                    //初始化
    SDKInitializer.initialize(getApplicationContext());
    setContentView(R.layout.activity_main);
    mMapView = (MapView) findViewById(R.id.bmapView);
    mBaiduMap = mMapView.getMap();
    mBaiduMap.setMapType(BaiduMap.MAP_TYPE_NORMAL);
    mBaiduMap.setMapStatus(MapStatusUpdateFactory.zoomTo(15));

    tv_Lon = findViewById(R.id.tv_Lon);
    tv_Lat = findViewById(R.id.tv_Lat);
    tv_Add = findViewById(R.id.tv_Add);
```

```
12
13          locService = new LocatingService(getApplicationContext());
14          LocationClientOption mOption = locService.getDefaultLocationClientOption();
15          mOption.setLocationMode(LocationClientOption.LocationMode.Hight_Accuracy);
16          locService.setLocationOption(mOption); //保存定位参数
17          locService.registerListener(listener);
18          locService.start();
19      }
```

(1) 第 2 行执行百度地图初始化，第 3 行显示布局文件定义的布局。这两个语句的顺序不可颠倒，因为在布局文件 activity_main 中引入地图作为背景，所以地图的初始化必须在先。

(2) 第 4～7 行对控件及地图对象初始化，包括地图对象的类型及缩放比例。百度地图类型可以使用 BaiduMap 中的 setMapType()方法来设置，地图类型有普通地图(MAP_TYPE_NORMAL)、卫星地图(MAP_TYPE_SATELLITE)和空白地图(MAP_TYPE_NONE)。

(3) 第 13～18 行初始化地图定位的服务对象，定义该对象对应的客户端对象，设置客户端对象定位模式为精确定位，获取该对象的默认设置属性值，对地图定位创建监听，最后启动服务。其中第 15 行设置定位模式为高精度(LocationMode.Hight_Accuracy)。百度地图支持三种定位模式：高精度定位模式(Hight_Accuracy)、低功耗定位模式(Battery_Saving)和仅设备定位模式(Device_Sensors)，这里的设备指 GPS。

另一个主要方法是定义地图定位的监听回调方法，在此方法中处理定位结果。

```
BDAbstractLocationListener listener = new BDAbstractLocationListener() {
    @Override
    public void onReceiveLocation(BDLocation location) {
        if (location != null
                    && (location.getLocType() == BDLocation.TypeGpsLocation
                     || location.getLocType() == BDLocation.TypeNetWorkLocation)) {
            if (location != null) {
                LatLng point = new LatLng(location.getLatitude(), location.getLongitude());
                tv_Lon.setText(location.getLongitude() + "");
                tv_Lat.setText(location.getLatitude() + "");
                tv_Add.setText(location.getAddrStr());

                //构建 Marker 图标
                BitmapDescriptor bitmap = null;
                bitmap = BitmapDescriptorFactory.fromResource(R.drawable.icon_map_mark);
                //构建 MarkerOption,用于在地图上添加 Marker
                OverlayOptions option = new MarkerOptions().position(point).icon(bitmap);
                //在地图上添加 Marker,并显示
                mBaiduMap.addOverlay(option);
                mBaiduMap.setMapStatus(MapStatusUpdateFactory.newLatLng(point));
            }
        }
    }
};
```

在该回调方法中主要实现两个功能：一是获取定位数据信息，二是使用图标标记定位点。在接收到定位信息后，获取定位点坐标的经度、纬度（第 6 行），并将该定位点的经度、纬度和地址串信息写入三个 TextView 对象中。第 15～18 行，执行创建一个地图覆盖物，将一个图标对象 bitmap 覆盖到地图之上，添加位置为定位点。这里，OverlayOptions 是地图覆盖物选型基类。

LocatingService. java 是地图定位属性设置、地图显示生命周期等管理服务类的实现代码文件，代码如下。

```
package ee.example.map_bdlocation;

import android.content.Context;

import com.baidu.location.BDAbstractLocationListener;
import com.baidu.location.LocationClient;
import com.baidu.location.LocationClientOption;

public class LocatingService {
    private static LocationClient client = null;
    private static LocationClientOption mOption;
    private Object objLock;

    public LocatingService(Context locationContext) {
        objLock = new Object();
        synchronized (objLock) {
            if (client == null) {
                client = new LocationClient(locationContext);
                client.setLocOption(getDefaultLocationClientOption());
            }
        }
    }

    //注册定位监听
    public boolean registerListener(BDAbstractLocationListener listener) {
        boolean isSuccess = false;
        if (listener != null) {
            client.registerLocationListener(listener);
            isSuccess = true;
        }
        return isSuccess;
    }

    public void unregisterListener(BDAbstractLocationListener listener) {
        if (listener != null) {
            client.unRegisterLocationListener(listener);
        }
    }

    //设置定位参数
    public static boolean setLocationOption(LocationClientOption option) {
        boolean isSuccess = false;
        if (option != null) {
            if (client.isStarted()) {
                client.stop();
```

```
46              }
47              client.setLocOption(option);
48          }
49          return isSuccess;
50      }
51
52      //设置默认定位选项
53      public LocationClientOption getDefaultLocationClientOption() {
54          if (mOption == null) {
55              mOption = new LocationClientOption();
56              mOption.setLocationMode(LocationClientOption.LocationMode.Hight_Accuracy);
                                        //设置定位模式为高精度
57              mOption.setCoorType( "bd09ll" );          //配合百度地图使用,建议设置为 bd09ll;
58              mOption.setScanSpan(3000);                //每隔 3 秒发起定位一次
59              mOption.setIsNeedAddress(true);           //设置需要地址信息
60              mOption.setIsNeedLocationDescribe(true);  //设置是否需要地址描述
61              mOption.setNeedDeviceDirect(false);       //设置不需要设备方向结果
62              mOption.setLocationNotify(false);         //设置不是 1 秒输出 1 次 GPS 结果
63              mOption.setOpenGps(true);                 //设置开启 GPS 定位
64              mOption.setIsNeedAltitude(false);         //设置定位时不需要海拔信息
65          }
66          return mOption;
67      }
68
69      public void start() {
70          synchronized (objLock) {
71              if (client != null && !client.isStarted()) {
72                  client.start();
73              }
74          }
75      }
76
77      public void stop() {
78          synchronized (objLock) {
79              if (client != null && client.isStarted()) {
80                  client.stop();
81              }
82          }
83      }
84
85  }
```

(1) 第 14～22 行,是 LocatingService 类的构造方法,对该类的成员对象进行初始化,与客户端的定位对象进行绑定。

(2) 第 25～32 行,注册定位监听器。当监听到百度地图的定位信息后,向客户端对象传递定位监听信息。

(3) 第 53～67 行,设置默认的定位选项参数。

setLocationMode()方法,设置定位的模式。有三种定位模式:高精度定位模式(Hight_Accuracy)、低功耗定位模式(Battery_Saving)和仅设备 GPS 定位模式(Device_Sensors)。其中 Hight_Accuracy 为默认模式。

setCoorType(String)方法,设置坐标类型。默认为 gcj02,设置返回的定位结果坐标系，如果配合百度地图使用，建议设置为 bd09ll。

setScanSpan(int)方法，设置发起连续定位请求的间隔时间，单位为毫秒(ms)。默认为 0，即仅定位一次，如果要不断地发起定位，一般要大于等于 1000ms 才是有效的。

setIsNeedAddress(Boolean)方法，设置是否需要地址信息。默认为 false，即不需要。

setIsNeedLocationDescribe(Boolean)方法，设置是否需要地址描述。默认为 false，即不需要。

setNeedDeviceDirect(Boolean)方法，设置是否需要设备方向结果。默认为 false，即不需要。

setLocationNotify(Boolean)方法，设置是否当 GPS 有效时按照每秒一次的频率输出 GPS 结果。默认为 false，即不是。

setOpenGps(Boolean)方法，设置是否开启 GPS 定位。默认为 false，即不开启。

setIsNeedAltitude(Boolean)方法，设置定位时是否需要海拔信息。默认为 false，即不需要，除基础定位版本都可用。

运行结果：本案例选择在真机上运行项目。因为百度地图无法从模拟器上获得反馈的地理位置信息，运行的结果往往达不到预期。

在运行项目之前，首先要接入手机，设置手机为运行设备。然后运行 Map_BDLocation 项目。在手机上看到的运行效果，如图 14-19(a)所示。用手势将地图放大，可以清楚地看到定位点及周边的建筑、道路等信息，如图 14-19(b)所示。

(a) 按初始比例显示定位点　　(b) 放大地图显示定位点

图 14-19　在百度地图上显示定位

2. POI 搜索

地图由多图层组成，可以以地图为底图开发多种应用。其中搜索兴趣点就是其中一种非常广泛的应用。百度地图和高德地图都提供了千万级别的 POI(Point Of Interest,兴趣点)服务,通过 POI 搜索,可以找到周边的学校、银行、餐馆、商店、医院、旅游景点等。这些兴趣点是覆盖在地图之上的另一个图层,通过本地搜索覆盖物 PoiOverlay 类进行管理。

下面利用高德地图 SDK,来开发对当前位置的周边 POI 搜索案例。

【案例 14.2】 使用高德地图 SDK,在地图上标注手机当前的位置,并以当前点为中心,搜索周边的兴趣点。要求当前点能跟随手机的位置变化,搜索兴趣点时,不仅在地图上标注兴趣点,而且单击每个兴趣点时能显示相应的名称和位置信息。

说明:在构建布局时,使用高德地图来作为背景。主要实现两大功能:一是定位,二是搜索 POI 点及相关信息。所以在本案例中,调用高德地图的二维地图功能、定位功能和搜索功能 SDK 来实现。

开发步骤及解析:过程如下。

1) 创建项目

在 Android Studio 中创建一个名为 Map_GaodeLocation 的项目。其包名为 ee.example.map_gaodelocation。

2) 生成 APK 签名

创建 APK 签名文件名为 GDMLApp.jks,存储在 E:\AndroidStudioProjects\Ch14\Map_GaodeLocation 路径下,并且记住启用该签名文件的密码。

3) 获取 Key 密钥

在高德开放平台控制台中新建项目,配置本 Android 项目的开发版和发布版 SHA1、本项目的包名 ee.example.map_gaodelocation。生成本项目的 Key 密钥为:dceceb6904bd9eb731fb6d439706a607。

4) 下载开发包

因为本案例要求在当前位置进行 POI 搜索,因此需要如下 JAR 包:Amap_2DMap_V6.0.0_20191106.jar 开发包,提供基于高德二维地图的相关功能开发;Amap_Location_V5.1.0_20200708.jar 开发包,提供高德地图与定位相关的功能开发;Amap_Search_V7.3.0_20200331.jar 开发包,提供与搜索相关的功能开发。

5) 配置 gradle

解压下载包,将下载的 Amap_2DMap_V6.0.0_20191106.jar 等三个文件都复制到本项目的 libs 目录下;然后将这些 JAR 文件添加到项目库中(右击目录中文件名,在弹出菜单中选择 Add As Library...);同步项目与 gradle 配置(依次选择菜单 File→Sync Project with Gradle Files),配置模块 build.gradle。执行完上述操作后,在项目的模块 build.gradle 文件中可以看到相关配置信息。

6) 声明权限与配置 Key 密钥

地图应用功能包含定位信息和 POI 搜索,在 AndroidManifest.xml 文件中,所需要添加的权限声明与案例 14.1 相同。配置高德地图开发 Key,在<application>元素内的<meta>子元素中添加,代码片段如下。

```
<meta-data
    android:name="com.amap.api.v2.apikey"
    android:value="dceceb6904bd9eb731fb6d439706a607" />
```

7）准备图片资源

事先准备标记兴趣点的多个图标以及界面中使用的一些图标，将它们复制到 res/drawable 目录中。

8）准备其他资源

颜色资源：编写 res/values 目录下的 colors.xml 文件，添加设置文字需要的颜色，如 darkgrey、grey、white、darkblue、greyblack 等。

密度资源：编写 res/values 目录下的 dimens.xml 文件，添加设置控件与边界距离密度，代码片段如下。

```
<dimen name="offset_title">5dp</dimen>
<dimen name="offset_contenttext">12dp</dimen>
```

9）设计布局

编写 res/layout 目录下的 activity_main.xml 文件。在布局文件中主要定义三部分内容。

(1) 引入 2D 高德地图控件作为背景图。声明高德地图控件的代码片段如下。

```
<com.amap.api.maps2d.MapView
    android:id="@+id/amapView"
    android:layout_width="match_parent"
    android:layout_height="match_parent"
    android:layout_alignParentLeft="true"
    android:layout_alignParentTop="true" />
```

(2) 使用相对布局设计兴趣点输入框。其中设置 TextView 控件的 android:clickable 为 true，即可代替按钮控件功能。在这里，先布局右侧作为“搜索”按钮的 TextView 控件，然后左侧剩余空间布局 EditText 控件。

(3) 使用相对布局来设计显示兴趣点详细说明层的容器。该层在初始时是不可见的，只有单击相应的兴趣点图标才显示。

10）开发逻辑代码

在 java/ee.example.map_gaodelocation 包下编写两个代码程序，一个是主逻辑控制代码程序 MainActivity.java 文件，另一个是提供主逻辑中所有需要的提示信息内容定义，程序文件为 ToastUtil.java。

在 MainActivity.java 文件中，需要实现 AMapLocationListener、LocationSource、OnMapClickListener、OnMarkerClickListener 和 OnPoiSearchListener 等接口。首先在 onCreate()方法中对高德地图控件进行初始化，自定义初始化 init()方法，代码片段如下。

```
@Override
protected void onCreate(Bundle savedInstanceState) {
    super.onCreate(savedInstanceState);
    ...
    setContentView(R.layout.activity_main);
    mapview = (MapView)findViewById(R.id.amapView);        //初始化地图控件
    mapview.onCreate(savedInstanceState);

    init();
```

```
10  }
11
12  //初始化 AMap 对象
13  private void init() {
14      if (mAMap == null) {
15          lp = new LatLonPoint(23.13615, 113.418260);
16          city = "广州市";
17          mAMap = mapview.getMap();
18          mAMap.setOnMapClickListener(this);
19          mAMap.setOnMarkerClickListener(this);
20
21          TextView searchButton = (TextView) findViewById(R.id.btn_search);
22          searchButton.setOnClickListener(this);
23          mSearchText = (EditText)findViewById(R.id.input_edittext);
24
25          locatingsetup();                              //定位当前标记点
26
27          mPoiDetail = (RelativeLayout) findViewById(R.id.poi_detail);
28          mPoiName = (TextView) findViewById(R.id.poi_name);
29          mPoiAddress = (TextView) findViewById(R.id.poi_address);
30      }
31      mAMap.moveCamera(CameraUpdateFactory.newLatLngZoom(
                new LatLng(lp.getLatitude(), lp.getLongitude()), 15));
32  }
```

(1) onCreate()方法主要完成三项工作：一是继承上级 onCreate()方法，二是检测定位权限，三是初始化布局操作。其中第 5 行显示布局文件 activity_main.xml，第 6 行初始化高德地图控件，第 7 行加载前一次应用被销毁前保存的每个实例的状态，该行语句与重写的 onSaveInstanceState(Bundle outState)方法相关。注意，这三行语句的顺序不可颠倒。

(2) init()是自定义方法。在该方法中，初始化地图对象及 Activity 界面上的控件对象。其中第 15、16 行是定位点的初始化。第 25 行调用自定义方法 locatingsetup()初始化定位标记点。第 27～29 行初始化 POI 详细信息 RelativeLayout 显示层。第 31 行设置高德地图的中心点为 lp，缩放比例为 15。

实现高德地图定位，需要实现 AMapLocationListener 定位接口，设置定位的监听方法 setLocationSource(this)，而该方法中的 this 监听对象指的是 LocationSource 接口。实现 LocationSource 接口，需要重写两个方法 activate()和 deactivate()。这两个方法是一对，activate()是定位按键单击时的触发事件，deactivate()是指定位按键结束事件，所以这两个方法内都会指向定位按键的监听器对象，需要重写 OnLocationChanged(AMapLocation location)定位回调方法。

```
//定位当前位置小蓝点
private void locatingsetup () {
    mAMap.setLocationSource(this);                          //设置定位监听
    mAMap.getUiSettings().setMyLocationButtonEnabled(true); //设置显示默认定位按钮
    mAMap.setMyLocationEnabled(true);                       //设置显示定位层并可触发定位
    //标注小蓝点
    locationMarker = mAMap.addMarker(new MarkerOptions()
                .anchor(0.5f, 0.5f)        //覆盖物的锚点比例，设置水平居中、垂直居中
                .icon(BitmapDescriptorFactory.fromBitmap(BitmapFactory.decodeResource(
```

```
                        getResources(), R.drawable.point_blue)))     //指定显示的图标
                    .position(new LatLng(lp.getLatitude(), lp.getLongitude())));
                                                                    //指定显示的位置
8   }
9   //激活定位
10  @Override
11  public void activate(OnLocationChangedListener listener) {
12      mListener = listener;
13      if (mlocationClient == null) {
14          mlocationClient = new AMapLocationClient(this);
15          mLocationOption = new AMapLocationClientOption();
16          //设置定位监听
17          mlocationClient.setLocationListener(this);
18          //设置为高精度定位模式
19          mLocationOption.setLocationMode(AMapLocationMode.Hight_Accuracy);
20          //设置客户端定位对象参数
21          mlocationClient.setLocationOption(mLocationOption);
22          //启动客户端定位对象
23          mlocationClient.startLocation();
24      }
25  }
26
27  //定位成功后回调函数
28  @Override
29  public void onLocationChanged(AMapLocation amapLocation) {
30      if (mListener != null && amapLocation != null) {
31          if (amapLocation != null
32                  && amapLocation.getErrorCode() == 0) {
33              mListener.onLocationChanged(amapLocation);  // 显示系统小蓝点
34              lp = new LatLonPoint(amapLocation.getLatitude(), amapLocation.getLongitude());
35              city = amapLocation.getCity();
36          }
37      }
38  }
39
40  //停止定位
41  @Override
42  public void deactivate() {
43      mListener = null;
44      if (mlocationClient != null) {
45          mlocationClient.stopLocation();
46          mlocationClient.onDestroy();
47      }
48      mlocationClient = null;
49  }
```

(1) locatingsetup()是自定义方法。在该方法中设置定位监听，显示默认的定位按钮，设置显示定位层并可触发。最后是初始化覆盖物(见第 7 行)，其中属性如下。

.anchor(0.5f, 0.5f)，设置覆盖物锚点比例，该设置意思是水平居中、垂直居中。

.icon(BitmapDescriptorFactory.fromBitmap(BitmapFactory.decodeResource(getResources(), R.drawable.point_blue)))，指定资源 R.drawable.point_blue 为覆盖物的图标。

.position(new LatLng(lp.getLatitude(), lp.getLongitude())),指定覆盖物的显示位置为 lp 定位点。

(2) activate()是 LocationSource 接口中的激活定位方法,在该方法中创建一个定位客户端对象 mlocationClient 和定位参数对象 mLocationOption,设置对定位建立监听。第 19 行设置 mLocationOption 为高精度定位模式。然后对 mLocationOption 设置定位参数,启动定位。

(3) 在定位成功后回调 OnLocationChanged()方法。在该方法中重新定位新的定位点,更新 lp 的经度、纬度数据,更新城市名称。

(4) 结束定位后,调用 deactivate(),在其中释放定位对象所占用的资源。

POI 应用需要标记 POI 点,将使用 2D 地图的 Marker 和 MarkerOptions 模块。并且单击这些 POI 的 marker 点会显示详细信息,单击地图的其他位置隐藏详细信息,所以要实现 OnMapClickListener 和 OnMarkerClickListener 监听接口。如果是 Android 5.2.1 以下版本,还需要高德地图的 InfoWindowAdapter 接口来定制 marker 的信息窗口,默认情况下,当单击某个 marker 时,如果该 marker 的 Title 和 Snippet 不为空,则会触发 getInfoWindow 和 getInfoContents 回调;自 5.2.1 版本开始,就不需要展示 InfoWindow 信息了,本书以 Android 11 为开发平台,所以开发时不需要实现 InfoWindowAdapter 接口。实现 OnMapClickListener 接口需要重写 onMapClick()方法,实现 OnMarkerClickListener 接口需要重写 onMarkerClick()方法,代码片段如下。

```
@Override
public void onMapClick(LatLng arg0) {
    whetherToShowDetailInfo(false);              //隐藏兴趣点的详细信息
    if (mlastMarker != null) {
        resetlastmarker();                       //重置之前被单击的 marker 状态
    }
}

@Override
public boolean onMarkerClick(Marker marker) {
    if (marker.getObject() != null) {
        whetherToShowDetailInfo(true);           //显示兴趣点的详细信息
        try {
            PoiItem mCurrentPoi = (PoiItem) marker.getObject();
            if (mlastMarker == null) {
                mlastMarker = marker;
            } else {
                //将之前被单击的 marker 置为原来的状态
                resetlastmarker();
                mlastMarker = marker;
            }
            detailMarker = marker;
            detailMarker.setIcon(BitmapDescriptorFactory
                        .fromBitmap(BitmapFactory.decodeResource(
                                getResources(), R.drawable.poi_marker_pressed)));
            setPoiItemDisplayContent(mCurrentPoi);
        } catch (Exception e) {
            e.printStackTrace();
```

```
27              }
28          }else {
29              whetherToShowDetailInfo(false);
30              resetlastmarker();
31          }
32          return true;
33      }
34
35      //确定是否显示兴趣点的详细信息
36      private void whetherToShowDetailInfo(boolean isToShow) {
37          if (isToShow) {
38              mPoiDetail.setVisibility(View.VISIBLE);
39          } else {
40              mPoiDetail.setVisibility(View.GONE);
41          }
42  }
43
44      //将之前被单击的 marker 置为原来的状态
45      private void resetlastmarker() {
46          int index = poiOverlay.getPoiIndex(mlastMarker);     //获取该 POI 的序号
47          if (index < 10) {
48              mlastMarker.setIcon(BitmapDescriptorFactory
                            .fromBitmap(BitmapFactory.decodeResource(
                                    getResources(),
                                    poiOverlay.markers[index])));
49          }else {
50              mlastMarker.setIcon(BitmapDescriptorFactory.fromBitmap(
51                      BitmapFactory.decodeResource(getResources(),
                                                R.drawable.marker_other_highlight)));
52          }
53          mlastMarker = null;
54      }
```

(1) 重写 onMapClick()方法，单击地图，即隐藏 POI 的详细信息。重写 onMarkerClick()方法，单击某个 POI，显示该点的名称及地址信息。是否显示该详细信息由 whetherToShowDetailInfo()方法确定；显示图标，由图标资源 R.drawable.poi_marker_pressed 提供(见第 23 行)。

(2) resetlastmarker()是自定义方法。该方法实现对于前 10 个 POI 点，使用带编号的图标标记，对于第 11 及之后的 POI 点，统一用图标 R.drawable.marker_other_highlight 标记。所以在还原 POI 的 marker 状态时要进行区分(见第 47～52 行)。

实现搜索 POI 功能，需要创建 PoiSearch 对象。OnPoiSearchListener 监听接口为 POI 搜索结果的异步处理回调接口，实现该接口需要重写 onPoiItemSearched()方法和 onPoiSearched()方法，前者为 POI 的 ID 搜索的结果回调，本案例该方法为空；后者返回 POI 搜索异步处理的结果，方法代码片段如下。

```
//开始进行 POI 搜索
protected void doSearchQuery() {
    currentPage = 0;
    query = new PoiSearch.Query(keyWord, "", city);   //创建一个 POI 查询对象
    query.setPageSize(20);                             //设置每页最多返回多少条 poiItem
```

```
    query.setPageNum(currentPage);                     //设置查第一页,首页的页码为 0

    if (lp != null) {
        poiSearch = new PoiSearch(this, query);
        poiSearch.setOnPoiSearchListener(this);
        poiSearch.setBound(new SearchBound(lp, 5000, true)); //
        //设置搜索区域为以 lp 点为圆心,其周围 5000 米范围
        poiSearch.searchPOIAsyn();                     // 异步搜索
    }
}

@Override
public void onPoiSearched(PoiResult result, int rcode) {
    if (rcode == AMapException.CODE_AMAP_SUCCESS) {
        if (result != null && result.getQuery() != null) {     //搜索 POI 的结果
            if (result.getQuery().equals(query)) {   //是否是同一条
                poiResult = result;
                poiItems = poiResult.getPois();   //取得第一页的 poiItem 数据,页数从 0 开始
                List<SuggestionCity> suggestionCities = poiResult.getSearchSuggestionCitys();
                            //当搜索不到 poiItem 数据时,会返回含有搜索关键字的城市信息
                if (poiItems != null && poiItems.size() > 0) {
                    whetherToShowDetailInfo(false); //清除 POI 信息显示
                    if (mlastMarker != null) {
                        resetlastmarker();           //并还原单击 marker 样式
                    }
                    if (poiOverlay != null) {
                        poiOverlay.removeFromMap(); //清理之前搜索结果的 marker
                    }
                    mAMap.clear();
                    poiOverlay = new myPoiOverlay(mAMap, poiItems);
                    poiOverlay.addToMap();
                    poiOverlay.zoomToSpan();

                    mAMap.addMarker(new MarkerOptions()
                                .anchor(0.5f, 0.5f)
                                .icon(BitmapDescriptorFactory
                                            .fromBitmap(BitmapFactory.decodeResource(
                                                    getResources(), R.drawable.point_blue)))
                                .position(new LatLng(lp.getLatitude(), lp.getLongitude())));

                } else if (suggestionCities != null && suggestionCities.size() > 0) {
                    showSuggestCity(suggestionCities);
                } else {
                    ToastUtil.show(this.getApplicationContext(),R.string.no_result);
                }
            }
        } else {
            ToastUtil.show(this.getApplicationContext(), R.string.no_result);
        }
    } else {
        ToastUtil.showerror(this.getApplicationContext(), rcode);
    }
}
```

(1) doSearchQuery()是自定义方法，该方法完成对输入的兴趣类关键词在定位点的周边进行搜索。其中，第 4 行，创建一个 POI 查询对象 query，PoiSearch. Query()方法有三个参数，第一个参数表示搜索字符串，第二个参数表示 POI 搜索类型，第三个参数表示 POI 搜索区域(空字符串代表全国)；第 5 行，设置查询的结果每页至多存放 20 条兴趣点信息；第 8～14 行，以 lp 为圆心、以 5000 米为半径进行 POI 异步搜索，并将搜索结果存放于 poiSearch 对象中，该对象与 query 对象绑定。

(2) 重写 onPoiSearched()方法，该方法是 OnPoiSearchListener 监听接口的回调方法，该方法中调用 poiOverlay 类的各种方法，来处理搜索 POI 的结果。如果 POI 没有搜索到数据，则调用 showSuggestCity()方法返回一些推荐城市的信息。

由于本案例中要求：前 10 个 POI 点的图标要标注编号，从第 11 个 POI 点后图标不用标编号，且使用另一种图标标注。所以采用自定义的 myPoiOverlay 类，而没有使用高德地图定义的 PoiOverlay 类。定义 myPoiOverlay 类的代码如下。

```
1   private class myPoiOverlay {
2       private AMap mamap;
3       private List < PoiItem > mPois;
4       private ArrayList < Marker > mPoiMarks = new ArrayList < Marker >();
5
6       public myPoiOverlay(AMap amap , List < PoiItem > pois) {
7           mamap = amap;
8           mPois = pois;
9       }
10
11      //添加 Marker 到地图中. @since V2.1.0
12      public void addToMap() {
13          for (int i = 0; i < mPois.size(); i++) {
14              Marker marker = mamap.addMarker(getMarkerOptions(i));
15              PoiItem item = mPois.get(i);
16              marker.setObject(item);
17              mPoiMarks.add(marker);
18          }
19      }
20
21      //去掉 PoiOverlay 上所有的 Marker. @since V2.1.0
22      public void removeFromMap() {
23          for (Marker mark : mPoiMarks) {
24              mark.remove();
25          }
26      }
27
28      //移动镜头到当前的视角. @since V2.1.0
29      public void zoomToSpan() {
30          if (mPois != null && mPois.size() > 0) {
31              if (mamap == null)
32                  return;
33              LatLngBounds bounds = getLatLngBounds();
34              mamap.moveCamera(CameraUpdateFactory.newLatLngBounds(bounds, 100));
35          }
36      }
37
```

```
    private LatLngBounds getLatLngBounds() {
        LatLngBounds.Builder b = LatLngBounds.builder();
        for (int i = 0; i < mPois.size(); i++) {
            b.include(new LatLng(mPois.get(i).getLatLonPoint().getLatitude(),
                    mPois.get(i).getLatLonPoint().getLongitude()));
        }
        return b.build();
    }

    //获取覆盖物相关信息
    private MarkerOptions getMarkerOptions(int index) {
        return new MarkerOptions()
                .position( new LatLng(mPois.get(index).getLatLonPoint()
                                .getLatitude(), mPois.get(index)
                                .getLatLonPoint().getLongitude()))
                    .title(getTitle(index)).snippet(getSnippet(index))
                    .icon(getBitmapDescriptor(index));
    }

    protected String getTitle(int index) {
        return mPois.get(index).getTitle();
    }

    protected String getSnippet(int index) {
        return mPois.get(index).getSnippet();
    }

    //从 Marker 中得到 POI 在 list 的位置. @since V2.1.0
    public int getPoiIndex(Marker marker) {
        for (int i = 0; i < mPoiMarks.size(); i++) {
            if (mPoiMarks.get(i).equals(marker)) {
                return i;
            }
        }
        return -1;
    }

    //返回第 index 的 POI 的信息. @since V2.1.0
    public PoiItem getPoiItem(int index) {
        if (index < 0 || index >= mPois.size()) {
            return null;
        }
        return mPois.get(index);
    }

    private int[] markers = {
        R.drawable.poi_marker_1,
        R.drawable.poi_marker_2,
        R.drawable.poi_marker_3,
        R.drawable.poi_marker_4,
        R.drawable.poi_marker_5,
        R.drawable.poi_marker_6,
        R.drawable.poi_marker_7,
        R.drawable.poi_marker_8,
```

```
            R.drawable.poi_marker_9,
            R.drawable.poi_marker_10
        };

        //标高覆盖物的图标
        protected BitmapDescriptor getBitmapDescriptor(int arg0) {
            if (arg0 < 10) {
                BitmapDescriptor icon = BitmapDescriptorFactory.fromBitmap(
                    BitmapFactory.decodeResource(getResources(), markers[arg0]));
                return icon;
            }else {
                BitmapDescriptor icon = BitmapDescriptorFactory.fromBitmap(
                                                BitmapFactory.decodeResource(getResources(),
                                                R.drawable.marker_other_highlight));
                return icon;
            }
        }
    }
```

ToastUtil.java程序文件定义本案例中所有Toast语句需要使用的提示信息集合，包括常见的异常处理代码的系统出错提示，代码如下。

```
package ee.example.map_gaodelocation;

import android.content.Context;
import android.util.Log;
import android.widget.Toast;

import com.amap.api.services.core.AMapException;

public class ToastUtil {

    public static void show(Context context, String info) {
        Toast.makeText(context, info, Toast.LENGTH_LONG).show();
    }

    public static void show(Context context, int info) {
        Toast.makeText(context, info, Toast.LENGTH_LONG).show();
    }

    public static void showerror(Context context, int rCode){
        try {
            switch (rCode) {
            //服务错误码
            case 1001:
                throw new AMapException(AMapException.AMAP_SIGNATURE_ERROR);
            case 1002:
                 throw new AMapException(AMapException.AMAP_INVALID_USER_KEY);
                 …
            case 4001:
                throw new AMapException(AMapException.AMAP_SHARE_FAILURE);
            default:
                Toast.makeText(context,"错误码:" + rCode , Toast.LENGTH_LONG).show();
                logError("查询失败", rCode);
```

```
33                  break;
34              }
35              } catch (Exception e) {
36                  Toast.makeText(context, e.getMessage(), Toast.LENGTH_LONG).show();
37                  logError(e.getMessage(), rCode);
38              }
39          }
40
41          private static void logError(String info, int errorCode) {
42              print(LINE); //start
43              print("                          错误信息                          ");
44              print(LINE); //title
45              print(info);
46              print("错误码: " + errorCode);
47              print("                                                            ");
48              print("如果需要更多信息,请根据错误码到以下地址进行查询");
49              print(" http://lbs.amap.com/api/android-sdk/guide/map-tools/error-code/");
50              print("若仍无法解决问题,请将全部 LOG 信息提交到工单系统,多谢合作");
51              print(LINE); //end
52          }
53
54          //log
55          public static final String TAG = "AMAP_ERROR";
56          static final String LINE_CHAR = "=";
57          static final int LENGTH = 80;
58          static String LINE;
59          static{
60              StringBuilder sb = new StringBuilder();
61              for(int i = 0;i<LENGTH;i++){
62                  sb .append(LINE_CHAR);
63              }
64              LINE = sb.toString();
65          }
66
67          private static void print(String s) {
68              Log.i(TAG,s);
69          }
70
71      }
```

(1) 第 19～39 行，分别对每个服务错误码抛出高德地图的错误提示信息。

(2) 第 41～52 行，定义一个错误信息在 LOG 中的显示格式。其中使用的自定义方法为 print()，其实现在第 67～69 行定义。

(3) 第 59～65 行，是一个结构体，在其中定义字符串 LINE 为一串由 80 个"="号构成的双横线。

运行结果：本案例选择在真机上运行项目，在运行项目之前，首先要接入手机，设置手机为运行设备。然后运行 Map_GaodeLocation 项目。在手机上看到的运行效果，显示的是输入搜索的 POI 关键词为"学校"，如图 14-20(a)所示。单击"搜索"按钮，即在当前定位点周边 5000 米范围内搜索学校，并标记图标，其中前 10 个学校的图标有序号数字，如图 14-20(b)所示。用手势将地图放大，可以清楚地看到定位点及周边 POI 的局部信息，如

图 14-20(c)所示。单击第 3 个 POI 点，在屏幕的下方显示学校名称和地址，如图 14-20(d)所示。在本案例中，单击任何一个 POI 标记图标，都会显示其学校名称和地址详细信息。然后单击地图，POI 点的详细信息便会消失。

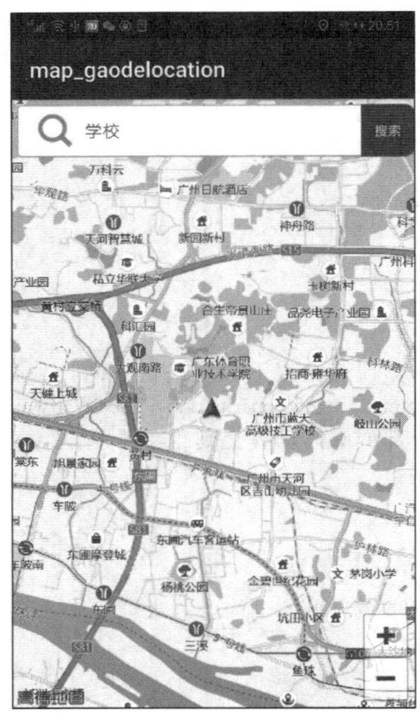

(a) 显示定位点并输入POI关键词

(b) 搜索定位点周边的POI

(c) 放大地图后POI的局部显示

(d) 单击其中的POI点显示详细信息

图 14-20　在高德地图上定位并搜索 POI

14.2　语音 SDK

现在，越来越多的 App 使用语音识别、语音播报等应用。虽然 Android 原生系统自身带有语音识别模块，但是需要在后台访问 Google 云服务器数据，在国内是没法正常使用的。在国内，有多个平台提供语音服务功能，可以很好地替代语音识别、在线语音合成等应用解决方案。常用的有讯飞语音识别、百度语音识别和微信语音服务等。本节主要介绍讯飞语音的语音识别和语音合成功能。

14.2.1　下载开发包

讯飞语音是科大讯飞推出的以语音交互技术为核心的人工智能开放平台，为开发者免费提供语音识别、语音合成等语音技术 SDK，以及人脸识别、声纹识别等。讯飞开发平台的网址为：https://www.xfyun.cn/。从平台上下载 SDK，需要首先注册平台账号，以开发者的身份在控制台页面创建新应用，新应用生成的 APPID 为 5f3c8d23，如图 14-21 所示。

图 14-21　在讯飞开放平台的控制台中创建新应用

单击该应用名称 SpeechActivity，进入应用的详细信息页面，如图 14-22(a)所示，左侧是相关语音应用的各项功能，往下拉网页，右侧有相关的 SDK 下载选项，在此单击下载 Android MSC，如图 14-22(b)所示。

14.2.2　配置开发环境

配置开发环境的步骤也是三步：一是复制开发包，二是配置模块的 build.gradle 文件，三是配置 AndroidManifest.xml。

1）步骤一

讯飞语音的一个应用对应一个 SDK，所以 Android 应用项目接入的 SDK 版本一定要与生成的 APPID 一致，下载的压缩包文件名尾部与 APPID 的 ID 号一致。解压下载的压缩包，解压后可以看到 libs 文件夹，其下包含两个文件夹 arm64-v8a、armeabiv-v7a 和一个

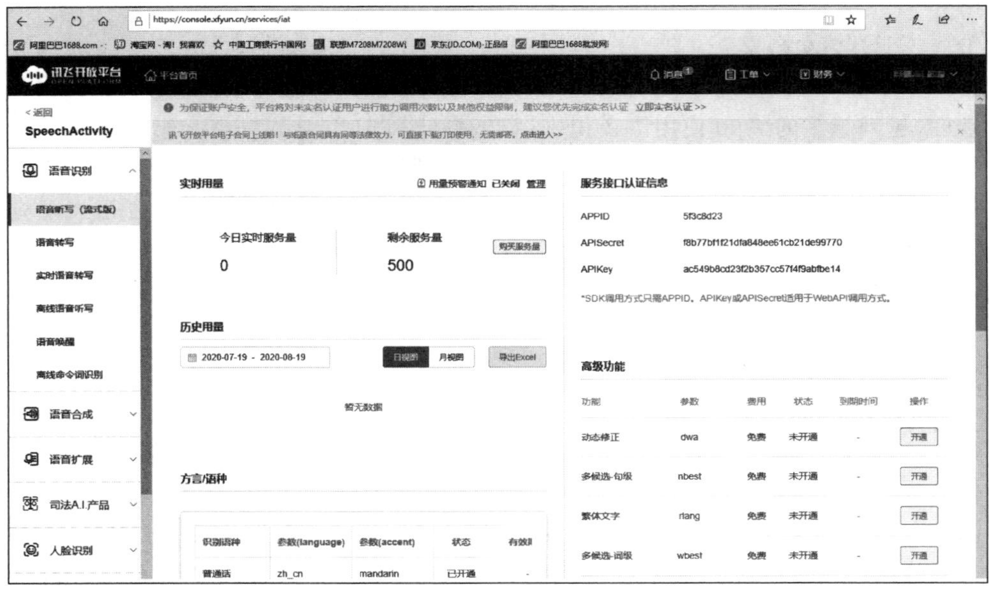

(a) SpeechActivity应用的服务接口认证信息

(b) SpeechActivity应用功能SDK下载列表

图 14-22 讯飞开放平台应用的详细信息

JAR 文件 Msc.jar，如图 14-23 所示。将它们全部复制到 Android 项目 app 下的 libs 目录中。

2）步骤二

arm64-v8a、armeabiv-v7a 两个目录中包含适应不同设备所需的.so 文件，在 Android 中默认复制到 jniLibs 目录下。为了便于管理，统一复制到 Android 项目的 libs 目录下，需要在模块级 build.gradle 文件的 android{}内手动添加指定目录的说明配置，代码如下。

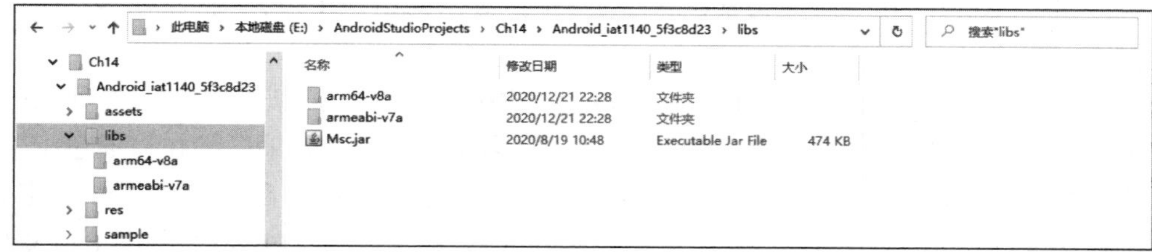

图 14-23　讯飞 SDK 解压包下的文件夹及文件信息

```
sourceSets {
    main {
        jniLibs.srcDir 'libs'
    }
}
```

复制 Msc.jar 到 libs 目录下，然后右击 Msc.jar，再选择 Add As Library...添加到项目库中。此时，在模块级 build.gradle 文件的 dependencies 中会自动增加如下依赖配置。

```
implementation files('libs/Msc.jar')
```

3）步骤三

在 AndroidManifest.xml 中，根据应用的需求加入必要的权限；要在应用的入口进行初始化；要设置讯飞 SDK 包的 APPID 等属性。

（1）添加权限。

在语音听写、识别应用时会使用手机上的录音、音量等控制，需要添加录音机权限；语音合成应用时会用到一些线上资源，需要添加与网络相关的权限声明；在语音应用中，语音与文字相互转换，以及一些设置参数信息需要使用 SharedPreferences 机制来保存，需要添加对外部存储设备的文件读、写权限声明。通常情况下，语音识别与合成应用需要的权限如下。

```
<!-- 获取手机录音机使用权限,听写、识别、语义理解需要用到此权限 -->
<uses-permission android:name="android.permission.RECORD_AUDIO"/>
<!-- 连接网络权限,用于执行云端语音能力 -->
<uses-permission android:name="android.permission.INTERNET"/>
<!-- 读取网络信息状态 -->
<uses-permission android:name="android.permission.ACCESS_NETWORK_STATE"/>
<!-- 获取当前 WiFi 状态 -->
<uses-permission android:name="android.permission.ACCESS_WIFI_STATE"/>
<!-- 允许程序改变网络连接状态 -->
<uses-permission android:name="android.permission.CHANGE_NETWORK_STATE"/>
<!-- 读取手机信息权限 -->
<uses-permission android:name="android.permission.READ_PHONE_STATE"/>
<!-- 外存储写权限,构建语法需要用到此权限 -->
<uses-permission android:name="android.permission.WRITE_EXTERNAL_STORAGE"/>
<!-- 外存储读权限,构建语法需要用到此权限 -->
<uses-permission android:name="android.permission.READ_EXTERNAL_STORAGE"/>
```

（2）应用入口初始化。

在程序开发中，专门编写一个初始化子类，用于纯粹定义初始化操作。在 Application

元素内使用 android:name 属性来指定该子类，作为应用的第一个被实例化的子类进行初始化操作，而且只被执行一次。

Application 元素的 android: name 属性用来指定应用在启动时，第一个被创建的 Application 子类实例，属性值是该子类的全名。全名的含义是：如果子类在当前包下，直接写入子类的名称；如果子类在当前包的下级包内，必须添加前面的路径。该属性值指定的子类就是该程序的入口。例如，在讯飞平台给出的案例中，入口初始化子类代码文件为 SpeechApp. java，位于项目目录的 voicedemo 子目录中，那么要求定义 Application 元素的 android:name 属性，代码片段如下。

```
<application
    ...
    android:name = ".voicedemo.SpeechApp"
    ...>
```

这个属性是可选的，大多数 App 都不需要这个属性。在没有这个属性的时候，Android 会启动 Application 元素中第一个< activity >所指类的实例。设置该属性是为了适用于某些应用要求初始化操作只做一次的限制。

（3）设置 APPID 属性。

如果新应用生成的 APPID 为 5f3c8d23，那么需要在< application >元素内添加< meta-data >子元素，代码片段如下。

```
<meta-data
    android:name = "IFLYTEK_APPKEY"
    android:value = "'5f3c8d23'" />
```

14.2.3 语音识别与合成应用

科大讯飞在语音应用方面属于专家级的，在其开放平台上可以看到很多强大的功能，并且提供了大量的功能应用案例。本节主要介绍语音识别与合成功能应用。

1. 语音识别

讯飞语音识别包括语音听写、语音默写、语音唤醒等功能。人们通常所说的语音识别主要是指讯飞的语音听写功能。

1）语音识别类及常用方法

开发语音识别应用，讯飞通过 SpeechRecognizer 类来实现。SpeechRecognizer 类常用的方法如下。

（1）createRecognixer()，创建语音识别对象。

（2）setParameter()，设置语音识别参数。

（3）startListening()，开始语音监听。参数为 RecognizerListener 对象，该对象为接口对象，需要重写 onBeginOfSpeech()、onError(SpeechError error)、onEndOfSpeech()、onResult()、onVolumeChanged()、onEvent()方法。

（4）stopListening()，结束监听语音。

(5) writeAudio(),把指定的音频流作为语音传给 SDK。

(6) cancel(),取消监听语音。

(7) destroy(),销毁语音识别对象,回收其占用的资源。

2) 初始化

初始化语音配置对象使用 SpeechUtility. createUtility()方法,只有初始化后应用才可以使用 Msc. jar 中提供的各项服务。建议使用一个单独的 Application 子类定义初始化,并作为应用项目的入口,即在 AndroidManifest. xml 中设置为 Application 的 android. name 属性值。初始化代码写在 Activity 的 onCreate()方法内,代码如下。

```
public class SpeechApp extends Application {
    @Override
    public void onCreate() {
        SpeechUtility.createUtility(SpeechApp.this, "appid = " + getString(R.string.app_id));
        super.onCreate();
    }
}
```

(1) SpeechApp 是继承于 Application 的子类。

(2) SpeechUtility. createUtility()初始化语音配置对象。其中 getString(R. string. app_id)是获得资源中记录的当前应用申请的 APPID 号,并且不要在'='与 APPID 号之间添加空格及空转义符。该 APPID 号必须和下载的 SDK 保持一致,否则会出现 10407 错误。比如 APPID 号为 5f3c8d23,初始化语句如下。

```
SpeechUtility.createUtility(SpeechApp.this, "appid = 5f3c8d23");
```

注意:此接口在非主进程调用会返回 null 对象,如需在非主进程使用语音功能,请增加参数:

```
SpeechConstant.FORCE_LOGIN + " = true"
```

3) 语音识别的方式

编写讯飞的语音识别应用,讯飞 SDK 提供了两种识别方式:一种是带 UI 界面的识别方式,另一种是无 UI 界面(自定义界面)的识别方式。

使用讯飞自带的 UI 界面,需要将下载的 SDK 解压包下的 assets 文件夹复制到 Android 项目的 app 目录下。因为 assets\iflytekh 文件夹下的所有文件是自带 UI 页面和相关其他服务的资源文件,比如语法文件、音频示例、词表等。在编写程序时,需要初始化听写 Dialog,如果只使用 UI 听写功能,无须创建 SpeechRecognizer 对象。初始化代码片段如下。

```
//初始化听写 Dialog,如果只使用 UI 听写功能,无须创建 SpeechRecognizer
mIatDialog = new RecognizerDialog(IatDemo.this, mInitListener);
//以下为 Dialog 设置听写参数
mIatDialog.setParams("xxx","xxx");
...
//开始识别并设置监听器
mIatDialog.setListener(mRecognizerDialogListener);
//显示听写对话框
mIatDialog.show();
```

不使用讯飞自带的 UI 界面，需要使用 SpeechRecognizer 对象，并且需要对 SpeechRecognizer 对象进行一系列的参数设置，根据回调消息来自定义界面，代码片段如下。

```
//初始化 SpeechRecognizer 对象
mIat = SpeechRecognizer.createRecognizer(IatDemo.this, mInitListener);

//设置语法 ID 和 SUBJECT 为空,以免因之前有语法调用而设置了此参数
mIat.setParameter( SpeechConstant.CLOUD_GRAMMAR, null );
mIat.setParameter( SpeechConstant.SUBJECT, null );
//或直接清空所有参数
//mIat.setParameter(SpeechConstant.PARAMS, null);

//设置返回结果格式,目前支持 Json、XML 以及 Plain 三种格式,其中 Plain 为纯听写文本内容
mIat.setParameter(SpeechConstant.RESULT_TYPE, "json");
//此处 engineType 为 cloud
mIat.setParameter( SpeechConstant.ENGINE_TYPE, engineType );
//设置语音输入语言,zh_cn 为简体中文
mIat.setParameter(SpeechConstant.LANGUAGE, "zh_cn");
//设置结果返回语言
mIat.setParameter(SpeechConstant.ACCENT, "mandarin");
//设置语音前端点:静音超时时间,单位 ms,即用户多长时间不说话则当作超时处理
//取值范围{1000～10000}
mIat.setParameter(SpeechConstant.VAD_BOS, "4000");
//设置语音后端点:后端点静音检测时间,单位 ms,即用户停止说话多长时间内即认为不再输入
//自动停止录音,范围{0～10000}
mIat.setParameter(SpeechConstant.VAD_EOS, "1000");
//设置标点符号,设置为"0"返回结果无标点,设置为"1"返回结果有标点
mIat.setParameter(SpeechConstant.ASR_PTT,"1");

//开始识别,并设置监听器
mIat.startListening(mRecogListener);
```

2. 语音合成

与语音听写相反，语音合成是将一段文字转换为语音。讯飞语音合成可根据需要合成出不同音色、语速和语调的声音，让机器像人一样开口说话。语音合成分在线语音合成和离线语音合成。

讯飞 SDK 语音合成功能主要通过 SpeechSynthesizer 类实现，SpeechSynthesizer 类常用的方法如下。

(1) createSynthesizer()，创建语音合成对象。

(2) setParameter()，设置语音合成参数。

(3) startSpeaking()，开始合成文本并播放音频。参数为 SynthesizerListener 对象，该对象为接口对象，需要重写 onSpeakBegin()、onSpeakPaused()、onSpeakResumed()、onBufferProgress(int progress, int beginPos, int endPos, java.lang.String info)、onSpeakProgress(int progress, int beginPos, int endPos)、onCompleted(SpeechError error)、onEvent()方法。

(4) synthesizeToUri()，合成到一个音频文件，不播放。使用此函数时，请考虑应用是

否有在设置的目录中保存文件的权限。

(5) pauseSpeaking(),暂停播放,仅在合成播放模式下有效。暂停播放,并不会暂停音频的获取过程,只是把播放器暂停。

(6) resumeSpeaking(),恢复播放。只有在暂停后,在当前暂停位置开始播放合成的音频。

(7) stopSpeaking(),停止合成。调用此函数,取消当前合成会话,并停止音频播放。调用此函数后,未合成的音频将不再返回,若已合成到文件模式,音频文件将不会被保存。

(8) destroy(),销毁已创建的语音合成对象。

关于讯飞语音应用的详细技术文档,可以参见讯飞开放平台提供的文档中心。网址为 https://www.xfyun.cn/doc/。

3. 应用案例

在讯飞开放平台上为各个功能的应用都提供有 Demo 案例。在下载的 SDK 压缩包内也有一个 sample 文件夹,包含一个可运行的 Demo 案例。本书给出一个简化版的 Demo 案例,便于初学者学习。

下面利用讯飞的语音识别和合成功能 SDK,来开发一个简单应用案例。案例代码主要来自讯飞平台的 speechDemo 样例,但是对界面进行了适当剪裁,突出语音识别与合成两项功能应用。

【案例 14.3】 使用讯飞语音 SDK,实现语音识别和语音合成两项功能。其中语音识别要求能识别中、英文语音;语音合成能实现中文普通话、部分中国地区方言、英文的语音合成播放。

说明:新建 Android 项目,从开放平台的样例中,把需要的类定义文件、布局文件和相关资源文件以及配置信息复制到新项目中,在此基础上加以修改。

本案例只实现无 UI 语音识别和在线语音合成功能。

开发步骤及解析:过程如下。

1) 创建项目

在 Android Studio 中创建一个名为 Speech_XfActivity 的项目。其包名为 ee.example.speech_xfactivity。

2) 下载开发包

从讯飞开放平台的控制台中创建新应用,然后选择语音识别下的语音听写项,下载 Android MSC 包,存储在 E:\AndroidStudioProjects\Ch14 路径下,文件名为 Android_iat1140_5f3c8d23.zip。该 SDK 包的 APPID 就是 5f3c8d23。

3) 复制并修改相关的代码文件

按照功能需要,从 speechDemo 样例中复制部分代码和资源文件。speechDemo 样例位于 E:\AndroidStudioProjects\Ch14\Android_iat1140_5f3c8d23\sample 文件夹下,复制的文件见项目的目录结构,如图 14-24 所示。

4) 配置开发环境

按照前面介绍的步骤,复制 SO 包、JAR 包文件到项目的 libs 目录中,在 Msc.jar 文件上右击,在弹出菜单上执行 Add As Library...菜单项,将该文件包添加到项目库中;配置模

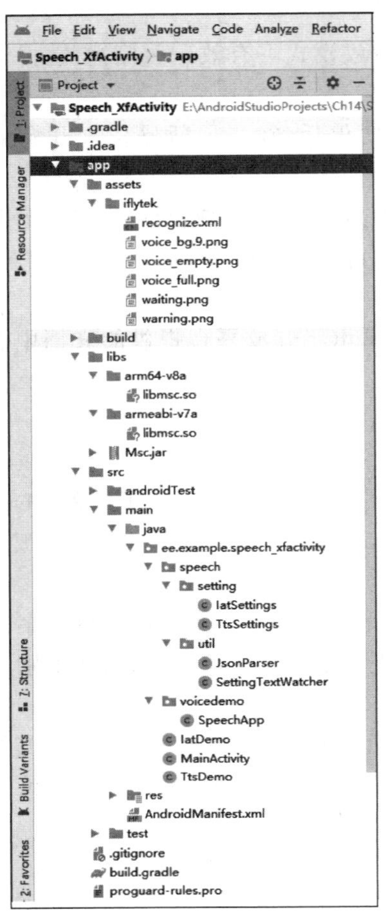

(a) 项目的类文件及库文件目录列表

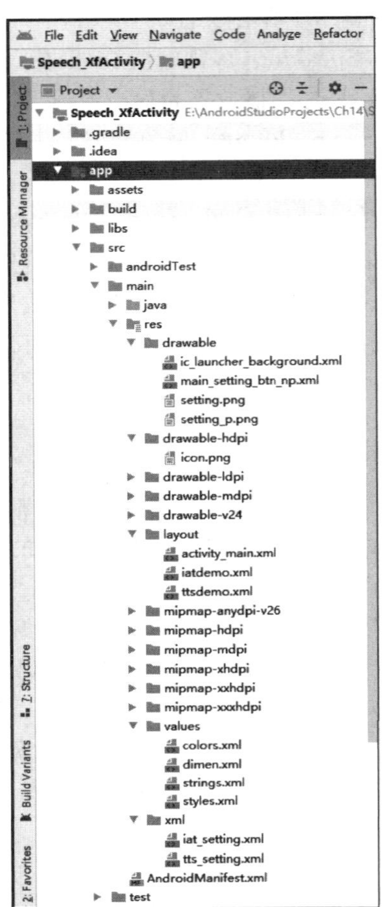

(b) 项目的布局文件和资源文件目录列表

图 14-24　语音应用项目的目录结构

块级的 build.gradle 文件，在 android {}内手动添加如下代码段。

```
sourceSets {
    main {
        jniLibs.srcDirs = ['libs']
    }
}
```

执行完上述操作后，建议执行同步 gradle 操作(File→Sync Project with Gradle Files)。

配置 AndroidManifest.xml 文件。根据本案例所用到的 Activity 来配置清单文件的 <activity>元素，权限声明可以从 speechDemo 的清单文件中复制，代码如下。

```
<?xml version = "1.0" encoding = "utf - 8"?>
<manifest xmlns:android = "http://schemas.android.com/apk/res/android"
    package = "ee.example.speech_xfactivity">

    <uses - permission android:name = "android.permission.INTERNET"/>
    <uses - permission android:name = "android.permission.RECORD_AUDIO"/>
```

```
7       <uses-permission android:name="android.permission.ACCESS_NETWORK_STATE"/>
8       <uses-permission android:name="android.permission.ACCESS_WIFI_STATE"/>
9       <uses-permission android:name="android.permission.CHANGE_NETWORK_STATE"/>
10      <uses-permission android:name="android.permission.READ_PHONE_STATE"/>
11      <uses-permission android:name="android.permission.WRITE_EXTERNAL_STORAGE"/>
12      <uses-permission android:name="android.permission.READ_EXTERNAL_STORAGE"/>
13
14      <application
15          android:icon="@drawable/icon"
16          android:name=".voicedemo.SpeechApp"
17          android:label="讯飞语音示例">
18          <activity
19              android:name=".MainActivity"
20              android:configChanges=
    "mcc|mnc|locale|touchscreen|keyboard|keyboardHidden|navigation|orientation|screenLayout|fontScale"
21              android:icon="@drawable/icon"
22              android:label="讯飞语音示例"
23              android:screenOrientation="fullSensor">
24              <intent-filter>
25                  <action android:name="android.intent.action.MAIN" />
26
27                  <category android:name="android.intent.category.LAUNCHER" />
28              </intent-filter>
29          </activity>
30
31          <activity android:name="ee.example.speech_xfactivity.speech.setting.TtsSettings" >
32          </activity>
33          <activity android:name="ee.example.speech_xfactivity.speech.setting.IatSettings" >
34          </activity>
35          <activity
36              android:name="ee.example.speech_xfactivity.IatDemo"
37              android:configChanges=
    "mcc|mnc|locale|touchscreen|keyboard|keyboardHidden|navigation|orientation|screenLayout|fontScale"
38              android:screenOrientation="fullSensor">
39          </activity>
40          <activity
41              android:name="ee.example.speech_xfactivity.TtsDemo"
42              android:configChanges=
    "mcc|mnc|locale|touchscreen|keyboard|keyboardHidden|navigation|orientation|screenLayout|fontScale"
43              android:screenOrientation="fullSensor">
44          </activity>
45
46          <meta-data
47              android:name="IFLYTEK_APPKEY"
48              android:value="'5f3c8d23'" />
49          <meta-data
50              android:name="IFLYTEK_CHANNEL"
51              android:value="Android_Demo" />
52      </application>
53
54  </manifest>
```

(1) 第 5～12 行，对本案例所需的权限进行声明。

(2) 第 16 行，是应用项目的入口配置，即<application>元素的 android.name 属性设置，此处与 speechDemo 的配置有些区别，一定要与项目中的 SpeechApp 子类的路径一致，否则程序无法正常执行。本案例的 SpeechApp 在当前包的子包 voicedemo 下。

（3）第 20、37、42 行，设置本 Activity 可以捕捉设备状态的变化，配置该属性，会直接调用 onCreate()方法中的 onConfigurationChanged()方法。

（4）第 46～51 行，为< application >附加 IFLYTEK_APPKEY 和 IFLYTEK_CHANNEL 的值，用来唯一标识本应用项目。

5）设计布局

编写 res/layout 目录下的 activity_main. xml 文件，设计两个按钮，单击后分别跳转到语音识别和语音合成页面，该布局为项目的主布局文件。语音识别和语音合成布局设计文件 iatdemo. xml 和 ttsdemo. xml 分别从 speechDemo 中复制过来，本案例对语音识别布局和语音合成布局做了部分删减修改，通过项目运行结果，可以清楚地知道删减内容，在此就不列出代码了。

另外，语音识别、语音合成的参数修改设置页面由列表布局文件 iat_setting. xml、tts_setting. xml 和 understand_setting. xml 设计，它们存放于 xml 目录下，一并要从 speechDemo 中复制过来。

6）开发逻辑代码

在 ee. example. speech_xfactivity 包及其下级包中定义了多个子类，它们各司其责，相互作用，共同实现语音识别和语音合成功能。每一个子类实现的功能及调用关系见表 14-1。

表 14-1　Speech_XfActivity 项目各子类的功能说明及调用关系

子 类 名	包　　名	说　　明
SpeechApp	ee. example. speech_xfactivity. voicedemo	Application 子类，初始化语音配置对象，赋予应用的 APPID 值，以便能调用讯飞语音 SDK 进行开发
MainActivity	ee. example. speech_xfactivity	Activity 子类，应用启动的第一个 Activity。在该类中完成应用需要的动态权限申请。 activity_main. xml 提供该 Activity 的页面布局设计
IatDemo	ee. example. speech_xfactivity	Activity 子类，实现无 UI 的语音识别功能，支持不同语言的语音识别功能。 调用 IatSettings 子类可以实现对语音识别的常用属性进行设置，并通过 SharedPreferences 对象保存参数。 调用 JsonParser 类获取 Json 结果中的 sn 字段。 iatdemo. xml 提供该 Activity 的页面布局设计
TtsDemo	ee. example. speech_xfactivity.	Activity 子类，实现线上语音合成功能，支持选择不同的发音人。 调用 TtsSettings 子类可以实现对语音合成的常用属性进行设置，并通过 SharedPreferences 对象保存参数。 ttsdemo. xml 提供该 Activity 的页面布局设计

续表

子类名	包名	说明
IatSettings	ee. example. speech _ xfactivity. speech. setting：	PreferenceActivity 子类，以列表的格式显示语言设置、前端点超时、后端点超时和标点符号的默认设置，并且可以修改之。 使用 PreferenceActivity 来显示这些设置列表，页面设置由 iat _setting. xml 文件提供
TtsSettings	ee. example. speech _ xfactivity. speech. setting	PreferenceActivity 子类，以列表的格式显示语速、音调、音量和音频流类型的默认设置，并且可以修改之。 使用 PreferenceActivity 来显示这些设置列表，页面设置由 tts_setting. xml 文件提供。 调用 SettingTextWatcher 子类可以限定参数设置的取值范围
SettingTextWatcher	ee. example. speech _ xfactivity. speech. util	TextWatcher 子类，定义部分输入框输入范围的约束控制，以保证输入数据的合理性
JsonParser	ee. example. speech _ xfactivity. speech. util	定义 JsonParser 类，对 Json 结果进行解析

在应用项目中，完成语音识别功能由 IatDemo. java 实现，其中对语音识别的属性设置通过调用 IatSettings. java 实现。IatSettings 是 PreferenceActivity 的子类，在子类的 onCreate()方法中调用 addPreferencesFromResource(R. xml. iat_setting)方法，从资源目录 xml 中取 iat_setting. xml 文件设置页面；属性值通过共享参数 SharedPreferences 对象保存到指定的文件中，调用 getPreferenceManager(). setSharedPreferencesName(PREFER_NAME)方法指定保存共享参数的文件名，其中的 PREFER_NAME 是静态常量，其所指字符串值即为文件名。

完成语音合成功能由 TtsDemo. java 实现，对语音合成的属性设置通过调用 IatSettings. java 实现。

运行结果：本案例选择在真机上运行项目，在运行项目之前，首先要接入手机，设置手机为运行设备。然后运行 Speech_XfActivity 项目。在手机上看到运行初始界面，如图 14-25(a)所示。

单击“语音听写”按钮，进入语音识别界面，单击“开始”按钮，就可以对着手机说一段诗词，如“床前明月光 疑是地上霜 举头望明月 低头思故乡”，句与句之间有短暂的停顿，片刻之后在识别区显示文字内容，并自动加上标点符号，如图 14-25(b)所示。单击屏幕下方的设置小语种处，出现语种设置选项列表，如图 14-25(c)所示。单击屏幕右上角齿轮状设置图标，进入识别配置参数浏览列表界面，可以看到当前的参数默认值，如图 14-25(d)所示。单击其中任一参数项，可以对该参数值进行修改，如图 14-25(e)所示。

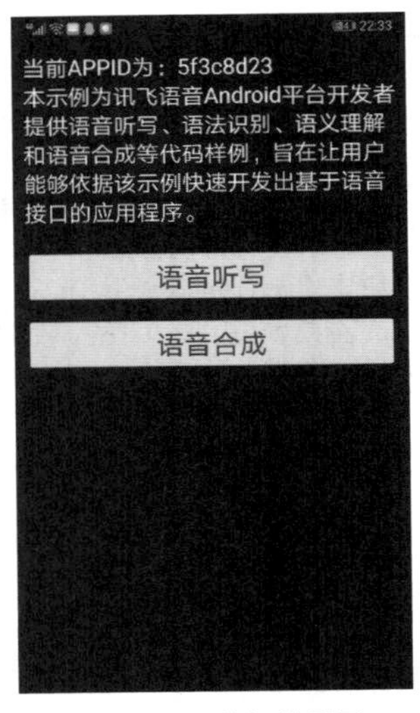

(a) 语音应用的初始界面

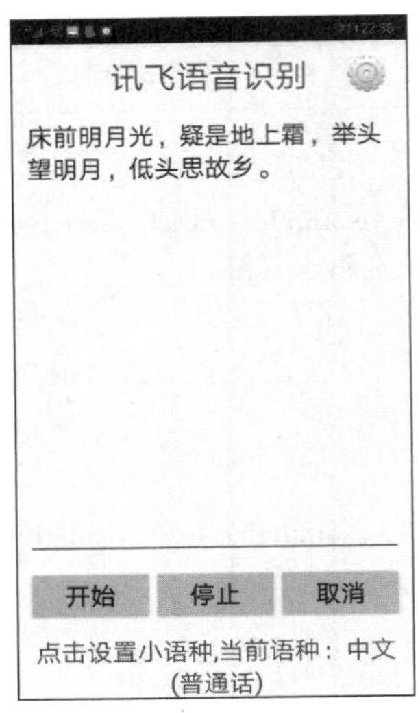

(b) 语音识别界面

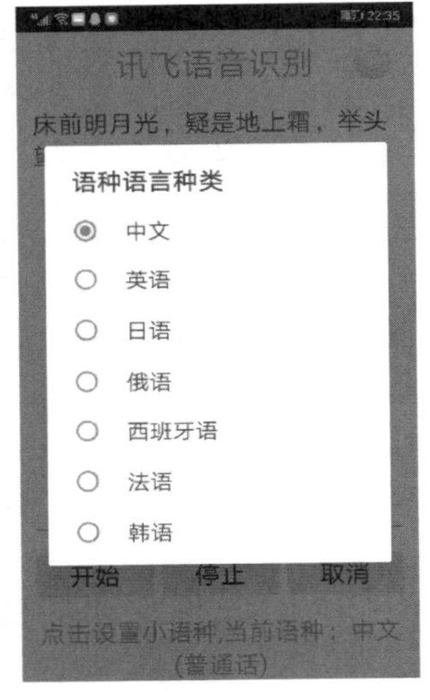

(c) 选择小语种列表

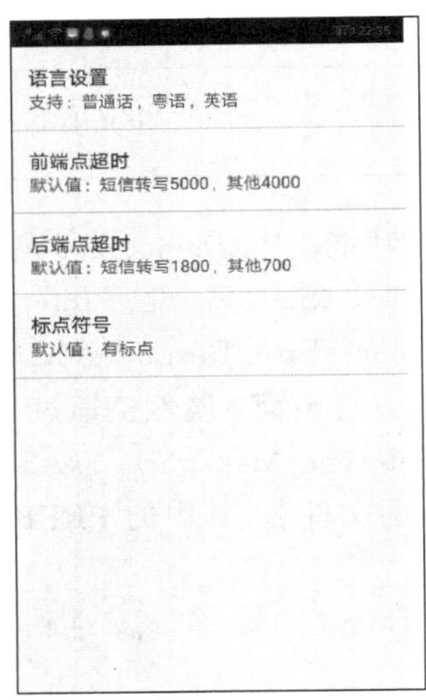

(d) 识别参数浏览列表

(e) 进入参数设置界面

图 14-25　首界面与语音识别应用运行界面

单击"语音合成"按钮，进入语音合成界面，如图 14-26(a)所示。单击"开始合成"按钮，就可以听到对语音合成区文字的语音播音，同时可以看到正在播报的文字段加上了底纹强调，如图 14-26(b)所示。此时播报的语音是在线选择的，默认是小燕的声音。单击"发言人"按钮，可以选择更多的人，如图 14-26(c)所示。如果选择了英文发言人，合成区的文字和语音将会以英文方式呈现。单击屏幕右上角齿轮状设置图标，进入语音合成配置参数浏览列表界面，可以看到当前的参数默认值，如图 14-26(d)所示。单击其中任一参数项，可以对该参数值进行修改，如图 14-26(e)所示。

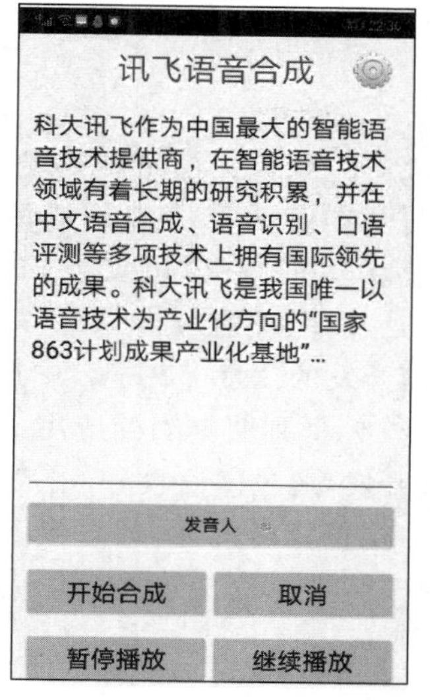

(a) 语音合成的初始界面

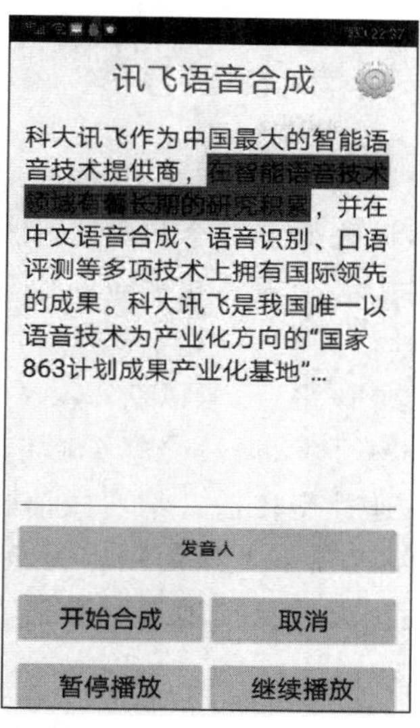

(b) 语音合成界面

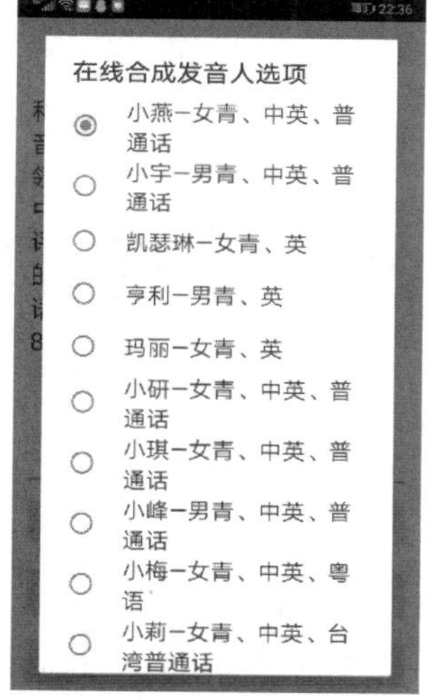

(c) 选择合成发音人列表

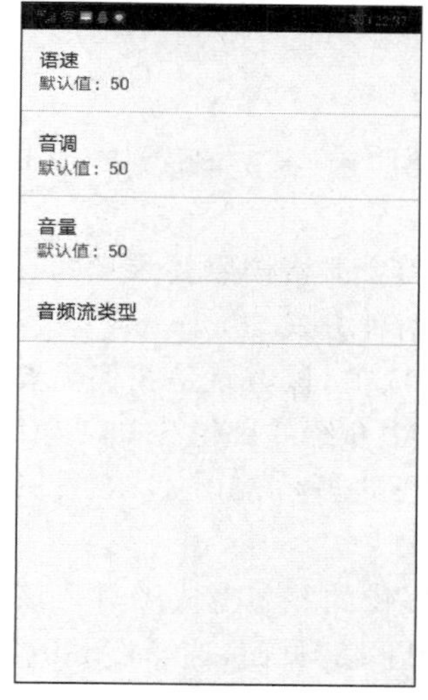

(d) 合成参数浏览列表

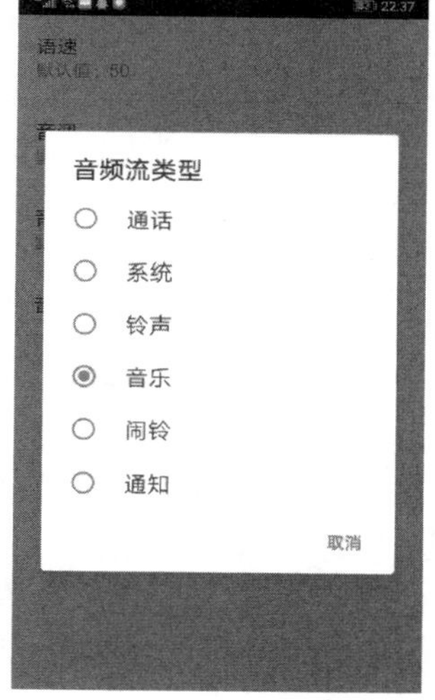

(e) 选择音频流参数界面

图 14-26　语音合成应用运行界面

14.3　社交 SDK

在中国，常用的社交平台是微信和 QQ。尤其是微信分享与收藏、QQ 分享与收藏、微信支付、支付宝支付等应用，越来越成为手机 App 中社交应用不可或缺的功能。在 App 中实现这些功能，可以根据第三方社交平台提供的 SDK 来快速实现。本节以开发 App 微信应用功能为例，简单介绍微信 SDK 的下载流程以及平台提供案例项目的 gradle 配置。

14.3.1 申请微信 APPID

微信开放平台网址为 https://open. weixin. qq. com/。首先要在该平台上注册申请成为开发者账号。申请账号有三个步骤，一是填写基本信息，二是邮箱激活，三是完善开发者资料。在完善资料中必须填写相关企业信息，可以填写所在公司或学校的信息，注册人信息中必须真实填写。

注册成功后登录该平台，可以获得关于微信开发的信息。在平台中选择“移动应用”，可以从中学习微信分享开发，通过分享给微信好友、分享到朋友圈的应用，使用 Android 移动应用被用户进行快速社交传播。学习微信收藏开发，通过微信收藏，用户可将移动应用的内容收藏到微信中，带来更多下次使用体验。学习微信支付开发，通过接入微信支付功能，用户可以在移动应用中方便快捷地通过微信支付来付款。完成这些微信应用开发，称为接入微信应用。

14.3.2 接入微信应用

1. 接入微信流程

接入微信流程分为三步：创建应用、提交审核、审核通过上线。

1）创建应用

在管理中心选择创建移动应用。严格按照要求填写、上传图片等信息。在第二页要求填写应用签名。获得应用签名有两种方式。

方式一：获取 Android 的 MD5 值，该值就是我们需要的应用签名。获取 MD5 值的方法参见本章 14.1.1 节之第 1 部分中介绍的获得 SHA1 值的方法。即在 Android Studio 中，打开 Gradle 窗格，展开项目的 Tasks→android，双击 signingReport 项，即可得到 MD5 的值。

方式二：利用微信开发平台提供的签名生成工具。该工具下载途径是：在微信开发平台上依次进入“资源中心”→“资源下载”页面，选择“Android 资源下载”页下面的“签名生成工具”，如图 14-27 所示。

下载该签名生成工具并安装运行，通过签名生成工具输入应用的包名，单击 Get Signature 按钮即可获取应用签名，单击 Copy to clipboard 按钮即可复制该签名。

2）提交审核

在填写完必填项目后，单击“提交审核”按钮，然后需要等待 7 天左右的审核时间，对于不合要求的填写申请，将不予以通过。

3）审核通过上线

审核通过后，开发者得到 APPID，可通过 APPID 进行微信分享、微信收藏等功能的开发，才能开发或发布应用。

2. 开发微信应用

微信开放平台既提供开发资源下载的平台，同时也是一个学习平台，几乎可以满足不同

图 14-27　微信开放平台 Android 资源下载页面

开发环境的需求，提供了大量的文档和示例代码。本书主要介绍 Android 应用开发，应该选择移动应用的资源。平台的移动应用包括如下。

(1) 微信分享与收藏。让用户可以实现从 Android App 中分享文字、图片、音乐、视频、网页、小程序至微信好友会话、朋友圈或添加到微信收藏。

(2) 微信支付。凡涉及支付功能，首先必须登记注册商户服务中心，提交相关资质材料，并通过审批，然后取得在线支付的资格，并签署在线协议，才能实现真正的支付操作。具体的申请流程见微信开放平台的微信 App 支付接入商户服务中心的申请流程指引文档。网址为 https://developers.weixin.qq.com/doc/oplatform/Mobile_App/WeChat_Pay/Vendor_Service_Center.html。

现在，平台只接受以公司主体的移动应用申请微信 App 支付权限，暂不支持个人开发的移动支付权限。移动支付包括支付、退款、对账等功能。

(3) 微信智能接口。通过微信接口开发图像识别、语音识别、语义理解和语音合成等功能。

还有微信登录、App 拉起微信小程序等应用功能。要实现这些微信应用功能开发，都需要为 App 申请专属的 APPID，并下载集成相应功能的 SDK，配置 build.gradle 和 AndroidManifest.xml 等开发环境。

如果要学习对微信的应用开发，可以在微信开放平台下载相关应用案例进行学习。网址为 https://developers.weixin.qq.com/doc/oplatform/Mobile_App/Resource_Center_Homepage.html。

小结

本章主要介绍了使用百度地图 SDK、高德地图 SDK 和科大讯飞语音 SDK 等常用第三方工具包的开发应用流程。如果要使用第三方开发包，操作流程都大同小异，归纳出以下 5 个步骤：其一，在第三方官网上注册开发者账户；其二，在相应官方网站上为应用项目获取开发密钥或 Key；其三，下载相关开发包和示例等资源；其四，Android 项目环境配置，包括添加相关库包到应用项目目录中，在 AndroidManifest.xml 文件中添加相关权限、属性等，设置 Gradle 相应配置及添加相关依赖等；其五，开始编程开发。

到本章为止，本书分门别类地介绍了关于 Android 应用开发的基础知识、开发技术和编程技巧，相信大家已具备了 Android 应用项目的开发能力，下一章将进行一个综合实例开发全过程的学习，让大家体会一下完整的应用项目的开发始终。

练习

利用百度地图 SDK 或高德地图 SDK，开发地图定位应用。要求：在地图上显示当前位置的定位点和经纬度等信息。

第 15 章 应用项目实例开发与发布

前面各章分别就相关控件、事件、存储、线程、服务、多媒体、网络通信、手机功能及第三方应用等主题进行学习。所有的案例也是为了突出学习内容、理解各控件作用、各机制控制逻辑而给出的。如果读者只是按章节的内容单一地学习开发技术，则体验不到一个完整项目的开发过程。本章以“我的音乐盒”项目为实训案例，学习实际项目的开发。

15.1 分析与设计

15.1.1 应用项目的需求分析

在网络上有许多音乐网站、娱乐网站、音乐社区、音乐资源库等。这些网站收集了海量的音乐、歌曲以及歌手的信息。但是对于个人而言，可能只是某个或某几个歌手的粉丝，只对部分歌曲及歌手感兴趣，要从海量的网络上找出自己喜欢的歌曲非常困难。那么，开发一套个性化的“我的音乐盒”应用项目十分必要。

“我的音乐盒”分为服务器端应用系统和客户端应用系统。

1. 服务器端系统

“我的音乐盒”服务器端应用系统主要完成对歌曲、音乐、歌手信息的收集及管理，对用户进行管理，负责提供向客户端传送歌曲等信息以及接收客户端请求和上传数据的接口。

服务器端应用系统使用 Java 开发，数据库使用 MySQL。

2. 客户端系统

“我的音乐盒”客户端应用系统主要完成对歌曲、音乐、歌手的播放和简单管理。使用 Android 进行开发，无论您身处何地，使用谁的手机，只要登录了客户端的 App，即可享用自己最爱的歌曲和歌手。主要的功能包括用户登录及欢迎界面、分类浏览与播放、相关信息管理，如图 15-1 所示。

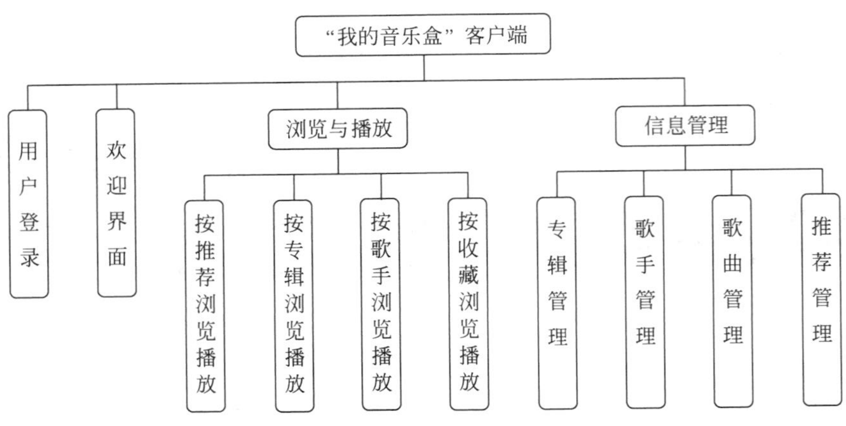

图 15-1　客户端功能模块

15.1.2　系统设计

1. 服务器端功能及数据库设计

“我的音乐盒”服务器端应用系统是 Web 管理后台程序，在浏览器上运行。服务器端应用分两大部分，一是基础管理，主要对用户和歌曲分类进行增、删、改、查等管理；二是音乐管理，主要对歌曲推荐、音乐、专辑和歌手进行管理。

根据应用的需求，服务器端数据库分别设计出用户信息表、音乐专辑表、歌曲信息表、歌手信息表和专辑类型表共 5 个表。表结构分别见表 15-1～表 15-5。

表 15-1　用户信息表 sys_user

字段名	数据类型	字段大小	是否主键	是否可为空	说　明
id	int		是	否	用户 ID 号，主键，自增量
account	varchar	20		否	账号
password	varchar	38		否	密码
name	varchar	50			用户名
mobile_phone	varchar	20			手机号

表 15-2　音乐专辑表 album

字段名	数据类型	字段大小	是否主键	是否可为空	说　明
id	int		是	否	专辑 ID 号，主键，自增量
name	varchar	20		否	专辑名称
picture_path	varchar	38			封面图片路径
music_count	int				歌曲数量
description	varchar	500			简介说明
publish_time	date				发行时间
singer_id	int				歌手 ID
theme_id	int				歌曲类型 ID

表 15-3 歌曲信息表 music

字段名	数据类型	字段大小	是否主键	是否可为空	说 明
id	int		是	否	歌曲 ID 号,主键,自增量
name	varchar	255			歌曲名称
album_id	int			否	专辑 ID
singer_id	int				歌手 ID
time_length	int				歌曲时长,单位为秒
file_path	varchar	255			文件链接
file_size	int				文件大小
description	varchar	500			歌曲简介
recommend_index	int			否	推荐排序。0 为未推荐,大于 0 为推荐的排序,按小到大排序

表 15-4 歌手信息表 singer

字段名	数据类型	字段大小	是否主键	是否可为空	说 明
id	int		是	否	歌手 ID 号,主键,自增量
name	varchar	20		否	姓名
picture_path	varchar	38		否	头像
birthday	varchar	50			生日
description	varchar	20			简介

表 15-5 专辑类型表 theme

字段名	数据类型	字段大小	是否主键	是否可为空	说 明
id	int		是	否	分类 ID 号,主键,自增量
code	varchar	20			代码
name	varchar	255			名称

2. 活动流程设计

根据客户端的功能,设计各活动的业务流程,如图 15-2 所示。

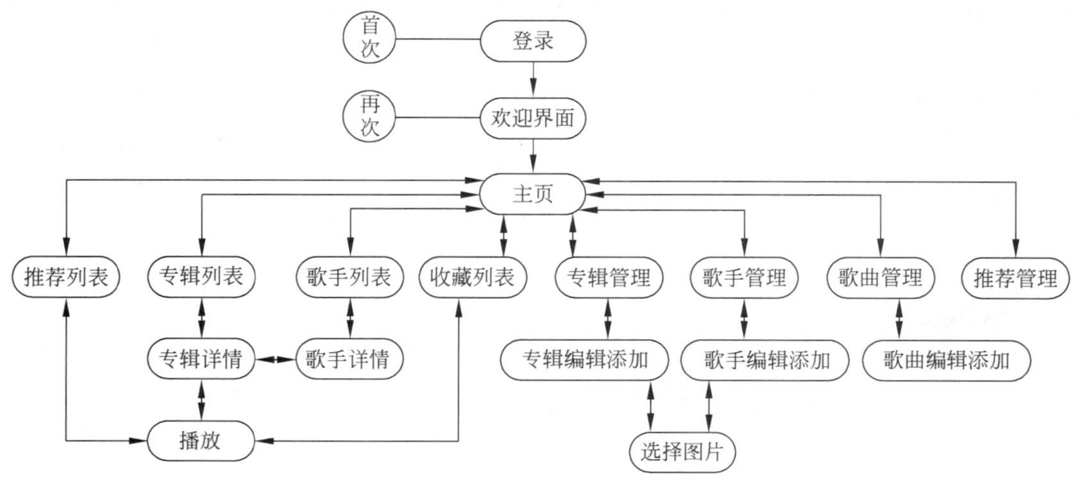

图 15-2 "我的音乐盒"客户端活动流程

(1) 用户登录。首次登录后，自动记住用户名和密码，再次打开 App 后自动跳过登录界面。登录成功进入主页。

(2) 欢迎界面。呈现系统应用名称，以及相关应用信息，数秒后消失，进入主页。

(3) 主页。设计导航栏和标签栏。标签栏包括推荐(默认)、专辑、歌手、收藏。导航栏中溢出的菜单包括专辑管理、歌手管理、歌曲管理、歌曲推荐、退出登录。

① 推荐。推荐歌曲信息列表，可进入歌曲播放 Activity。

② 专辑。专辑信息列表，可进入专辑详细信息 Activity，由专辑详情 Activity 进入歌曲播放 Activity。

③ 歌手。歌手信息列表，可进入歌手详细信息 Activity，由歌手详情 Activity 进入专辑详情 Activity，由专辑详情 Activity 进入歌曲播放 Activity。

④ 收藏。如果有收藏歌曲显示收藏歌曲信息列表，可进入歌曲播放 Activity。

⑤ 专辑管理。专辑信息列表 Activity，可删除专辑项，可进入添加或编辑专辑项 Activity，在添加 Activity 中可进入选择图片 Activity。

⑥ 歌手管理。歌手信息列表 Activity，可删除歌手项，可进入添加或编辑歌手项 Activity，在添加 Activity 中可进入选择图片 Activity。

⑦ 歌曲管理。歌曲信息列表 Activity，可删除歌曲项，可设置推荐标志，可进入添加或编辑歌曲项 Activity。

⑧ 歌曲推荐。推荐歌曲排序列表 Activity，可取消推荐，可上移、下移调整推荐顺序。

15.2 服务器端 Web 管理程序的部署说明

"我的音乐盒"由服务器端的 Web 管理程序和客户端的 App 两部分构成。开发流程是先完成 Web 管理程序的开发、部署，再完成 App 的开发。本书以学习 Android 应用开发为主，在此省略 Web 管理程序开发过程。我们已经开发好服务器端 Web 管理程序并打包。读者可从 github 网站上下载，下载地址为 https://github.com/zhanggx/music-store-web.git。在 github 上用户不用登录 github，直接在这里单击 Code 旁的下拉按钮，选择 Download ZIP 下载压缩文件就可以了，如图 15-3 所示。或者从 gitee 网站上下载，下载地址为 https://gitee.com/zhangproject/music-store-web。在 gitee 上需要注册登录后才能完成下载，但是下载速度要比 github 快。

图 15-3　在 github 上下载服务器端 Web 管理程序压缩包

我们已在网络服务器端部署好了 Web 管理程序。但是，为防止服务器租用到期或其他因素导致网络服务器无法访问，建议读者在本地计算机上部署 Web 管理程序。

"我的音乐盒"服务器端 Web 管理程序安装配置在本地机上，需要本地机的软件环境满足以下几个基本条件。

(1) Windows 7/Windows 10 操作系统，建议使用 64 位的操作系统。

(2) 安装 Java SDK，建议安装 Java 11 或者 Java 12。

(3) 安装数据库管理系统 MySQL。

(4) 安装集成开发环境 IntelliJ IDEA。

(5) 安装打包工具 Maven。

(6) 安装服务器 Tomcat 8.5。

下面就软件的下载安装、开发环境的配置以及本案例服务器端程序的打包部署分别进行简要说明。

15.2.1 安装 Java SDK

下载网址为 https://www.oracle.com/java/technologies/。需要安装 Java 8 以上版本，建议安装 Java 11 或者 Java 12。

15.2.2 安装 MySQL

下载 MySQL 社区版 8.0.X 版本，下载网址为 https://dev.mysql.com/downloads/installer/。下载 mysql-installer-web-community-8.0.X.0.msi 文件。(请注意，MySQL 各版本部分特性存在一些差异，请不要使用其他版本。)

在 Windows 7 下安装 MySQL 8.0.X 版本需要安装.NET Framework，启动 MySQL 的安装程序时会有提示，如图 15-4 所示。

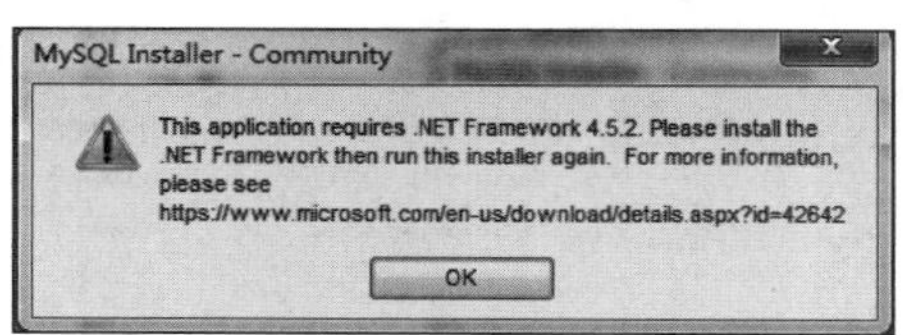

图 15-4 MySQL 8.0.X 版本要求安装.NET Framework 提示信息

可以到微软网站下载.NET Framework 应用程序。

运行 MySQL 安装程序 mysql-installer-web-community-8.0.X.0.msi，在安装对话框中选择安装类型时，选择 Developer Default，如图 15-5 所示。

安装程序在安装过程中会自动下载依赖包，如图 15-6 所示。

下载完成后单击 Next 按钮，在账户和角色对话框，输入 MySQL 默认 root 用户的密码，如图 15-7 所示。这里设置的用户名为 root，密码为 123456。注意，请记住您设置的密码，在设置应用程序配置文件时需要使用。

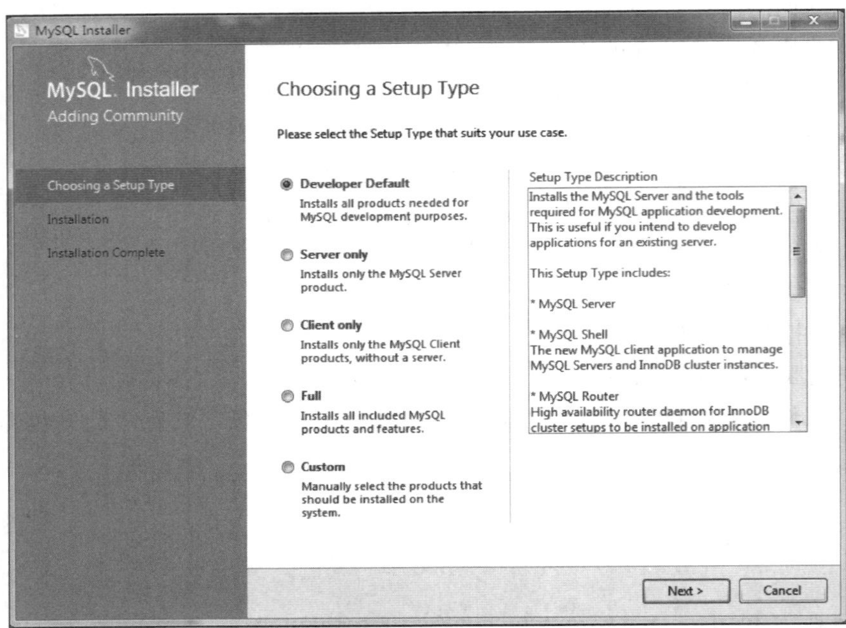

图 15-5　选择安装类型对话框

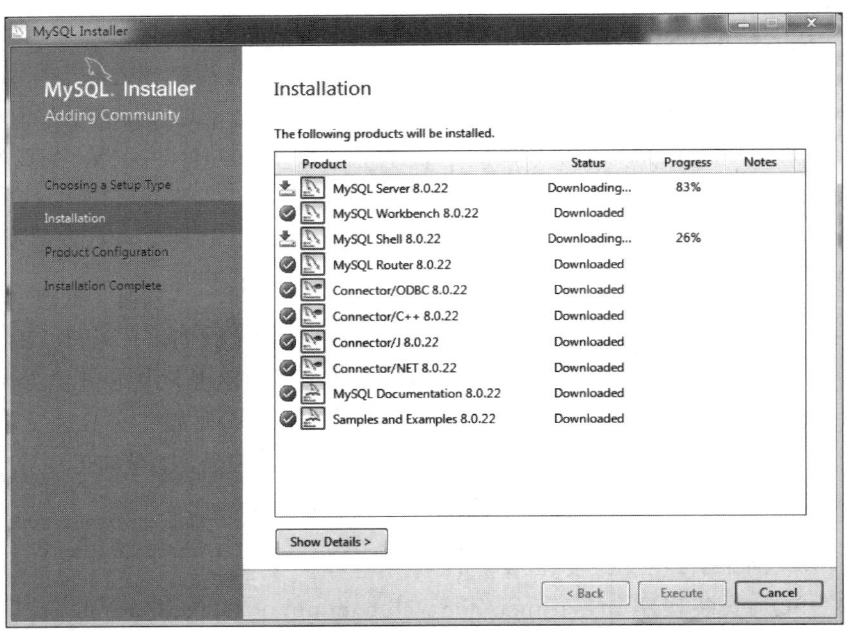

图 15-6　自动下载依赖包对话框

下一步配置 Windows 服务，保持默认配置，直接单击 Next 按钮即可，如图 15-8 所示。

接下来是应用配置(Apply Configuration)，单击 Execute 按钮执行。完成后单击 Finish 按钮。在最后的 Connect To Server 页面，可以输入 MySQL 的用户名和密码(分别为 root 和 123456)，单击 Check 按钮测试连接，如图 15-9 所示。

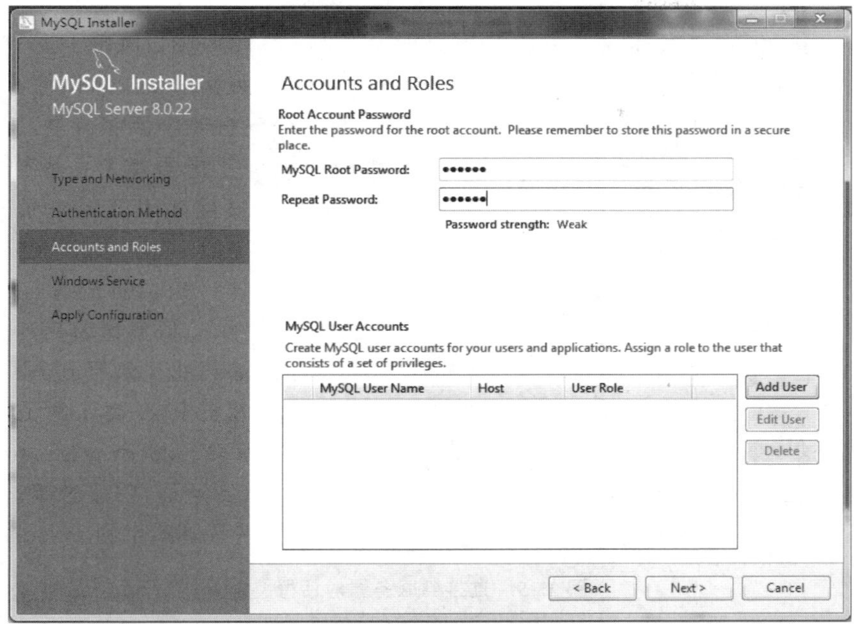

图 15-7　账户与角色对话框

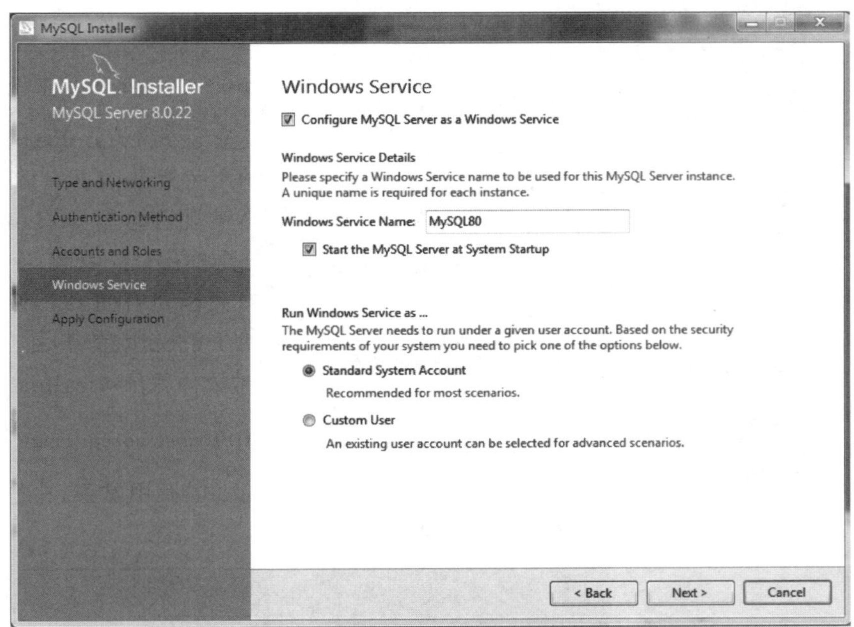

图 15-8　配置 Windows 服务对话框

检测完成后单击 Next 按钮进入下一步，执行 Apply Configuration。最后安装完成，如图 15-10 所示。

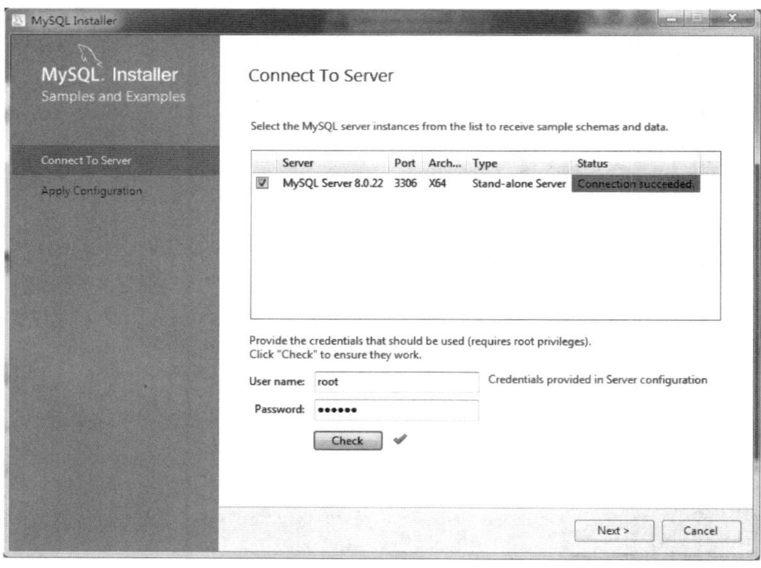

图 15-9　连接到服务器对话框

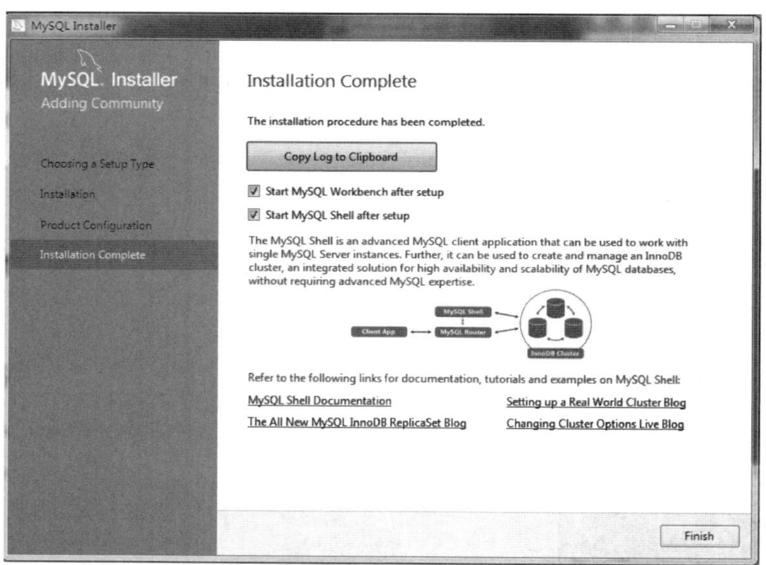

图 15-10　安装完成对话框

15.2.3　安装数据库

打开 MySQL Workbench，在 MySQL Connections 下面，会自动出现刚才安装的数据库。选择刚创建的数据库 Local instance MySQL80/localhost:3306，输入密码(123456)，如图 15-11 所示。

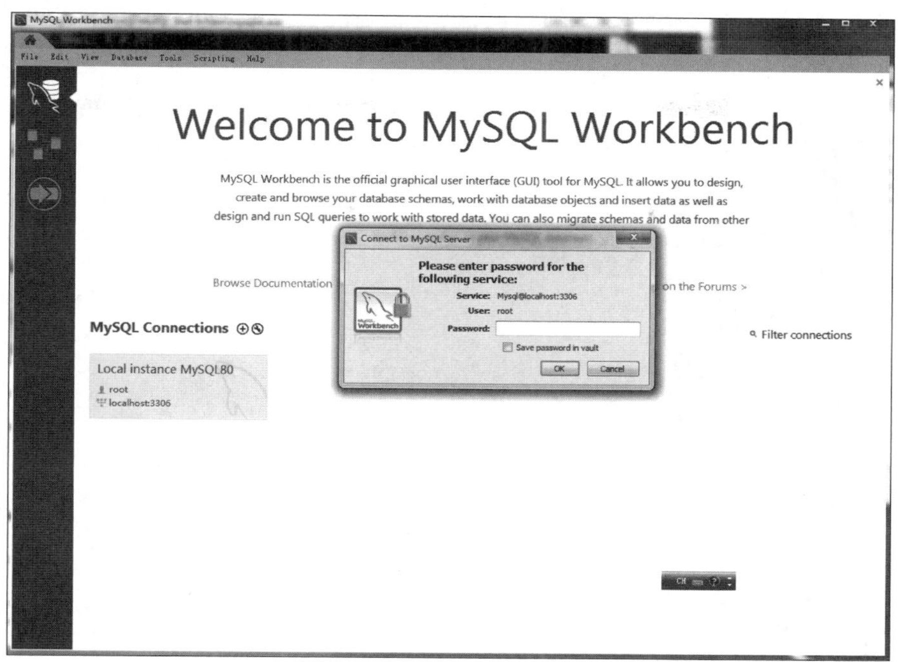

图 15-11 启动 MySQL

进入 MySQL 管理主界面，在左侧的 Navigator 栏目底部选择 Schemas，单击右键弹出菜单，选择 Create Schema，输入名称 music_store，字符集选择 utf8mb4 和 utf8mb4_genera 即可。输入完成后，单击 Apply 按钮，如图 15-12 所示。

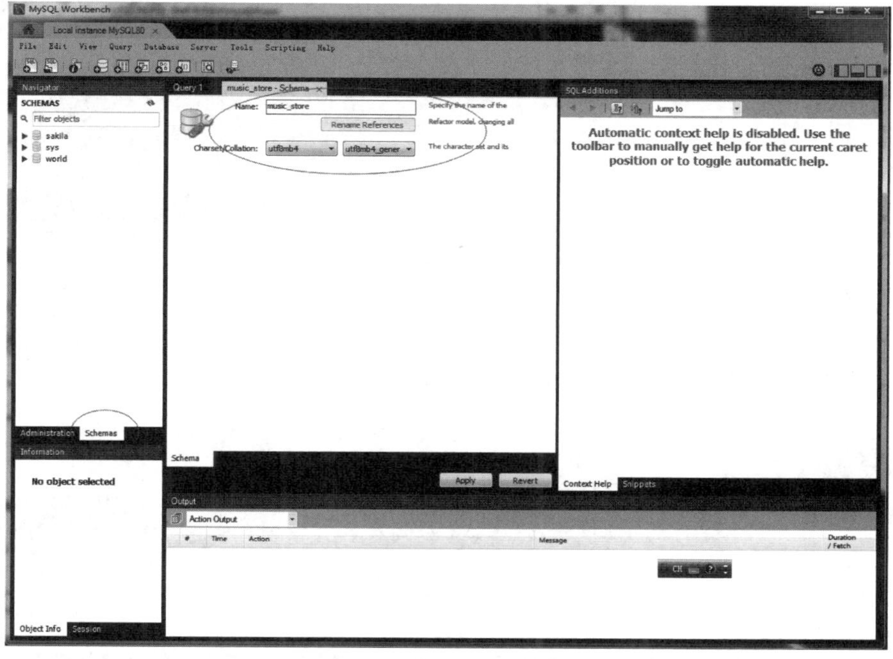

图 15-12 MySQL 管理界面

接下来会出现一个让您检查的窗口，直接单击 Apply 按钮，进入创建数据库界面，如图 15-13 所示。

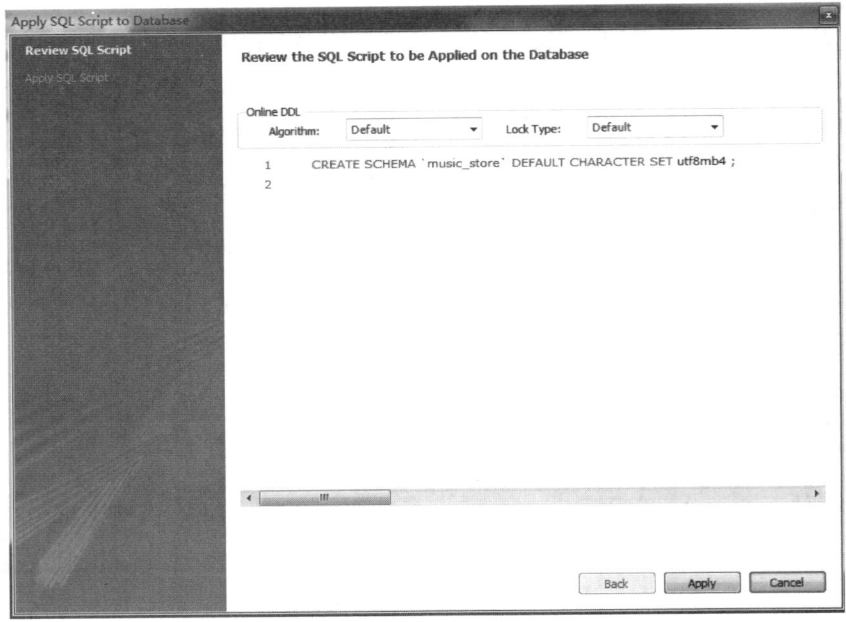

图 15-13　创建数据库

创建 Schema 之后，在 Query 窗口打开 music_store. sql 脚本，或者将该脚本文件的内容复制过去。单击 Query 窗口的“运行”按钮。运行完毕后，在左侧选择 music_store 的 schema，在其上单击右键，选择菜单 Refresh All 刷新，就可以看到相应的表已创建好，如图 15-14 所示。

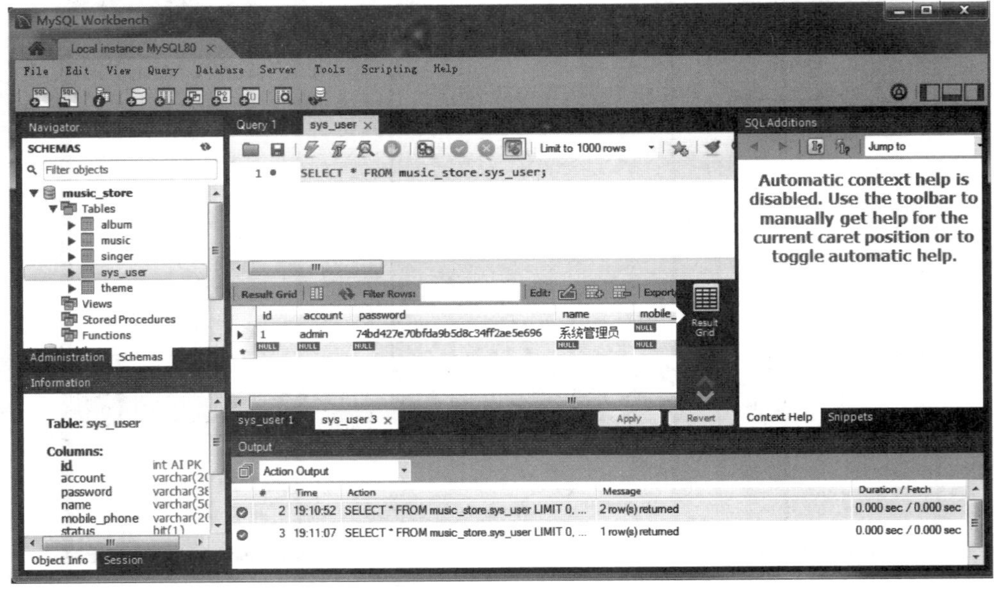

图 15-14　创建数据表

在左侧导航栏中的 Schemas 中选择 music_store 下的 sys_user 表，选择右击菜单 Select Rows，可以打印当前 sys_user 表的数据。

您也可以安装第三方数据库管理工具 Navicat 来创建和管理 MySQL，使用更方便。

15.2.4　安装 IDE 并配置项目开发环境

我们已经打包了服务器端的 Web 管理程序，打包文件名为 musicstore.war，该文件可在下载的服务器端程序 music-store-web-master 的目录下找到。如果读者想了解 Web 的程序源代码，或想在源代码基础上做二次开发，可以学习本节内容。否则可跳过本节和 15.2.5 节直接学习 15.2.6 节。

1. 下载并安装 IDE

在 jetbrain 网站下载 IntelliJ IDEA 的社区版，或者 Ultimate 版本（Ultimate 可以试用 30 天），下载网址为 https://www.jetbrains.com/idea/download/#section=windows。下载后安装。

2. 配置项目的 SDK

在 IntelliJ IDEA 中打开 music-store-web，第一次加载项目需要下载依赖包，可能需要花费较长时间。首先需要配置项目的 SDK，在左侧的项目导航栏右击菜单 Open Module Setting，弹出项目结构设计对话框，然后在左侧的菜单中选择 Project Settings-Project，请务必确认项目的 SDK 和 15.2.1 中安装的 Java SDK 要保持一致，如图 15-15 所示。

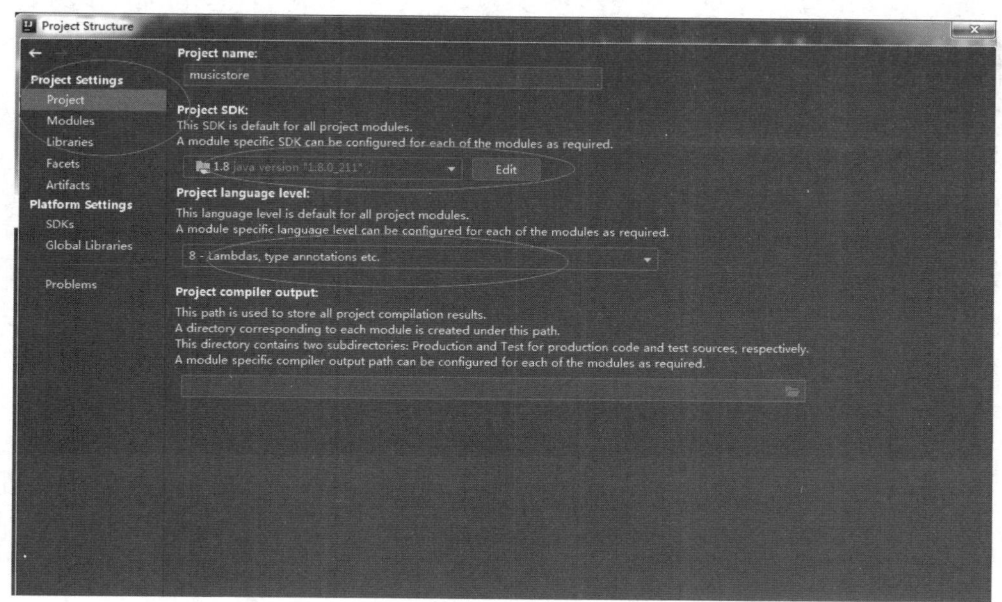

图 15-15　项目中的 SDK 配置

3. 修改配置文件

在本项目的 src/main/resources 目录下有若干配置文件。需要确认其中的一些配置信息，必要时需要修改。

（1）确认 application. properties 文件中的 profile 使用的是 dev 配置。

```
spring.profiles.active = dev
server.port = 18086            //为程序运行的端口,可根据需要修改此端口
```

（2）修改 application-dev. properties 文件中的相关配置。

① 确认 spring. datasource. url 配置中的数据库名称是否正确。

② 根据数据库的用户名和密码修改 spring. datasource. username 和 spring. datasource. password 的值。

③ 根据当前项目的路径修改 musicstore. file. path 的值。请注意目录的斜杠应该为"/"，如 D:/Tools/Tomcat 8.5/webapps/upload。

④ 根据当前计算机的 IP 地址和应用端口修改 musicstore. image. http. url 和 musicstore. music. http. url 的值。

（3）创建运行配置。

单击 Add Configuration，出现运行配置窗口（如果有 unknown 的项目，请删除），单击"+"号，选择 Application，在 Main Class 项目中选择 MusicStoreApplication，如图 15-16 所示。

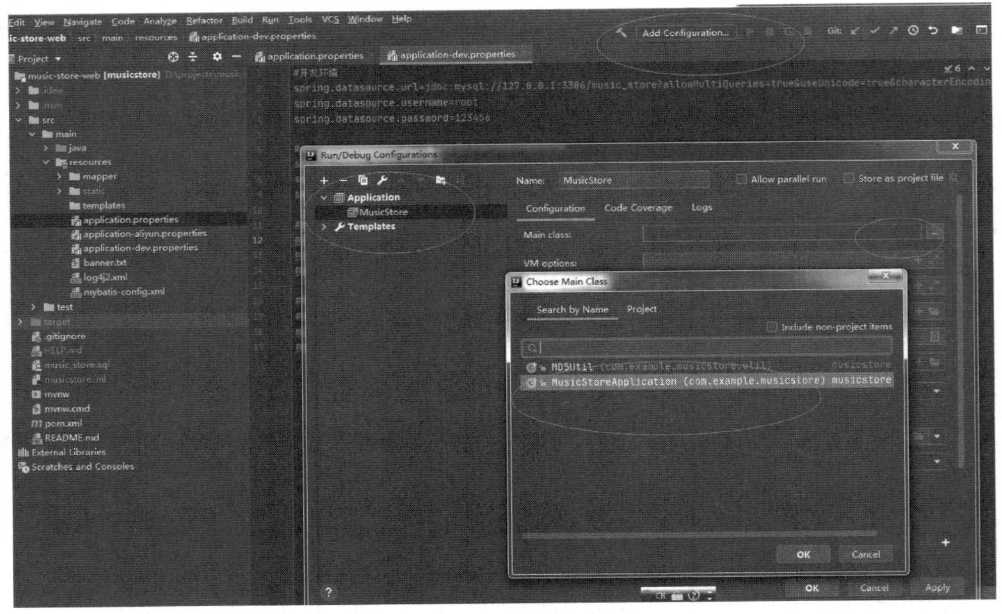

图 15-16 创建运行配置

4. 运行程序

在浏览器输入 http://localhost:18086，打开服务器端的 Web 管理程序。也可以输入 http://IP 地址:18086 测试。在运行 Web 管理程序时，需要注意下列事项。

1）防火墙设置处理

在联机测试环境中，请务必关闭本机的防火墙，或者将 18086 端口设置为允许。

2）客户端上传文件生效处理

在 IntelliJ IDEA 中运行后台应用程序，客户端上传的文件不会实时更新，客户端上传文件后，需要重新启动程序，才会生效。

3）创建多个 profile

可以根据自己的需要，创建多个 profile。例如，将 application-dev. properties 复制成 application-release. properties，修改其中的配置，然后将 application. properties 中的 spring. profiles. active 修改为 release，等等。

15.2.5　打包 WAR

打包需要使用 maven 打包工具，下载网址 http://maven. apache. org/download. cgi，下载 maven 3. X 版本。如果读者想利用已生成的打包文件进行部署，可以跳过本节，直接学习 15.2.6。

1. 下载打包工具

下载 apache-maven-3. X. X-bin. zip，解压该压缩包文件，并将解压后的路径配置到系统的 PATH 环境变量中。

2. 编辑 pom. xml 文件

打开 pom. xml 文件，在 dependency 下的 org. springframework. boot 下面需要增加如下代码片段。

```
<exclusions>
    <exclusion>
        <groupId>org.springframework.boot</groupId>
        <artifactId>spring-boot-starter-tomcat</artifactId>
    </exclusion>
</exclusions>
```

注意：在原来的 pom. xml 文件中已经增加了这段代码，只是处于注释状态，如图 15-17 所示。这是因为 spring boot 自带了一个 Tomcat，在 IDE 中运行时，不需要配置自己的 Tomcat。但是在打包时，需要去掉这里的注释符号<! --　-->，否则打包时会出错。

修改 pom. xml 文件后，右上角会出现一个刷新按钮，请务必刷新。或者在 IntelliJ IDEA 左侧导航栏的 pom. xml 文件上单击右键，选择 Maven→Reload project 刷新。

3. 开始打包

在命令行窗口中，更改当前目录为本项目的根目录，然后输入如下命令开始打包。

```
mvn clean package
```

注意：使用该命令，是在<maven 解压目录>\bin 路径已经添加到系统的 PATH 环境变量中的前提下。如果未添加，只能在 maven 解压目录下运行：<maven 解压目录>\bin\mvn clean package。

```
</dependency>
<dependency>
    <groupId>org.springframework.boot</groupId>
    <artifactId>spring-boot-starter-web</artifactId>
    <!--忽略内嵌tomcat，打包部署到tomcat。注*本地运行的时候要把这一段忽略引入个注释掉，要不然项目启动不了-->
    <!--
    <exclusions>
        <exclusion>
            <groupId>org.springframework.boot</groupId>
            <artifactId>spring-boot-starter-tomcat</artifactId>
        </exclusion>
    </exclusions>
    -->
</dependency>
<dependency>
    <groupId>org.mybatis.spring.boot</groupId>
```

图 15-17　打包配置项

第一次打包需要下载依赖，可能需要花费较长时间。

打包完毕后在当前项目的 target 目录下生成了打包文件 musicstore. war，该文件可用于部署。

15.2.6　部署 WAR

1. 下载并安装 Tomcat 8.5

在 https://tomcat. apache. org/download-80. cgi 中下载 Tomcat 8.5。对于 Windows 操作系统，可下载 32-bit/64-bit Windows Service Installer 安装文件。

安装 Tomcat 8.5 时，Tomcat 的默认端口为 8080，请确认端口是否与其他程序冲突，如果有冲突需要修改端口号。Tomcat 的默认安装路径为 C:\Program Files\Apache Software Foundation\Tomcat 8.5，您也可以选择安装在其他地方，安装完成后，会启动 Tomcat 的服务。

2. 复制打包文件

将 musicstore. war 包复制到 Tomcat 安装目录下的 webapps 目录中（进入 Tomcat 目录可能会提示权限问题，确定即可），Tomcat 运行时，会自动解压这个文件。

3. 修改配置文件

在 Windows 的文件资源管理器中进入 musicstore 文件夹下的\WEB-INF\classes，可以看到配置文件 application. properties 和 application-dev. properties。请按 15.2.4 节中第 3 部分所描述的方法对这两个文件进行修改。

4. 配置文件路径

在 Tomcat 的安装路径下的 webapps 文件夹中创建一个 upload 子文件夹，如图 15-18 所示。在 upload 下面再创建 images 和 music 两个子文件夹。

然后打开 musicstore\WEB-INF\classes 下的 application-dev. properties 文件，将 musicstore. file. path 配置为 upload 的路径，如图 15-19 所示。

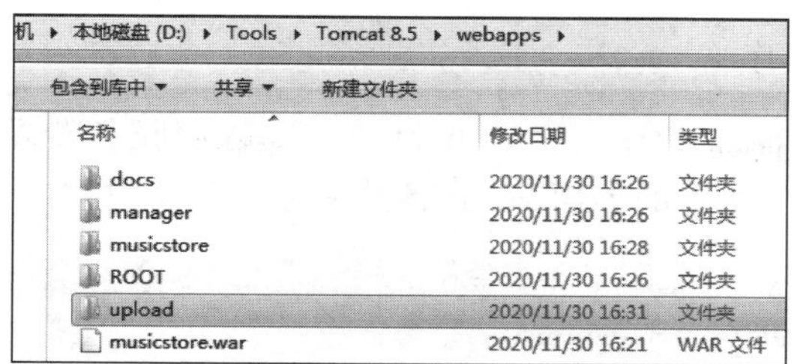

图 15-18　在 webapps 中创建 upload 目录

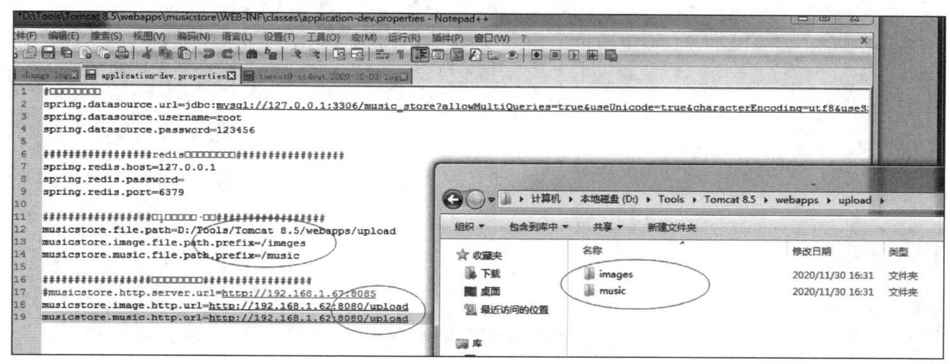

图 15-19　配置 application-dev. properties 文件中的路径

注意：在 musicstore. image. http. url 和 musicstore. music. http. url 后面请务必加上 upload 子文件夹。

5. 重启 Tomcat 服务

配置完成之后需要重启 Tomcat 服务。在 Tomcat 的管理窗口点击 Stop 按钮关闭服务，如图 15-20 所示，再打开服务；或者进入 Windows 服务中重启 Tomcat 服务。

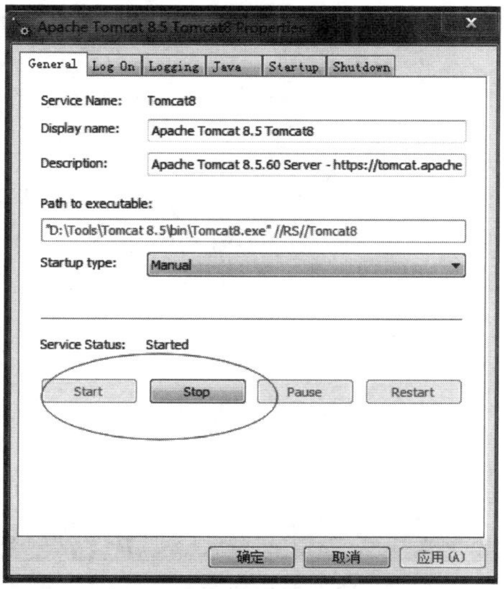

图 15-20　Tomcat 的管理窗口

6. 启动服务器端程序

在浏览器地址栏上输入 http://localhost:8080，启动服务器端程序，验证安装是否正确。如果出现如图 15-21 所示的界面，表示部署成功。

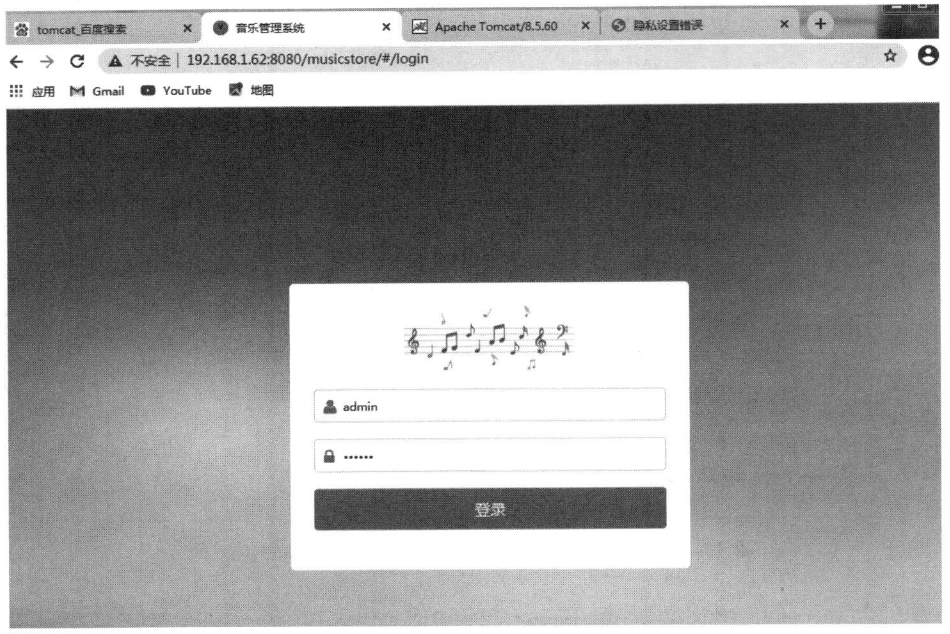

图 15-21　启动服务器端管理系统的登录界面

单击“登录”按钮进入管理系统首页。在页面的左侧是系统的功能菜单，单击“音乐管理”展开全部菜单项，如图 15-22 所示。

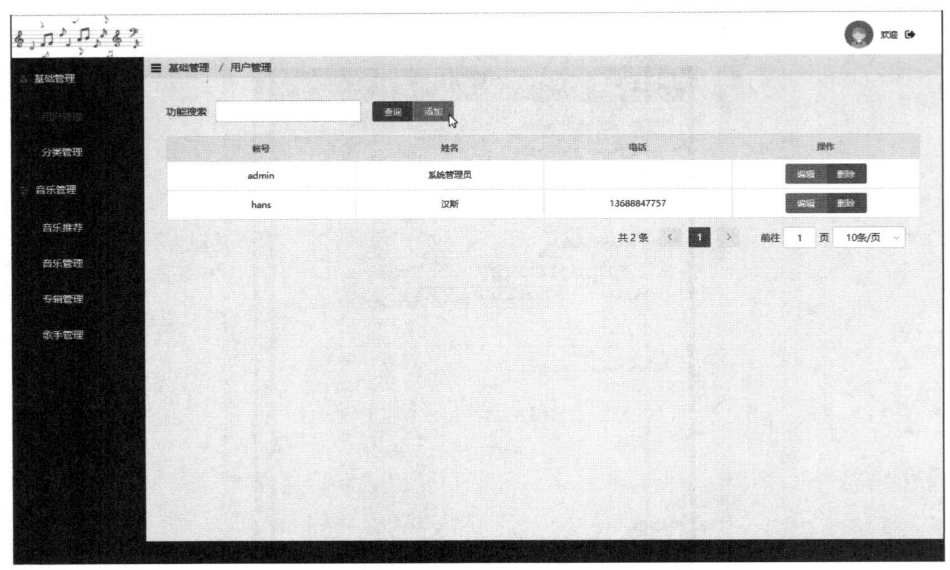

图 15-22　服务器端管理系统的首页界面

15.3　客户端 App 实现

"我的音乐盒"的客户端，所有的数据来自服务器端，通过服务器管理系统提供的接口，实现与服务器数据库的通信。在客户端创建一个本地的数据库表，存放歌曲、歌手、专辑、推荐等简要信息。关于用户名与密码、歌曲的播放状态可使用 SharedPreferences 存储。本书从第 4 章开始，已经陆续在各章后面的练习中涉及"我的音乐盒"相关的 Activity 开发设计。本章向读者提供客户端 App 的完整实现源代码。读者可从 github 网站上下载，下载地址为 https://github.com/zhanggx/music-store-android.git。或者从 gitee 网站上下载，下载地址为 https://gitee.com/zhangproject/music-store-android。

15.3.1　目录结构规划

在客户端的 App，除了实现各个 Activity 类外，还有一些适配器类、辅助类、服务类、数据处理类和工具类等。在创建应用项目后，需要对各个类的 Java 实现代码进行划分，分不同的目录来组织存放。本项目的 Java 代码文件，目录结构如图 15-23 所示。

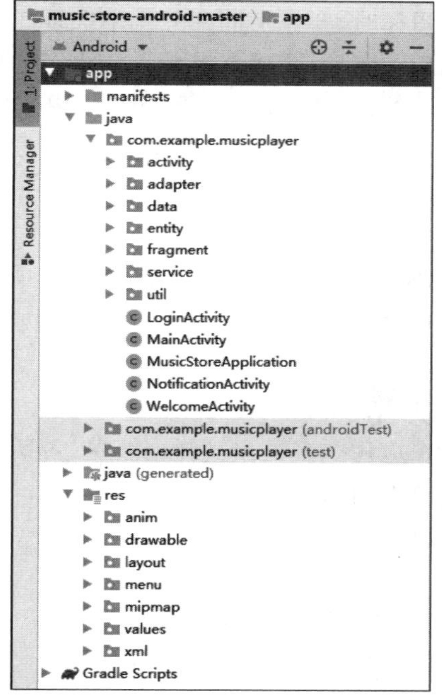

图 15-23　项目的目录结构

1. 根目录

在根目录中，存放 App 的入口 Activity，以及登录界面、欢迎界面、主页界面、通知界面的 Activity 实现代码文件。

2. activity 目录

在本目录中，存放 App 除了根目录之外的所有 Activity 中的实现代码文件。

3. adapter 目录

在本目录中，存放 App 中所有的适配器类的实现代码文件。

4. data 目录

在本目录中，存放 App 中所有的数据处理类的实现代码文件。

5. entity 目录

在本目录中，存放 App 中所有的实体类的实现代码文件。

6. fragment 目录

在本目录中，存放 App 中所有的 Fragment 类的实现代码文件。

7. service 目录

在本目录中，存放 App 中所有的涉及播放音乐等服务类的实现代码文件。

8. util 目录

在本目录中，存放 App 中的所用常量的定义类、上下文操作的辅助类、登录信息的数据存储类、网络服务请求通信类、通知创建的辅助类、屏幕相关数据计算与获取类的实现代码文件。

15.3.2 素材准备

开发任何一个实际的 Android 应用项目，都需要考虑在不同的 Android 设备上运行的 UI 效果，良好的用户体验是 App 的生命线。构成 UI 的元素包括图片、菜单、样式、文字表达和适当动画等，在开发中都需要精心地设计与准备。

1. 图片

当前智能手机的常见分辨率一般有 hdpi、xhdpi、xxhdpi，以 xhdpi、xxhdpi 居多，所以在准备图标时，要按不同的分辨率准备 2～3 套图片、图标。这样，在不同分辨率设备中运行时所见到的图片、图标在屏幕上的比例保持不变，用户界面得以良好地呈现。

另外，在 drawable 目录中，使用图形描述文件定义了一些控件的背景图。例如按钮、下拉框、输入框、编辑框等控件的背景图形。

2. 菜单

在主页面、二级页面和部分三级页面的导航栏中有设计菜单项，有些列表项也设计有弹出菜单，因此，需要在 res 目录中创建 menu 目录，编写菜单项的描述文件。

3. 动画

在应用中多处出现弹出菜单、对话框，当它们出现和消失时适当设计进入和退出动画；在播放歌曲时，为了让界面表现得更加丰富，设计一些简单动画。以增强用户的体验感。因此，需要在 res 目录中创建 anim 目录，编写动画描述文件。

4. 样式

对应用项目中的欢迎界面样式、输入框样式、对话框动画样式进行定义，使得 UI 呈现的效果与系统默认不同，更有个性化。

5. 字符串

对于项目中会出现的文字内容进行声明，使得文字表达规范统一，易于维护。

15.3.3 开发实现

客户端 App 的逻辑实现由 com. example. musicplayer 包下的所有 Java 代码协作完成。说明如下。

包含 17 个 Activity 类的实现代码。它们分别为：App 入口 Activity，通知栏展开 Activity，登录 Activity，欢迎界面 Activity，主页 Activity，专辑详情 Activity，歌手详情 Activity，播放 Activity，专辑管理 Activity，歌手管理 Activity，歌曲管理 Activity，推荐管理 Activity，专辑编辑/添加 Activity，歌手编辑/添加 Activity，歌曲编辑/添加 Activity，选择图片 Activity 以及图片预览 Activity。在这些 Activity 类的实现代码中，部分使用了视图绑定技术。在实际的应用项目开发中，有些 UI 中呈现的内容较多，换言之，布局文件中设计的控件数量较多。为了避免在代码中写入大量的 findViewById()方法，在本项目中使用了 ViewBinding 对视图中的控件进行绑定。

包含 5 个适配器类实现代码。AlbumArrayAdapter. java 和 AlbumListAdapter. java 用于专辑列表的适配；SingerListAdapter. java 用于歌手列表的适配；SongListAdapter. java 用于歌曲列表的适配；ItemObjectArrayAdapter. java 用于项目中下拉列表框的适配。

包含 4 个 Fragment 类的实现代码。AlbumListFragment. java 实现专辑列表 Fragment 类；SingerListFragment. java 实现歌手列表 Fragment 类；RecommendListFragment. java 实现推荐列表 Fragment 类；CollectListFragment. java 实现收藏列表 Fragment 类。

还包含一些对实体数据类、后台服务类、数据存储类、网络通信等工具类的实现代码。分别介绍如下。

1. 实体类

存放对实体类的数据规范，其中包括实体类的 Parcelable 接口、自定义 ItemObject 接口、网络通信请求与应答的 Bean 数据类、用户数据类的实现。这里的自定义 ItemObject 接口是对 Parcelable 接口中方法的补充。ItemObject 接口的代码如下。

```
public interface ItemObject{
    int getId();
    String getName();
}
```

在本应用项目中存在大量的 Activity 之间传递对象数据的情形，例如歌曲、歌手、专辑、推荐等。为了方便两个 Activity 之间传递实体对象，需要将对象进行序列化。在 Android 中序列化一个对象有两种方式，一种是实现 Serializable 接口，另一种就是实现 Parcelable 接口，实现 Parcelable 接口的效率要高于实现 Serializable 接口。本项目使用的是 Parcelable。

使用 Parcelable 的步骤如下。

(1) 实现 Parcelable 接口。

需要定义一个 Parcelable 的子类，用于实现 Parcelable 接口。

（2）实现接口的两个方法。

两个方法的代码如下。

```
public int describeContents();      //内容接口描述，默认返回 0
public void writeToParcel(Parcel dest, int flags);  //将对象序列化为一个 Parcel 对象，
                                                    //即将对象存入 Parcel 中
```

（3）实例化静态内部对象 CREATOR，实现接口 Parcelable. Creator。实例化 CREATOR 时要实现其中的两个方法：writeToParcel()和 createFromParcel()方法。也就是说，先利用 writeToParcel()方法写入对象，再利用 createFromParcel()方法从 Parcel 中读取对象。因此这两个方法中的读写顺序必须一致，否则会出现数据紊乱。

以 Singer.java 为例，其代码如下。

```
public class Singer implements Parcelable, ItemObject {
    private int id;
    private String name;;
    private String picturePath;
    private String pictureUrl;
    private String birthday;
    private String description;
    private String timeStamp;                  //创建时间
    //构造方法
    public Singer(){ }
    //带 Singer 参数的构造方法
    public Singer(Singer singer){
        id = singer.id;
        copyFrom(singer);
    }
    //带 Parcel 参数的构造方法
    protected Singer(Parcel in) {
        id = in.readInt();
        name = in.readString();
        picturePath = in.readString();
        pictureUrl = in.readString();
        birthday = in.readString();
        description = in.readString();
        timeStamp = in.readString();
    }
    //将对象存入 Parcel 中，即序列化为 Parcel 对象
    @Override
    public void writeToParcel(Parcel dest, int flags) {
        dest.writeInt(id);
        dest.writeString(name);
        dest.writeString(picturePath);
        dest.writeString(pictureUrl);
        dest.writeString(birthday);
        dest.writeString(description);
        dest.writeString(timeStamp);
    }
    //内容接口描述，返回 0
    @Override
    public int describeContents() {
```

```
            return 0;
        }
        //实例化静态内部对象 CREATOR,实现接口 Parcelable.Creator
        public static final Creator<Singer> CREATOR = new Creator<Singer>() {
            @Override
            public Singer createFromParcel(Parcel in) {
                return new Singer(in);
            }

            @Override
            public Singer[] newArray(int size) {
                return new Singer[size];
            }
        };
        //实现相关的 get...()和 set...()方法
        public int getId() { return id; }
        public void setId(int id) { this.id = id; }
        public String getName() { return name; }
        public void setName(String name) { this.name = name; }
        public String getPictureUrl() { return pictureUrl; }
        public void setPictureUrl(String pictureUrl) { this.pictureUrl = pictureUrl; }
        public String getBirthday() { return birthday; }
        public void setBirthday(String birthday) { this.birthday = birthday; }
        public String getDescription() { return description; }
        public void setDescription(String description) { this.description = description; }
        public String getTimeStamp() { return timeStamp; }
        public void setTimeStamp(String timeStamp) { this.timeStamp = timeStamp; }
        public String getPicturePath() { return picturePath; }
        public void setPicturePath(String picturePath) { this.picturePath = picturePath; }
        //实现成员方法
        public void copyFrom(Singer singer) {
            name = singer.name;
            picturePath = singer.picturePath;
            pictureUrl = singer.pictureUrl;
            birthday = singer.birthday;
            description = singer.description;
            timeStamp = singer.timeStamp;
        }
    }
```

2. 服务类

存放对音乐播放的相关类和接口,包括如下几个。

PlayMusicInfo.java 是对音乐播放的状态实现。

PlayServiceCallBack.java 是定义播放操作回调方法的接口,将在 activity 目录下的 PlayActivity.java 中实现,作用于播放进度条对音乐播放进度的跟踪。

PlayServiceBinder.java 是定义播放操作方法的接口。

PlayService.java 是一个服务子类,用于实现播放的各种操作、状态信息的获取、播放的广播发送、创建更新播放通知等。

3. 本地数据库管理

本项目使用到的音乐、歌手、分类、专辑及用户等数据都来源于网络数据库。但是对于在 App 上执行收藏时，有关的收藏信息存放于本地。于是，需要在本地创建一个数据表，以及对这个数据表进行操作和管理。

在 data 目录中，MusicDataUtils.java 就是这个本地数据表的管理类实现，其创建的数据表代码片段如下。

```
Log.d(TAG, "onCreate");
String create_sql = "CREATE TABLE IF NOT EXISTS " + TABLE_NAME + " ("
                    + "id INTEGER PRIMARY KEY NOT NULL,"
                    + "name VARCHAR NOT NULL,"
                    + "albumId INTEGER NOT NULL,"
                    + "albumName VARCHAR NOT NULL,"
                    + "singerId INTEGER NOT NULL,"
                    + "singerName VARCHAR NOT NULL,"
                    + "timeLength INTEGER NOT NULL,"
                    + "timeLengthText VARCHAR NOT NULL,"
                    + "fileUrl VARCHAR NOT NULL,"
                    + "filePath VARCHAR NOT NULL,"
                    + "fileSize INTEGER NOT NULL,"
                    + "fileSizeText VARCHAR NOT NULL,"
                    + "description VARCHAR NOT NULL,"
                    + "albumPictureUrl VARCHAR NOT NULL,"
                    + "timeStamp VARCHAR,"
                    + "collect_timeStamp LONG"
                    + ");";
Log.d(TAG, "create_sql:" + create_sql);
db.execSQL(create_sql);
```

4. 网络通信实现

1）使用 Bean 类

使用 Bean 类定义与网络数据通信时向服务器端发送的请求数据和接收服务器端返回的结果数据的通用属性和方法。这里定义的 Bean 类即 JavaBean，在 MVC 设计模型中是模型层，在一般的程序中，称它为数据层，该数据层用来设置数据的属性和一些行为，提供获取属性和设置属性的 get()/set()方法。JavaBean 是一种 Java 语言写成的可重用组件，所以，一般利用 Bean 类来存放一些特定的属性或行为，而不存放值，这样就能多次调用 Bean 类中的属性并赋值使用，实现重复使用的功能。

在项目的 entity 目录中，定义了这些类。

RequestBean.java，主要实现向服务器发送用户信息的 Bean 数据。

ResultBeanBase.java、ResultBean.java，主要实现从服务器端接收返回结果的 Bean 数据。

对于返回结果 ResultBean 类，为了使得返回结果的类型的通用性，使用统一代码规范 ResultBean < T >。这里的< T >是泛型参数类型，可以代表 String、Map、List 等数据类型。例如 ResultBean.java 的代码片段如下。

```
public class ResultBean < T > extends ResultBeanBase {
```

```
    private T data;
    public ResultBean(){ }
    public ResultBean(JsonObject jsonObject){
        super(jsonObject);
    }
    public T getData() {
        return data;
    }
    public void setData(T data) {
        this.data = data;
    }
}
```

2）使用 Json 和 Gson 技术

Json(JavaScript Object Natation)是一种文本形式的、轻量级的数据交换格式，与 XML 相似，比 XML 更为轻量，是被广泛采用的客户端和服务端交互的解决方案。Json 的解析和生成的方式很多，在 Android 平台上最常用的类库有 Gson 和 FastJson 两种，这里使用的是 Gson 类。

Gson 类位于 com.google.gson.Gson 包中。Gson 类提供了 toJson()和 fromJson()两个方法。

toJson(<对象>)方法，实现序列化，即将对象转换为 Json 字符串。

fromJson(<Json 字符串>,<对象>)方法，实现反序列化，即将 Json 字符串转换为对象。

在进行序列化与反序列化操作前，首先实例化一个 Gson 对象，然后才能使用 toJson()和 fromJson()方法，并且 Gson 借助 TypeToken 获取泛型参数类型的数据，可以方便地进行客户端与服务器端种类数据的通信。

例如，本项目首次用户登录时需要向服务器端上传用户名和密码，待服务器管理系统验证登录信息正确，返回 success 的成功结果后，才能通过登录进入 App 主页。为实现这部分网络通信，在 util 目录下的 NetworkRequestUtils.java 中有代码片段如下。

```
public static ResultBean<User> login(String account, String password){
    try {
        RequestBean requestBean = new RequestBean();
        requestBean.setAccount(account);
        requestBean.setPassword(password);
        String json = gson.toJson(requestBean);
        String result = postJsonData(LOGIN_URL, json);
        if (!TextUtils.isEmpty(result)){
            ResultBean<User> resultBean =
                    gson.fromJson(result, new TypeToken<ResultBean<User>>() { }.getType());
            return resultBean;
        }
    }catch(Throwable tr){
            tr.printStackTrace();
    }
    return null;
}
```

在项目根目录下 LoginActivity.java 接收服务器端的返回结果并执行相应操作，代码片段如下。

```
@Override
protected ResultBean<User> doInBackground(Void... voids) {
    LoginActivity loginActivity = loginActivityWeakReference.get();
    if (loginActivity == null){
        return null;
    }
    try {
        ResultBean<User> resultBean = NetworkRequestUtils.login(account,password);
        if (resultBean!= null&&resultBean.isSuccess()){
            loginActivity.loginUtils.setLoginUser(resultBean.getData().getAccount(),
                                            resultBean.getData().getName());
        }
        return resultBean;
    }catch(Throwable tr){
        tr.printStackTrace();
    }
    return null;
}
```

例如，进入主页的推荐页面，即可看到推荐歌曲列表。这些歌曲信息来自服务器端的数据库中。那么，在 NetworkRequestUtils.java 中有如下代码片段。

```
public static ResultBean<List<Music>> getRecommendMusicList(){
    try {
        String result = get(RECOMMEND_MUSIC_URL);
        if (!TextUtils.isEmpty(result)){
            ResultBean<List<Music>> resultBean
                = gson.fromJson(result, new TypeToken<ResultBean<List<Music>>>()
                        { }.getType());
            return resultBean;
        }
    }catch(Throwable tr){
        tr.printStackTrace();
    }
    return null;
}
```

在 fragment 目录中，负责推荐页面显示的 AlbumListFragment.java 有如下代码片段。

```
@Override
protected ResultBean<List<Album>> doInBackground(Void... voids) {
    AlbumListFragment albumListFragment = fragmentWeakReference.get();
    if (albumListFragment == null){
        return null;
    }
    return NetworkRequestUtils.getAlbumList();
}
```

5. 配置清单文件 AndroidManifest.xml

本项目的配置清单文件 AndroidManifest.xml 应用声明本项目应用到的相关权限，代码片段如下。

```
<uses-permission android:name="android.permission.INTERNET" />
<uses-permission android:name="android.permission.READ_EXTERNAL_STORAGE" />
<uses-permission android:name="android.permission.WRITE_EXTERNAL_STORAGE" />
<uses-permission android:name="android.permission.FOREGROUND_SERVICE" />
```

在<application>元素节点中，需要添加一些特定的属性以保证项目正常运行。

首先要指定应用项目的入口 Activity，该入口 Activity 用于实现对项目进行全局的初始化操作，尤其是对于那些引用了第三方应用包的项目。本项目的入口 Activity 为根目录下的 MusicStoreApplication.java，设置属性 android:name=".MusicStoreApplication"。

其次，Android 11(API 30)为外部存储设备上的应用和用户数据提供了更好的保护，执行分区存储。由于本应用涉及对 SD 卡中图片文件的读写访问操作，需要停用与分区存储相关的变更，需要添加属性 android:requestLegacyExternalStorage="true"。

最后，本应用项目使用明文网络流量，例如明文 HTTP。在 Android 8.1(API 27)或更低版本中，应用程序的默认值为 true，而面向 Android 9(API 28)或更高级别的应用默认为 false。因此，需要设置属性 android:usesCleartextTraffic="true"。

在<application>元素节点下，需要声明所有使用到的 Activity、Service 和 Provider，代码片段如下。

```
<activity android:name=".WelcomeActivity"
    android:theme="@style/WelcomeTheme" android:screenOrientation="portrait">
    <intent-filter>
        <action android:name="android.intent.action.MAIN" />
        <category android:name="android.intent.category.LAUNCHER" />
    </intent-filter>
</activity>
<activity android:name=".LoginActivity"
    android:theme="@style/WelcomeTheme" android:screenOrientation="portrait">
</activity>
<activity android:name=".MainActivity"
    android:launchMode="singleTask" android:screenOrientation="portrait">
</activity>
<activity android:name=".activity.AlbumActivity" android:screenOrientation=
                          "portrait"> </activity>
<activity android:name=".activity.SingerActivity" android:launchMode="singleTop">
                          </activity>
<activity android:name=".activity.PlayActivity" android:screenOrientation=
                          "portrait"> </activity>
<activity android:name=".NotificationActivity"> </activity>
<activity android:name=".activity.AlbumManageActivity" android:screenOrientation=
                          "portrait"> </activity>
<activity android:name=".activity.AlbumEditActivity"
    android:windowSoftInputMode="stateVisible|adjustResize" android:
                                  screenOrientation="portrait">
</activity>
<activity android:name=".activity.SingerManageActivity"> </activity>
<activity android:name=".activity.SingerEditActivity"
    android:windowSoftInputMode="stateVisible|adjustResize" android:
                                  screenOrientation="portrait">
</activity>
<activity android:name=".activity.SongManageActivity"> </activity>
```

```
<activity android:name = ".activity.SongEditActivity"
    android:windowSoftInputMode = "stateVisible|adjustResize" android:screenOrientation
                                  = "portrait">
</activity>
<activity android:name = ".activity.RecommendManageActivity"
    android:screenOrientation = "portrait">
</activity>
<activity android:name = ".activity.PicturePreviewActivity" android:screenOrientation = "
                         portrait"> </activity>
<service android:name = ".service.PlayService" android:exported = "false"> </service>
<provider
    android:name = "androidx.core.content.FileProvider"
    android:authorities = "com.example.musicplayer.fileprovider"
    android:exported = "false"
    android:grantUriPermissions = "true">
    <meta-data
        android:name = "android.support.FILE_PROVIDER_PATHS"
        android:resource = "@xml/file_paths" />
</provider>
```

15.4 项目调试与测试

在项目开发过程中，开发人员每完成一段 Java 程序，都需要运行一下，以检验运行结果是否达到预期。如果运行过程或结果出现问题，称为程序 Bug。要排除 Bug，可以使用 Android Studio 提供的调试工具。

15.4.1 调试程序

1. 调试程序的步骤

调试程序在 Android Studio 中进行，调试步骤如下。

(1) 在有可能出问题的代码处设置断点。在哪里设置断点需要有一定的分析与判断。

(2) 单击工具 Debug app 按钮，按钮图标为，开始调试。

(3) Debug app 按钮会重启应用程序，需要等待较长时间。如果程序已经运行，可以单击 Attach To Process 按钮，会弹出一个选择进程的对话框，选择 com.example.musicplayer。

(4) 使用调试工具，监测代码执行过程。调试程序如图 15-24 所示。

2. 调试工具说明

调试信息出现在 Android Studio 的下方窗格，如图 15-25 所示。对于调试窗格中的一些常用工具按钮简要说明如下。

(1) Step Over(F8)，程序向下执行一行，如果当前行有方法调用，不进入该方法，待该方法执行完毕返回，继续进入下一行执行。

(2) Step Into(F7)，程序向下执行一行，如果当前行有用户自定义方法(非官方类库方法)调用，则进入该方法执行。

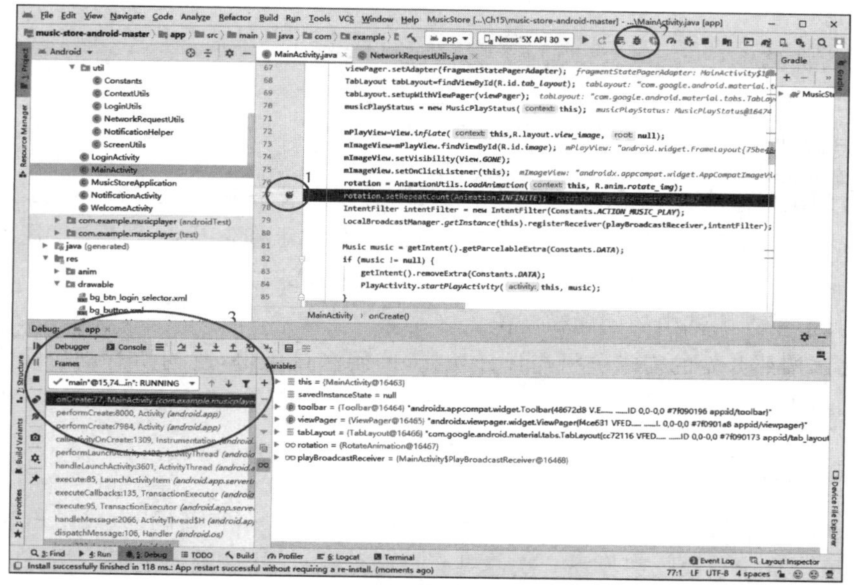

图 15-24　调试程序运行界面

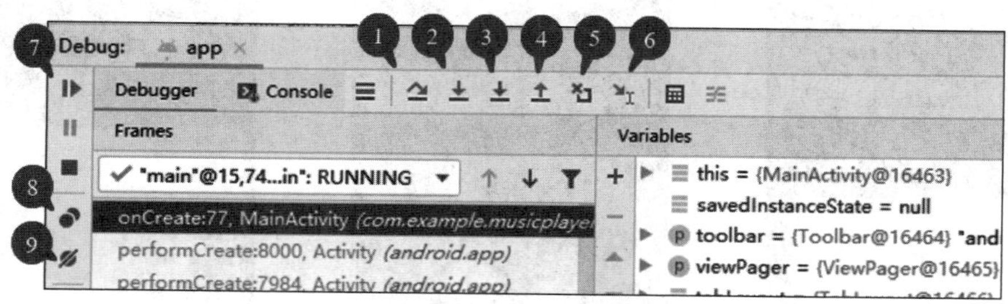

图 15-25　调试程序运行界面

(3) Force StepInto(Alt+Shift+F7)，程序向下执行一行，如果当前行有方法调用，则进入该方法执行。

(4) Step Out(Shift+F8)，如果在调试的时候进入了一个方法，并认为该方法没有问题，可以使用 Step Out 跳出该方法，返回到该方法被调用处的下一行语句执行。注意，这时该方法已执行完毕。

(5) Drop Frame，返回到当前方法的调用处重新执行，并且所有上下文变量的值也回到重新执行时刻。只要调用链中还有上级方法，可以跳到其中的任何一个方法。

(6) Run toCursor(Alt+F9)，一直运行到光标所在的位置。

(7) Resume Program(F9)，一直运行程序直到碰到下一个断点。

(8) View Backpoints(Ctrl+Shift+F8)，查看设置过的所有断点并可以设置断点的一些属性。在单击了 View Backpoints 按钮后，会出现一个断点属性窗口，可以对断点进行一些更高级的设置。

(9) Mute Backpoints，选中后所有的断点被设置成无效状态。再次点击可以重新设置所有断点有效。

15.4.2 测试

1. 项目测试

对项目的编码结束后，需要对项目进行测试，查找系统漏洞、Bug、项目各功能是否按设计要求实现，收集用户试用的反馈意见，完善系统。测试工作包括：设计测试用例，进行功能测试、性能测试和用户体验测试。对于测试出的 Bug，要进行 Bug 定位并修改，修改后要进行回归测试，直到测试中的问题被解决。

功能测试按照开始的需求分析与设计功能，使用测试用例，跑遍所有的应用页面，观察记录测试的结果。

2. 项目运行界面

运行“我的音乐盒”应用项目，首次运行时需要输入用户名和密码，完成用户登录。再次打开 App，可以直接进入欢迎界面。欢迎界面会停留数秒，然后进入主页。主页上有 4 个标签，默认显示推荐歌曲列表，分别如图 15-26～图 15-28 所示。

图 15-26　用户登录

图 15-27　欢迎界面

图 15-28　主页界面

在推荐歌曲列表中，每一栏列出了歌名、歌唱者的头像和姓名、专辑名、歌曲的时长，还有一个播放按钮。单击播放按钮，开始播放该栏的歌曲。在播放页面上端有一个导航栏，左侧是一个返回导航按钮，右侧是一个收藏按钮；歌手的头像在页面中央，以圆形呈现并顺时针旋转；下面有歌曲播放的进度条，以及暂停和停止播放按钮；如果单击了暂停按钮，会出现播放按钮。在播放歌曲时从上往下滑，可以看到展开的通知信息，里面有正在播放的歌曲信息，这里的播放通知栏是自己设计的显示布局，如图 15-29 所示。

在专辑歌曲列表中，每行有两个专辑项，单击专辑项后进入专辑详情页，详情页显示图

(a) 推荐歌曲列表

(b) 播放歌曲界面

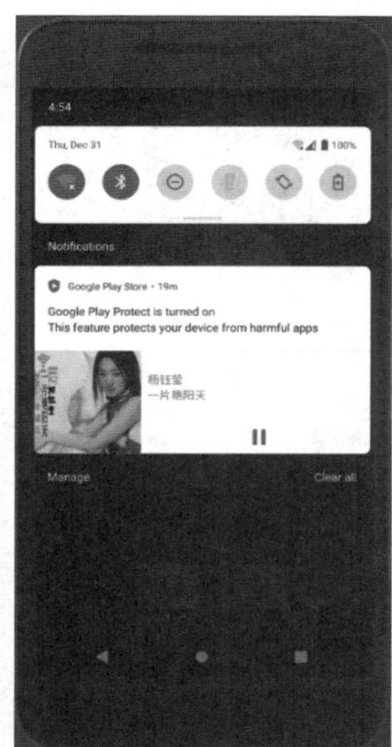

(c) 查看通知信息

图 15-29 推荐模块的运行效果

片、专辑内容、专辑下的歌曲列表。单击歌曲列表栏上的播放按钮，开始播放该栏的歌曲，如图 15-30 所示。

(a) 专辑表格列表

(b) 专辑详情界面

(c) 播放专辑音乐

图 15-30 专辑模块的运行效果

在歌手列表中，显示歌手的信息。单击每行右边的进入按钮，进入歌手的详情页，在详情页显示歌手的图片、歌手的专辑歌曲列表，单击专辑进入专辑详情。在专辑详情页可以单击歌曲列表栏上的播放按钮，开始播放该栏的歌曲，如图 15-31 所示。

(a) 歌手列表

(b) 歌手详情界面

(c) 专辑详情界面

图 15-31　歌手模块的运行效果

在歌曲播放页的右上角都有一个收藏按钮。当遇到特别欣赏的歌曲，可以单击收藏。新装的 App 在收藏页是没有信息的，只有单击了收藏，才会在这里显示收藏歌曲列表。当 App 正在播放歌曲时，如果跳转到其他页面，可在导航栏看到一个不停旋转的图标，如图 15-32 所示。

(a) 收藏歌曲列表

(b) 播放歌曲在后台运行

图 15-32　收藏模块的运行效果

在主页的导航栏右侧，有“…”溢出菜单按钮，单击该按钮，弹出菜单项，如图 15-33 所示。

图 15-33 溢出菜单项

在专辑管理页面显示专辑列表，每一栏有专辑代表图片、专辑名称、类型以及删除按钮；在导航栏左侧有返回上一页导航按钮，右侧是添加菜单项。单击删除按钮会出现删除确认框，单击添加菜单项会进入添加专辑页；单击专辑栏，可进入专辑编辑页，如图 15-34 所示。

(a) 专辑项列表　(b) 删除专辑项确认框　(c) 添加新的专辑　(d) 编辑专辑信息

图 15-34 专辑管理模块的运行效果

在歌手管理页面显示歌手列表，每一栏有歌手的图片、姓名、出生日期、简介信息以及删除按钮；在导航栏左侧有返回上一页导航按钮，右侧是添加菜单项。单击删除按钮会出现删除确认框，单击添加菜单项会进入添加歌手页；单击歌手栏，可进入歌手编辑页，如图 15-35 所示。

(a) 歌手项列表

(b) 编辑歌手信息

图 15-35　歌手管理模块的运行效果

在歌曲管理页面显示歌曲列表，每一栏有歌曲的图片、歌名、演唱歌手姓名、专辑名、歌曲时长以及一个菜单图标；在导航栏左侧有返回上一页导航按钮，右侧是添加菜单项。单击菜单图标按钮会弹出菜单，有两个菜单项：删除、推荐歌曲。单击添加菜单项会进入添加歌曲页，如图 15-36 所示。

(a) 歌曲项列表

(b) 添加歌曲信息

图 15-36　歌曲管理模块的运行效果

在歌曲推荐管理页面显示已被推荐歌曲列表，每一栏有歌曲的图片、歌名、演唱歌手姓名、专辑名、歌曲时长以及一个菜单图标；在导航栏左侧有返回上一页导航按钮，右侧是添加菜单项。单击菜单图标按钮会弹出菜单，有三个菜单项：取消推荐、上移、下移推荐歌曲排序，如图 15-37 所示。

如果在溢出菜单中选择了“退出登录”菜单项，会弹出确认对话框，如图 15-38 所示。

图 15-37 歌曲推荐管理

图 15-38 退出登录确认框

在测试时应注意以下两点。

(1) 在测试运行时，可以看到，每当要执行删除操作和退出操作时，都会出现确认对话框。这是对于危险操作的一种常用处理模式。也是我们在测试中应该检测的功能点。

(2) 不仅在模拟器上进行测试，还需要在真机上进行测试。并且要使用不同屏幕分辨率、不同配置型号、不同 Android 版本的多台真机进行测试。

15.5 打包发布

在应用项目开发完成并通过测试后，就可以发布了。发布的前提是打包。

15.5.1 打包

对应用项目进行签名、打包、生成 APK 文件，才能安装在手机上运行。下面简单介绍本项目在 Android Studio 中的打包流程。

第一步：在 Android Studio 的 Build 菜单下选择 Generate Signed Bundle/APK...，进入 Generate Signed Bundle or APK 对话框。选择 APK 单选项，然后单击 Next 按钮。

第二步：如果首次进行打包，需要创建签名文件；再次进行打包，则单击 Choose

existing...按钮，选择已生成的签名文件，然后跳转到第四步。在此选择 Create new...按钮，进入创建签名文件对话框，如图 15-39 所示。单击 Key store path：项输入框内的图标，选择签名文件的存储地址，并为签名文件命名，如图 15-40 所示。然后单击 OK 按钮返回创建签名文件对话框。

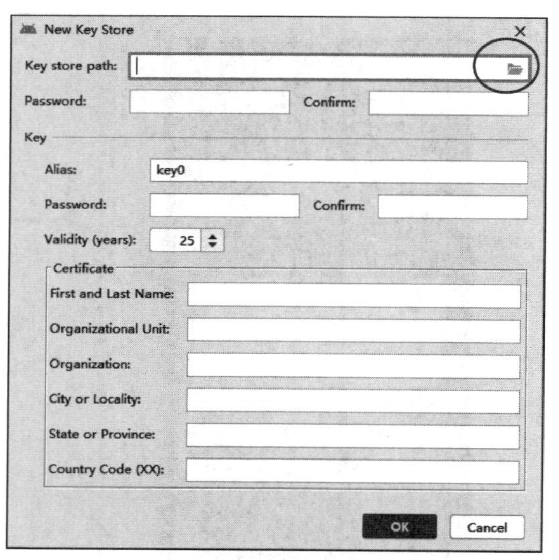

图 15-39　创建签名文件对话框

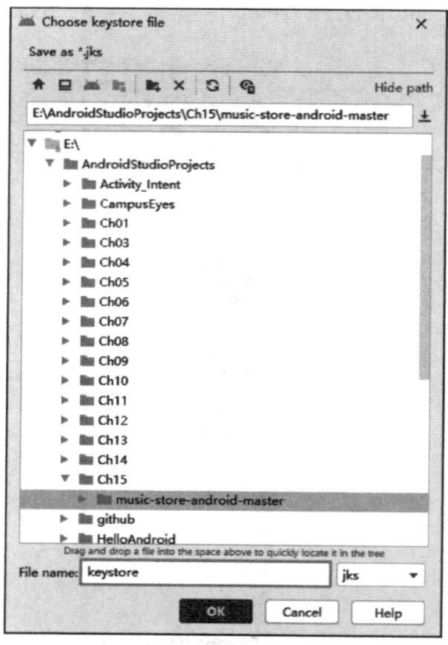

图 15-40　指定签名文件名及路径

第三步：在创建签名文件对话框中，设置数字证书文件密码、别名及密码、有效年限和开发者的相关验证信息，如图 15-41 所示。填写完毕单击 OK 按钮，即完成签名文件的创建。

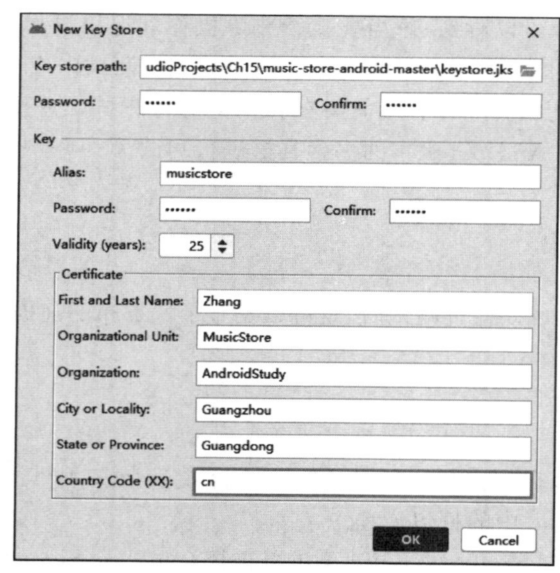

图 15-41　输入签名文件相关信息

第四步：确认签名文件及别名的输入密码，并设置和记住密码，如图 15-42 所示。然后单击 Next 按钮。

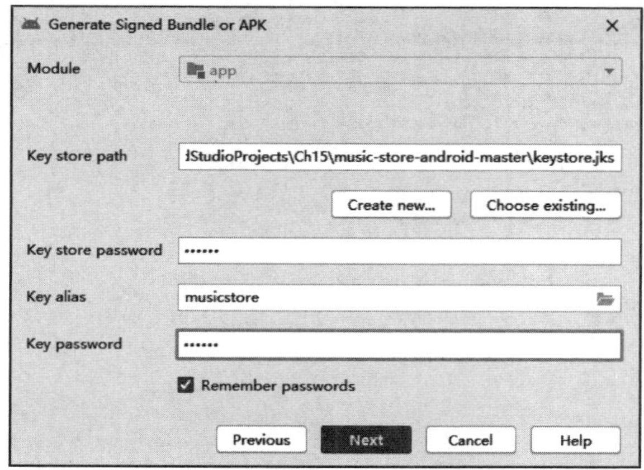

图 15-42　确认签名文件密码信息

第五步：接下来是生成打包文件。在对话框中的 Build Variants 内选择 release，在下面的签署版本中选择 V2 项。最后单击 Finish 按钮，如图 15-43 所示。

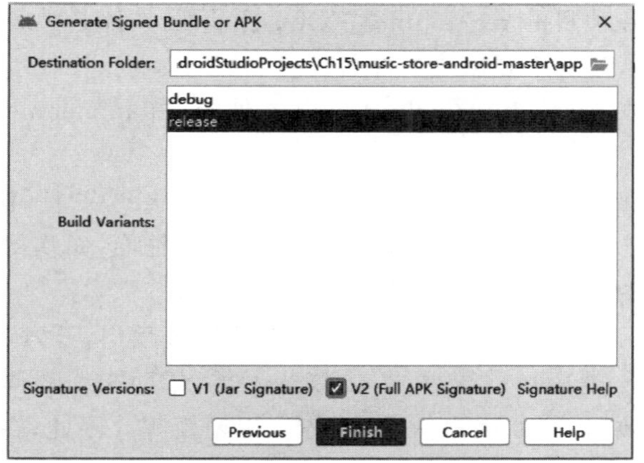

图 15-43　选择打包文件相关信息

至此，完成了打包操作流程。可以在指定的文件夹下看到新生成的打包文件。该文件存储在指定签名文件路径下的 app\release 文件夹下，文件名为 app-release. apk，如图 15-44 所示，可以修改该文件名为 musicstore. apk。

15.5.2　发布上线

1. 准备工作

App 发布上线，需要做好上线前的准备工作，包括设置 App 图标、名称、版本号；把开发模式切换为上线模式，除了代码的切换外，还需要修改 AndroidManifest. xml；对关键业

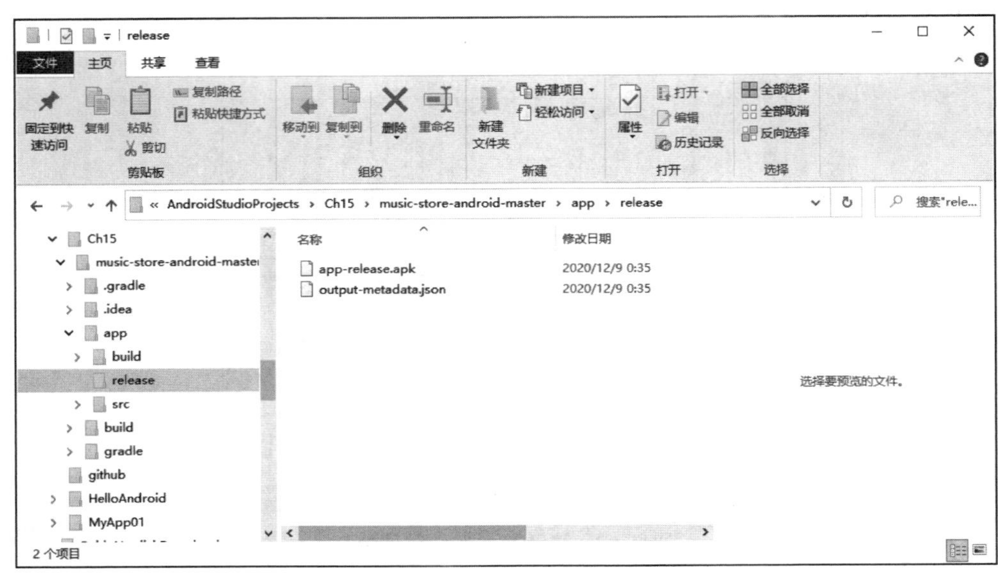

图 15-44　打包文件的存储地址

务数据进行加密处理等。

App 的图标保存在项目的 res/mipmap-XXX 目录下，文件名为 ic_launcher. png。如果需要更换，可在此替换，注意不要修改文件名。

App 的名称保存在 res/values/strings. xml 文件中，由 app_name 项定义。如果需要更换，可在此修改为新名称。

App 的版本号保存在项目的模块级 build. gradle 文件中，由 versionCode 和 versionName 两个参数确定。versionCode 必须为整数。注意，每次 App 升级，versionCode 和 versionName 要同时更改，版本号只能递增，不能减小。

上线模式，就是去掉多余的开发调试代码。在开发过程中，开发者为了调试方便，常常在代码中添加一些 Log 类、Toast 类和 AlertDialog 类，帮助在编程调试时发现 Bug。有些调试信息会涉及敏感数据，例如用户信息、业务流程逻辑等。因此，在 App 打包前，需要去掉多余的调试信息，形成上线模式。

在清单文件 AndroidManifest. xml 中，也要修改部分属性用于上线版。主要涉及 2 点修改：

(1) <application>元素节点中的属性 android:debuggable 是调试模式的开关属性。默认为 false，表示上线模式；如果设置为 true，则表示调试模式。如果项目测试通过后，为保险起见，要设置属性 android:debuggable="false"。

(2) App 发布后，所有的 activity 和 service 默认是对外部开放的。如果不希望 activity 和 service 开放，则需要在<activity>和<service>元素节点下添加属性 android:exported="false"。

2. 选择上线平台

要上线 App，需要选择合适的应用市场线上平台。当前国内 Android 应用市场分为第三方市场和手机厂商市场。第三方应用市场主要有应用宝、360 手机助手、百度、豌豆荚(目前被阿里收购了)等;厂商应用市场有华为、小米、VIVO、OPPO、魅族、三星等。

第三方市场对App类型的限制比较小,容易审核通过。

厂商市场是具有硬件制造能力的厂商,他们在自己品牌的手机里默认安装应用商店。厂商市场对应用的上架把控比较严格,对应用质量也要求较高,不容易通过审核。目前,厂商市场的应用分发量已超过第三方市场。在应用分发量排行中,依次为华为、OPPO、VIVO、小米、第三方市场。

另一方面,正是由于应用市场中的App数量巨大,我们发布的App能否被发现并得到用户量,是需要App的质量保障和推广宣传的。

3. 上线发布

上线到应用市场,首先要注册开发者账号。不同的应用市场,其上线的操作流程也有些差异,其实可以在各自开发者平台官网上找到相关操作流程。不同类型的App在不同的应用市场需要提供的证书会有所不同,需要上传哪些证明材料,特别是软件著作权证明或免责函等都会有所不同。在申请上线时要严格按照平台要求填报。只要资料齐全,是可以通过审核而上架属于自己的App的。

小结

本章较全面地介绍了开发一个Android综合应用项目的实战案例,从应用项目的需求分析、系统功能模块设计、数据库设计、服务器部署,到客户端App开发、测试、打包,直至项目发布全过程,给读者一个开发全貌体验。重点掌握Android客户端的开发过程,学习Android应用的编程技巧和开发技术。通过对全书上、下册的学习,相信大家已具备了Android应用项目的开发能力。

参考文献

[1] 张冬玲,杨宁. Android应用开发教程[M]. 北京：清华大学出版社,2013.

[2] Ed Burnette. Android基础教程[M]. 张波,高朝勤,杨越,等译. 北京：人民邮电出版社,2009.

[3] 欧阳燊. Android Studio开发实战[M]. 北京：清华大学出版社,2017.